RESCEU International Symposium Series

Universal Academy Press, Inc. *Tokyo, Japan*
ISSN 1341-9757

RISS is a series of books dedicated to the publication of symposium proceedings sponsored by the Research Center for the Early Universe (RESCEU), University of Tokyo.

The Research Center for the Early Universe (RESCEU), founded in the School of Science at the University of Tokyo in July 1995, is a "Center of Excellence" funded by Monbusho (the Ministry of Education, Science, Sports and Culture). Its objective is to study the origin of the Universe from both observational and theoretical perspectives, and we aim to make it a worldwide base for the research of the early Universe.

"How did our Universe begin, and how was its present structure formed?" — this question is as old as human history. The Standard Big Bang Model based on the theory of general relativity has been around for a while, but studies to give concrete substance to the model and to verify it are still in their early stages. We need to explore the early Universe to get the whole picture of the evolution of the Universe from its creation up to the present. Fortunately, we can observe the early Universe by observing the distant Universe because of the finite speed of light. Understanding the early Universe requires systematic observations not only of the full spectrum of light from gamma rays to radio waves but also of particle rays. In addition, we need theoretical models built on the latest results in nuclear and particle physics. Computation is also indispensable to make quantitative comparisons between results from observations and those from theories. Thus it is crucial to establish an organisation which brings together the results from these different fields and can promptly launch a project motivated by them.

No. 1 (RISS-1)
Edited by K. Sato, T. Suginohara, and N. Sugiyama
Cosmological Constant and the Evolution of the Universe
1996 ISBN 4-946443-32-0

No. 2 (RISS-2)
Edited by M. Minowa
Dark Matter in the Universe and its Direct Detection
1997 ISBN 4-946443-39-8

No. 3 (RISS-3)
Edited by K. Sato, T. Yanagida, and T. Shiromizu
Particle Cosmology
1998 ISBN 4-946443-46-0

No. 4 (RISS-4)
Edited by K. Sato and M. Kawasaki
Birth and Evolution of the Universe
2001 ISBN 4-946443-57-6

No. 5 (RISS-5)
Edited by K. Sato and T. Shiromizu
New Trends in Theoretical and Observational Cosmology
2002 ISBN 4-946443-73-8

New Trends in Theoretical and Observational Cosmology

Proceedings of the 5th RESCEU International Symposium on "New Trends in Theoretical and Observational Cosmology" held on November 13–16, 2001, at The University of Tokyo, Japan

RESCEU INTERNATIONAL SYMPOSIUM SERIES NO. 5

Edited by

K. Sato
Department of Physics, School of Science, The University of Tokyo
and
Research Center for the Early Universe (RESCEU),
School of Science, The University of Tokyo

T. Shiromizu
Research Center for the Early Universe (RESCEU),
School of Science, The University of Tokyo

Universal Academy Press, Inc.
Tokyo, Japan

New Trends in Theoretical and Observational Cosmology

Proceedings of the 5th RESCEU International Symposium on "New Trends in Theoretical and Observational Cosmology" held on November 13–16, 2001, at The University of Tokyo, Japan

Edited by K. SATO and T. SHIROMIZU

RESCEU International Symposium Series No. 5 (RISS-5)
ISSN 1341-9757

Published by
Universal Academy Press, Inc.

Postal Address: C.P.O. Box 235, Tokyo 100-8691, JAPAN
Address for Visitors: BR-Hongo-5 Bldg., 6–16–2, Hongo, Bunkyo-ku, Tokyo 113-0033, JAPAN
Telephone — + 81 3 3813 7232
Facsimile — + 81 3 3813 5932
E-Mail — general@uap.co.jp
WWW URL — http://www.uap.co.jp

ISBN 4-946443-73-8

Printed in Japan

Preface

The International Symposium on New Trends in Theoretical and Observational Cosmology was held from 13–16 November 2001 at the Sanjo Conference Hall, the University of Tokyo. This was the fifth Symposium of RESCEU, the Research Center for the Early Universe. The total number of the participants was about 180, coming from 10 countries, and 34 people were from abroad.

RESCEU was established by the COE program of MEXT, Ministry of Education, Culture, Sports, Science and Technology in 1995, and the research subjects of RESCEU are very wide, from the quantum creation of the universe, Inflation, brane-world cosmology, galaxy formation, large-scale structure, star formation, chemical evolution, origin of elements, supernova and neutron stars and etceteras. This is the last year for COE program RESCEU, but two years ago, ten years extension was approved. Formally it is not extension. So we have to say that the new research center, RESCEU-II, was established by the MEXT rule.

The purpose of this conference was to bring out a state -of -the -art discussion and assessment of cosmology by putting together the latest observational data against theoretical ideas on the birth and evolution of the Universe.

This proceedings includes about 17 reviews by distinguished researchers and 83 contributed papers by active researchers. We hope this proceedings is useful for people who want to know recent development in theoretical and observational cosmology.

Finally the editors would like to express deepest gratitude to MEXT for the support on behalf of all participants.

For the Organizing Committee,
Katsuhiko Sato
Tetsuya Shiromizu

February 2002

Congratulatory Address

Good morning. I am Junko Kawamura and I am very happy to be here today to offer you this greeting from the Ministry of Education, Culture, Sports, Science and Technology Monbukagaku-sho—which is sponsoring this COE International Symposium. I am very grateful that all of you are here today attending this symposium. I especially thank those scientists who have come all the way to Japan in spite of your very tight schedule as well as quite uncertain situation.

This symposium is organized by COE Program Research Center for the Early Universe.

First, I would like to talk briefly about COE program. COE programs started seven years ago to establish Centers of Excellence" which would serve as outstanding research bases in Japan. Our ministry actively supports research institutions and units with the potential to conduct research at a high international level with a view to raising them to COEs. We have selected several projects every year based on the recommendation of the Science Council. There are 37 on-going projects.

I am pleased to say that this "Research Center for the Early Universe headed by Professor Sato was selected as one of the first COE programs started in 1995. We chose 6 projects in that year, from almost 200 excellent proposals.

I understand the purpose of RESCEU is to quest for the origin and evolution of the universe through physical law and observations. "Where did we come from? How has our universe begun and how the present beautiful and fruitful structure of our universe was formed?" These are the questions which we, humankind, have been asking since the beginning of our history.

This COE program was originally supposed to continue for five years. During those years RESCEU produced many brilliant scientific results including the first clear identification of cosmic antiprotons and measurement of the antiproton spectrum by the group led by late Professor Orito. RESCEU also built a submillimeter-wave telescope at the summit of Mt. Fuji and an optical telescope, and has operated them since 1998 and 2000, respectively. These prominent results made us support RESCEU for two more years expecting further achievement.

I hear a stellar show –the Leonid meteor shower– will take place in the nighttime sky in a week. I believe that this symposium will also be a stellar event with many presentations of new excellent and exciting scientific results by RESCEU.

Last but not least, I should like to express our appreciation to the organizers of the Research Center for Early Universe, The University of Tokyo. I wish you all the best for a productive and enjoyable meeting.

Thank you very much.

Junko Kawamura
Director
Scientific Research Aid Division
Ministry of Education, Culture,
Sports, Science and Technology

November 2001

Contents

Primordial Perturbations from Inflation

David WANDS
Relativity and Cosmology Group, School of Computer Science and Mathematics, University of Portsmouth, Portsmouth PO1 2EG, United Kingdom

Abstract

I review the standard analysis of adiabatic scalar and tensor perturbations produced by slow-roll inflation driven by a single scalar field, before going on to discuss recent work on the role of non-adiabatic modes during and after inflation. Isocurvature perturbations correlated with adiabatic modes may be produced during multi-field inflation and would give valuable information about the physics of the early universe. A significant contribution from correlated isocurvature perturbations is not ruled out by present data, but should be either detected or ruled out by future observations.

1. Introduction

In recent years a standard model has emerged for the origin of the large-scale structure of our Universe [1]. Observations of the cosmic microwave background (CMB) reveal primordial anisotropies on the surface of last scattering of the CMB photons. Structure can form from these initial perturbations about a Friedmann-Robertson-Walker (FRW) universe by gravitational instability to form the galaxies, clusters of galaxies and superclusters observed in large-scale surveys. These observations are consistent with an almost scale-invariant initial spectrum of adiabatic density perturbations at last-scattering [2]. Inflation is a dynamical model of the early universe which can explain the origin of perturbations on arbitrarily large scales from small-scale vacuum fluctuations of light fields.

Observational data are increasingly being used to constrain the cosmological parameters of the background FRW model of the universe since last-scattering. But this can only be done in the context of some model for the nature of the primordial perturbations. Constraints on the form of the primordial perturbations yield constraints on the dynamics and high energy physics driving inflation in the very early universe.

2. Single-field inflation

The simplest models of inflation are driven by a scalar field, ϕ, slowly rolling down its potential, $V(\phi)$, in a spatially flat FRW spacetime with scale factor a. The classical evolution is determined by the Klein-Gordon equation

$$\ddot{\phi} + 3H\dot{\phi} = -V_\phi \,, \tag{1}$$

(where V_ϕ denotes $dV/d\phi$) coupled to the Friedmann equation for the Hubble expansion ($H \equiv \dot{a}/a$)

$$H^2 = \frac{8\pi}{3M_P^2}\left(V + \frac{1}{2}\dot{\phi}^2\right) . \tag{2}$$

This gives inflation (i.e., accelerated expansion, $\ddot{a} > 0$) for $V > \dot{\phi}^2$.

The slow-roll approximation truncates these to a first-order system

$$3H\dot{\phi} \simeq -V_\phi \tag{3}$$

$$H^2 \simeq \frac{8\pi}{3M_P^2} V \, . \tag{4}$$

This assumes the evolution is potential-dominated ($V \gg \dot{\phi}^2$) and over-damped ($3H|\dot{\phi}| \gg |\ddot{\phi}|$) but can give a useful approximation to the growing mode solution when the dimensionless slow-roll parameters [1] are small

$$\epsilon \equiv \frac{M_P^2}{16\pi}\left(\frac{V_\phi}{V}\right)^2 \ll 1 \quad , \quad |\eta| \equiv \frac{M_P^2}{8\pi}\left|\frac{V_{\phi\phi}}{V}\right| \ll 1 \, . \tag{5}$$

Linear perturbations of a massless field in an FRW background obey the wave equation

$$\ddot{\delta\phi} + 3H\dot{\delta\phi} - \nabla^2\delta\phi = 0 \, . \tag{6}$$

Arbitrary inhomogeneities can be decomposed into spatial harmonics, such as Fourier modes. Each mode with fixed comoving wavenumber k has two characteristic timescales

- oscillation period (determined by the physical wavelength) a/k
- damping timescale (determined by the Hubble expansion) H^{-1}

The evolution of each mode is thus naturally split into two regimes

- small scales ($a/k < H^{-1}$) under-damped *oscillations*
- large scales ($a/k < H^{-1}$) over-damped, or *frozen-in*

In a conventional (non-inflationary) matter- or radiation-dominated universe the comoving Hubble length $H^{-1}/a = \dot{a}^{-1}$ increases with time so that modes are frozen-in ($k < aH$) at early times and only come within the Hubble length at late times ($k > aH$). But in an inflationary era the comoving Hubble length decreases and modes that begin as under-damped oscillators on small scales are stretched by the accelerated expansion beyond the Hubble length. Thus initial zero-point fluctuations of the quantum vacuum on small-scales (where the effects of the cosmological expansion is negligible) leads to a spectrum of overdamped perturbations on large-scales. Linear evolution ensures that the perturbations are described by a Gaussian random field at all times.

Perturbations of a massive or self-interacting field have an additional oscillation timescale set by the effective mass, $m^2 \equiv V_{\phi\phi}$. Massive fields ($m^2 \geq 9H^2/4$) remain underdamped even on large-scales and effectively no perturbations are generated. But any light field ($m^2 < 9H^2/4$) acquires a spectrum perturbations $\langle \delta\phi^2 \rangle \simeq (H/2\pi)^2$ at Hubble-crossing. In particular, the inflaton must be light ($|\eta| \ll 1$) during slow-roll inflation.

3. Cosmological perturbations

Arbitrary perturbations of an FRW cosmology can be decomposed into two types

- *adiabatic perturbations* perturb the solution along the same trajectory in phase-space as the background solution. Thus perturbations in any scalar x can be described by a unique perturbation in expansion with respect to the background

$$H\delta t = H\frac{\delta x}{\dot{x}} \quad \forall \quad x \,, \tag{7}$$

  e.g., the adiabatic density perturbation $\delta\rho/\rho \propto H\delta\rho/\dot{\rho}$.

- *entropy perturbations* perturb the solution off the background solution

$$\frac{\delta x}{\dot{x}} \neq \frac{\delta y}{\dot{y}} \quad \text{for some } x \text{ and } y \,. \tag{8}$$

  One example is an *isocuravture* perturbation of the baryon-photon ratio $S = \delta(n_B/n_\gamma) = (\delta n_B/n_B) - (\delta n_\gamma/n_\gamma)$.

Although the amplitude of adiabatic perturbations (such as the density perturbation) is notoriously gauge-dependent, the adiabaticity condition (7) is not. Moreover, from this definition it is clear that purely adiabatic perturbations (along the background trajectory) must remain adiabatic on large scales and cannot generate entropy perturbations (off that trajectory) [5, 3].

If the perturbed expansion, $H\delta t$, is evaluated in the spatially-flat gauge [4] then it coincides with the gauge-invariant scalar curvature perturbation, ζ, on uniform-density hypersurfaces [5], first introduced by Bardeen, Steinhardt and Turner [6]. This is a particularly useful quantity as it remains constant for adiabatic perturbations in the large-scale limit.

In single-field inflation the over-damped perturbations of the inflaton field on large scales are adiabatic perturbations of the growing mode solution and entropy perturbations vanish in the large-scale limit [3]. Thus single-field inflation models predict a primordial adiabatic perturbation on large-scales whose amplitude can be calculated in terms of the scalar field perturbations at Hubble-crossing ($k = aH$) during inflation. This in turn can be related to the inflaton potential $V(\phi)$ in the slow-roll approximation [7]

$$A_S^2 = \left\langle \left(\frac{H\delta\phi}{\dot{\phi}} \right)^2 \right\rangle_{k=aH} \simeq \frac{32}{75} \frac{V}{\epsilon M_P^4} \tag{9}$$

The time-dependence of the inflaton field during inflation leads to a scale dependence of the resulting spectrum, which in the slow-roll approximation is determined by the slow-roll parameters

$$n_S - 1 \equiv \frac{d \ln A_S^2}{d \ln k} \simeq -6\epsilon + 2\eta \,. \tag{10}$$

In the extreme slow-roll limit this yields the scale-invariant Harrison-Zel'dovich spectrum, $n_S - 1$.

Gravitational waves, corresponding to tensor metric perturbations independent of the scalar perturbations to linear order [8], are another massless degree of freedom excited during inflation and frozen-in on large scales to give [7]

$$A_T^2 = \left\langle \left(\frac{H}{M_P} \right)^2 \right\rangle_{k=aH} \simeq \frac{32}{75} \frac{V}{M_P^4} \,, \tag{11}$$

with spectral tilt

$$n_T \equiv \frac{d \ln A_T^2}{d \ln k} \simeq -2\epsilon \,. \tag{12}$$

A_S^2 and A_T^2 describe the contribution of scalar and tensor perturbations to the the CMB anisotropies on large angular scales. A prediction of single-field slow-roll models of inflation is that there should be a consistency condition relating the scalar-tensor ratio to the tensor tilt [7]:

$$\frac{A_T^2}{A_S^2} \simeq -\frac{1}{2} n_T \,. \tag{13}$$

Unfortunately there is no guarantee that the tensor contribution is large enough to ever be detectable (let alone its tilt measurable). Currently favoured hybrid-type inflation models generally occur at low energies with $\epsilon \ll 1$ and $A_T^2 \ll A_S^2$ [30].

4. Non-adiabatic effects

4.1. Single field

In single-field models the Hubble damping generally causes the decaying mode solution to the Klein-Gordon equation to rapidly decay, suppressing any non-adiabatic perturbations on large scales. Nonetheless, even in single-field inflation it is possible to significantly alter the curvature perturbation on finite, but super-Hubble scales. Non-adiabatic (decaying mode) and/or gradient terms may have an effect if $z \equiv a\dot{\phi}/H$ (which is a monotonic increasing function of time in the slow-roll approximation) decreases back below its value at Hubble-crossing [9]. This is possible in some models of inflation where the slope of the inflaton potential decreases abruptly [10] and the inflaton field enters a transient friction-dominated 'fast-roll' regime [11] described by $\ddot{\phi} \simeq -3H\dot{\phi}$, leading to $z \propto a^{-2}$ for a finite period.

Another case in which the instantaneous value of the scalar curvature perturbation calculated at horizon-crossing may not be a good estimate of the final value on large scales is when the inflaton stops [12], $\dot{\phi} = 0$. In this case the apparent divergence of $\zeta = H\delta\phi/\dot{\phi}$ is transient if at the same time $V_\phi \neq 0$ and, ironically, the slow-roll approximation $\zeta \simeq V^{3/2}/V'$ can give a better estimate of the final value, giving as it does an estimate of the perturbation in the growing-mode solution [9].

4.2. Two fields

If more than one light scalar field exists during inflation then there is more than one allowed phase-space trajectory for FRW cosmologies. A spectrum of entropy perturbations (off the background trajectory) will be generated on large scales from initial vacuum fluctuations on small scales.

In the case of two canonical light fields φ_1 and φ_2 the background trajectory is described by $\dot{\varphi}_1$ and $\dot{\varphi}_2$ and arbitrary field perturbations can be decomposed along and orthogonal to the background trajectory [3]:

$$\delta\sigma \equiv \cos\theta\delta\varphi_1 + \sin\theta\delta\varphi_2\,, \quad \delta s \equiv -\sin\theta\delta\varphi_1 + \cos\theta\delta\varphi_2 \tag{14}$$

where the angle of the trajectory in field-space is given by $\tan\theta = \dot{\varphi}_2/\dot{\varphi}_1$. At any instant, the adiabatic perturbation $\delta\sigma$ determines the scalar curvature perturbation $R = H\delta\sigma/\dot{\sigma}$, while the entropy perturbation determines the isocurvature perturbation $S \propto \delta s$ at that time.

The coupled evolution equations for $\delta\sigma$ (in arbitrary gauge) and δs (which is automatically gauge invariant) were derived in Ref. [3] (for a different treatment of the same equations see [13]). In the slow-roll limit, and on large scales, the equations can be reduced to [3, 14]

$$
\begin{aligned}
3H\dot{\delta\sigma} + \left(V_{\sigma\sigma} - \dot{\theta}^2\right)\delta\sigma &= 2\left(\dot{\theta}\delta s\right) - 2\left(\frac{V_\sigma}{\dot{\sigma}} + \frac{\dot{H}}{H}\right)\dot{\theta}\delta s\,, \\
3H\dot{\delta s} + \left(V_{ss} + 3\dot{\theta}^2\right)\delta s &= 0
\end{aligned} \tag{15}
$$

For $\dot{\theta} = 0$ the equations are decoupled and reduce to the standard slow-roll equations for canonical field perturbations. But for a curved trajectory the entropy perturbation δs appears as an extra driving term for $\delta\sigma$. As a result the curvature perturbation is no longer constant on large scales. This additional contribution to the spectrum of scalar curvature perturbations at the end of inflation weakens the consistency condition (13) of single-field inflation to an inequality [15, 16, 17]

$$
\frac{A_T^2}{A_S^2} \leq -22n_T\,. \tag{16}
$$

Langlois [18] was the first to point out that another consequence of this coupling is that any residual isocurvature perturbation after multi-field inflation will in general be correlated with the curvature perturbation. The adiabatic and entropy field perturbations at horizon crossing are, by their construction (14), independent random fields

$$
\langle\delta\sigma^2\rangle_{k=aH} = \langle\delta s^2\rangle_{k=aH} \simeq (H/2\pi)^2\,, \quad \langle\delta\sigma\delta s\rangle_{k=aH} = 0\,. \tag{17}
$$

The subsequent evolution on large scales can be parameterised by a transfer matrix, which has the general form [19]

$$
\begin{pmatrix} R \\ S \end{pmatrix} = \begin{pmatrix} 1 & T_{RS} \\ 0 & T_{SS} \end{pmatrix} \begin{pmatrix} R \\ S \end{pmatrix}_{k=aH} \tag{18}
$$

The form of this matrix is determined by the general definition of adiabatic and entropy perturbations [5] introduced in section 3. and the evolution of two field perturbations during inflation in Eq. (15) provides just one example.

The two coefficients T_{RS} and T_{SS} are model-dependent transfer functions which determine the final power spectra

$$
\langle R^2\rangle \propto 1 + T_{RS}^2\,, \quad \langle S^2\rangle \propto T_{SS}^2\,, \quad \langle RS\rangle \propto T_{RS}\,. \tag{19}
$$

For example, if all the particle species present after inflation are in thermal equilibrium determined by a single temperature then there is a unique phase-space trajectory for the FRW cosmologies and only adiabatic perturbations are possible, $T_{SS} = 0$. The other extreme is that one species remains completely decoupled after Hubble-crossing corresponding to $T_{RS} = 0$, leaving uncorrelated isocurvature perturbations. But the general case is that non-zero isocurvature perturbations survive and are correlated with the curvature perturbations.

5. Curvaton model for the origin of structure

Taking into account the effect upon the large-scale curvature perturbations of entropy perturbations possible in multi-field or multi-fluid cosmological models leads to the realisation that the curvature perturbation calculated at Hubble-crossing during inflation only gives a *lower bound* for the curvature perturbation at the start of the epoch of structure formation. In a recent paper with David Lyth [20], I addressed the question of whether it is possible to have a viable model of structure formation if there is effectively no curvature perturbation produced on large scales during inflation. The answer is yes!

In our model the *curvaton* is supposed to be a light field during inflation ($m_s^2 \equiv V_{ss} \ll H^2$) which does not affect the dynamics during inflation, i.e., is not the inflaton, and thus its perturbations correspond to isocurvature perturbations. If this field is decoupled from other inflation or matter fields then the long-wavelength perturbations are effectively frozen-in until the Hubble rate H in the expanding universe drops to below the mass m_s. At this point the curvaton field begins to oscillate and its energy density redshifts as $\rho_s \propto a^{-3}$. This grows relative to the radiation density $\rho_\gamma \propto a^{-4}$ and will come to dominate the energy density of the universe unless the field decays. This is the well-known Polonyi (or moduli) problem associated with massive weakly-coupled fields. But a late-decaying field, so long as it decays before nucleosynthesis, may not be a bad thing. Late entropy release can dilute the abundance of other dangerous relics and can be used as a model for baryogenesis or leptogenesis. Crucially, its perturbations can also produce a large-scale curvature perturbation [23, 20, 21].

As an example consider a complex scalar field $\phi = |\Sigma| e^{is/v}$ whose modulus is fixed, $|\Sigma| \sim v$, by a mexican-hat potential with large effective mass $m_{|\Sigma|} \sim v$, but whose U(1) symmetry is only broken by non-renormalisable terms so that s has a small mass $m_s \sim v^2/M_P$. The radial vev is stabilised, but the pseudo-Goldstone boson s will acquire an almost scale-invariant spectrum of isocurvature fluctuations during a period of inflation, with Hubble rate H, if we have

$$\frac{v^2}{M_P} \ll H \ll v\,. \tag{20}$$

Assuming that after inflation s decays only with gravitational strength, $\Gamma \sim m_v^3/M_P^2$ then we naturally have $\rho_s \sim \rho_\gamma$ at the decay time and hence [20]

$$\langle \zeta^2 \rangle \sim \langle (\delta s/s)^2 \rangle \sim (H/v)^2\,. \tag{21}$$

If the decay products thermalise completely then the isocurvature perturbation during inflation, δs, is converted into an adiabatic curvature perturbation,

ζ. It is then indistinguishable from conventional inflaton models for structure formation, other than violating the consistency condition (13), but respecting the inequality (16).

An interesting alternative possibility is that, because the curvaton decay can occur relatively late, some particle species have already dropped out of equilibrium when the curvaton decays. These species would then have have an isocurvature perturbation relative to the radiation produced by the curvaton decay, but one that would be 100% correlated with the curvature perturbation. A recent example of a late-decaying field that could produce correlated curvature and isocurvature perturbations is the sneutrino field in leptogenesis models [22].

A similar model for the origin of large-scale structure from initially isocurvature axion perturbations has been recently proposed by Enqvist and Sloth [23] in the context of the pre big bang scenario [24]. In this case the rapid increase in the Hubble rate during the pre big bang phase must be compensated by the rapidly growing dilaton coupling to yield a scale-invariant spectrum of axion perturbations [25]. This appears to be the only possible origin of cosmological structure in pre big bang type models where essentially no curvature perturbation is produced on large scales during the 'inflationary' phase [24].

6. Observational data

The form of the primordial perturbation spectra is increasingly being constrained by astronomical observations, especially cosmic microwave background experiments. The overall amplitude of the anisotropies on large scales is still set by the COBE data [26] which gives an amplitude $A^2 = 1.9 \times 10^{-5} \pm 10\%$. A recent compilation comparing data against FRW models with adiabatic scalar and tensor perturbations by Wang, Tegmark and Zaldarriage [2] gives a constraint on the spectral index $0.80 < n_S < 1.03$, and an upper limit on the contribution from gravitational waves, $A_T^2/A_S^2 < 0.008$, with the spectral index n_T being unbounded.

Most studies of non-adiabatic perturbation spectra to date have considered only uncorrelated isocurvature perturbations which tend to only give an additional source of anisotropies on large angular scales and hence their contribution to the CMB anisotropies is severely constrained [27]. However I have emphasized that it is natural in inflation models to consider isocurvature perturbations *correlated* with the standard adiabatic mode [28, 29, 19]. In particular the cross-correlation can decrease the CMB anisotropies on large angular scales and a significant contribution from isocurvature modes cannot be ruled out from current data. Indeed considering correlated CDM-isocurvature modes (with scale-invariant correlation angle Δ) the best-fit to the current CMB data [19] has $n_S = 0.8$, $\cos\Delta = 1$ and a similar contribution to large-angle anisotropies from adiabatic and isocurvature modes. Future CMB data from the MAP satellite would certainly be able to distinguish between such a model and a purely adiabatic spectrum.

7. Conclusions

It is quite possible that the primordial perturbations observed by future CMB experiments will remain consistent with the scale-invariant Gaussian spectrum of adiabatic density perturbations proposed more than thirty years ago by Harrison and Zel'dovich. In this case we would be able to extract little about the physics of inflation and would only be able to place bounds on allowed deviations from the extreme slow-roll limit.

If we are to learn more about the dynamical history of the early universe we need to find deviations from scale-invariance or traces of either tensor perturbations or non-adiabatic effects which, though not evident in current observations, could be detected by future experiments.

Acknowledgements

I am grateful to RESCEU at the University of Tokyo for their kind hospitality. My work is supported by the Royal Society.

References

[1] A. R. Liddle and D. H. Lyth, *Cambridge, UK: Univ. Pr. (2000) 400 p.*
[2] X. Wang, M. Tegmark and M. Zaldarriaga, arXiv:astro-ph/0105091.
[3] C. Gordon, D. Wands, B. A. Bassett and R. Maartens, Phys. Rev. D **63**, 023506 (2001) [arXiv:astro-ph/0009131].
[4] H. Kodama and M. Sasaki, Prog. Theor. Phys. Suppl. **78**, 1 (1984).
[5] D. Wands, K. A. Malik, D. H. Lyth and A. R. Liddle, Phys. Rev. D **62**, 043527 (2000) [arXiv:astro-ph/0003278].
[6] J. M. Bardeen, P. J. Steinhardt and M. S. Turner, Phys. Rev. D **28**, 679 (1983).
[7] J. E. Lidsey, A. R. Liddle, E. W. Kolb, E. J. Copeland, T. Barreiro and M. Abney, Rev. Mod. Phys. **69**, 373 (1997) [arXiv:astro-ph/9508078].
[8] J. M. Bardeen, Phys. Rev. D **22**, 1882 (1980).
[9] S. M. Leach, M. Sasaki, D. Wands and A. R. Liddle, Phys. Rev. D **64**, 023512 (2001) [arXiv:astro-ph/0101406].
[10] A. A. Starobinsky, JETP Lett. **55**, 489 (1992) [Pisma Zh. Eksp. Teor. Fiz. **55**, 477 (1992)].

[11] S. M. Leach and A. R. Liddle, Phys. Rev. D **63**, 043508 (2001) [arXiv:astro-ph/0010082].
[12] O. Seto, J. Yokoyama and H. Kodama, Phys. Rev. D **61**, 103504 (2000) [arXiv:astro-ph/9911119].
[13] J. c. Hwang and H. Noh, Phys. Lett. B **495**, 277 (2000) [arXiv:astro-ph/0009268].
[14] N. Bartolo, S. Matarrese and A. Riotto, Phys. Rev. D **64**, 083514 (2001) [arXiv:astro-ph/0106022]; Phys. Rev. D **64**, 123504 (2001) [arXiv:astro-ph/0107502].
[15] D. Polarski and A. A. Starobinsky, Phys. Lett. B **356**, 196 (1995) [arXiv:astro-ph/9505125].
[16] J. Garcia-Bellido and D. Wands, Phys. Rev. D **53**, 5437 (1996) [arXiv:astro-ph/9511029].
[17] M. Sasaki and E. D. Stewart, Prog. Theor. Phys. **95**, 71 (1996) [arXiv:astro-ph/9507001].
[18] D. Langlois, Phys. Rev. D **59**, 123512 (1999) arXiv:astro-ph/9906080.
[19] L. Amendola, C. Gordon, D. Wands and M. Sasaki, arXiv:astro-ph/0107089.
[20] D. H. Lyth and D. Wands, arXiv:hep-ph/0110002.
[21] T. Moroi and T. Takahashi, Phys. Lett. B **522**, 215 (2001) [arXiv:hep-ph/0110096]; arXiv:hep-ph/0112335.
[22] K. Hamaguchi, H. Murayama and T. Yanagida, arXiv:hep-ph/0109030; T. Yanagida, *in these proceedings*.
[23] K. Enqvist and M. S. Sloth, arXiv:hep-ph/0109214.
[24] J. E. Lidsey, D. Wands and E. J. Copeland, Phys. Rept. **337**, 343 (2000) [arXiv:hep-th/9909061].
[25] E. J. Copeland, R. Easther and D. Wands, Phys. Rev. D **56**, 874 (1997) [arXiv:hep-th/9701082].
[26] E. F. Bunn, A. R. Liddle and M. J. White, Phys. Rev. D **54**, 5917 (1996) [arXiv:astro-ph/9607038].
[27] E. Pierpaoli, J. Garcia-Bellido and S. Borgani, JHEP **9910**, 015 (1999) [arXiv:hep-ph/9909420]; K. Enqvist, H. Kurki-Suonio and J. Valiviita, Phys. Rev. D **62**, 103003 (2000) [arXiv:astro-ph/0006429].
[28] M. Bucher, K. Moodley and N. Turok, Phys. Rev. D **62**, 083508 (2000) [arXiv:astro-ph/9904231].
[29] R. Trotta, A. Riazuelo and R. Durrer, Phys. Rev. Lett. **87**, 231301 (2001) [arXiv:astro-ph/0104017].
[30] D. H. Lyth and A. Riotto, Phys. Rept. **314**, 1 (1999) [arXiv:hep-ph/9807278].

Brane-World Cosmology and the 5D Big Bang

Jaume Garriga
Departament de Física Fonamental, Universitat de Barcelona, Diagonal 647, 08028 Barcelona, Spain
Yukawa Institute for Theoretical Physics, Kyoto University, Kyoto 606-8502, Japan

Abstract

The possibility of having gravity localized on a brane which moves in a higher dimensional space has recently stimulated the search for alternatives to the standard inflationary paradigm, where the Big Bang would result from the collision of 4-dimensional extended objects. In particular, Bucher has suggested that our universe may arise in the collision of two vacuum bubbles which nucleate in a five-dimensional bulk. Here, I review this proposal, with an emphasis on the issue of primordial density perturbations.

1. Introduction

Our understanding of the origin of the universe and its large scale structure is based on the inflationary paradigm, which was proposed some twenty years ago and whose predictions agree very well with the most recent cosmological data. It is quite remarkable that no compelling alternatives to inflation have emerged so far, and in view of the successes of the theory one might just take the attitude that none are really needed. Over the past two years, however, much has been learned about gravity in theories with "branes" which move in spacetimes with extra dimensions [2, 3]. In particular, it has been pointed out that gravity may appear to be four dimensional even if spacetime has more than four non-compact dimensions, a phenomenon which has come to be known as the localization of gravity on a brane. Such developments have opened up the possibility of an entirely different picture for the birth and evolution of our universe, and a few ideas have been put forward as alternative theories of initial conditions [4, 1, 5]. Some of them have little or nothing to do with inflation, and on this account they may be worth exploring.

In particular, Bucher [1] has proposed that the Big Bang hypersurface is the locus of collision of two vacuum bubbles which nucleate in a five-dimensional bulk. As the bubbles expand towards each other, their surface develops small quantum fluctuations which slightly deform the bubble shape. This leads to density perturbations on the Big Bang surface, and one is led to speculate whether these may be the seeds for the observed cosmological density perturbations. In what follows, I will briefly summarize Bucher's model, after which I will present in some detail the mechanism for the generation of cosmological perturbations in this scenario. The following discussion is based on work done in collaboration with Takahiro Tanaka [6]. Similar calculations have been done by J.J. Blanco-Pillado and M. Bucher [7], who have also considered the case of bubbles nucleating in an initial de Sitter phase.

2. A Big Bang from bubble collision

In Bucher's model one starts with a metastable (or "false") vacuum in five dimensions, which is flat space or a very mildly expanding de Sitter space. This decays through bubble nucleation into an anti-de Sitter (AdS) phase, where the five dimensional cosmological constant is negative. The model requires the existence of degenerate discrete AdS vacua. Upon collision of two bubbles corresponding to different vacua, a domain wall forms, where gravity is localized à la Randall-Sundrum [2]. The "Big Bang" hypersurface is the locus where the worldsheets of the two bubbles meet, and the domain wall, or "local brane", is the place where we are supposed to live.

Let us use coordinates $X^A = (X^i, W, T)$ in the original 5D Minkowski space, where a pair of bubbles of radius R_0 nucleate at $T = 0$, separated by a distance $2d$ along the W axis. The radius R_0 is a fixed parameter of the theory, which is related to the bubble wall tension σ and to the energy gap ϵ between the original Minkowski phase and the final AdS phases [8] (this relation, however, will not be needed in the following discussion.) Let us choose the origin of coordinates to be at the center of one of the bubbles. After nucleation, the bubble wall expands with constant acceleration, following a hyperbolic trajectory which can be parametrized as

$$T = R_0 \sinh \beta, \qquad (1)$$

$$R = R_0 \cosh \beta, \qquad (2)$$

where $R \equiv (|\vec{X}|^2 + W^2)^{1/2}$, and β is the boost parameter of the bubble wall

$$v = dR/dT = \tanh \beta.$$

The collision takes place on the plane $W = d$, along the hyperboloid

$$T^2 - |\vec{X}|^2 = d^2 - R_0^2. \qquad (3)$$

To the future of this hyperboloid, and on the plane $W = d$, a domain wall forms where gravity is four dimensional [2, 9]. We shall asume that, locally, the collision process is almost instantaneous compared with the lengthscales of our interest, and that a fixed fraction of the energy of the collision is channeled into the degrees of freedom which live on the local brane. We shall also assume the usual fine tuning between the tension of the local brane and the AdS radius [2], so that the effective 4D cosmological constant is sufficiently small.

In the unperturbed setup, the 4D Minkowski space on the plane $W = d$ is matched into an open FRW model along the hypersurface (3). Along this hyperboloid, the metric is continuous, although its derivative is not (this jump in the extrinsic curvature corresponds to the jump in the Hubble rate accross the surface). It is convenient to use Milne coordinates to describe the 4D Minkowski space before the collision,

$$^{(4)}ds^2 = -dt^2 + t^2 d\Omega_H^2, \tag{4}$$

where the $t = const.$ surfaces have the hyperbolic geometry described by $d\Omega_H^2$. In terms of Minkowski coordinates, Milne time is given by $t = (T^2 - |\vec{X}|^2)^{1/2}$. This form of the metric is valid for $t < R_0 \cosh^{-1}(d/R_0)$.

For $t > R_0 \cosh^{-1}(d/R_0)$, the 4D metric on the "local brane" is given by

$$^{(4)}ds^2 = -dt^2 + a^2(t) d\Omega_H^2.$$

The local brane separates the interiors of the two vacuum bubbles, which have already collided and continue to expand. Due to the $O(3,1)$ symmetry of the process, the bulk metric on both sides of the local brane is given by Schwarzschild-AdS [9, 10]. The evolution of the scale factor after collision is given by the usual Friedmann equation

$$H^2 \approx \frac{\kappa^2}{3}\rho + \frac{1}{a^2}. \tag{5}$$

Here, $H = \dot{a}/a$, $\kappa^2 = 8\pi G$ is the effective 4D gravitational coupling and ρ is the matter energy density which is deposited on the local brane after collision. This will also include some "dark radiation"[9], because a significant fraction of the energy of the colliding bubbles may not stick to the brane; it may just fly into the bulk contributing to the effective mass of the Schwarzschild-AdS. In (5) we have also neglected "brany" corrections in the right hand side, proportional to $\kappa^4\rho^2\ell^2$. These will be unimportant provided that the AdS radius ℓ in the bulk is sufficiently small.

From (3), the spatial curvature scale of the open universe at the moment of collision is given by $(d^2 - R_0^2)^{1/2}$, so the flatness problem is solved provided that d is sufficiently large. Of course, the homogeneity problem is also solved, due to the residual $O(3,1)$ symmetry of the colliding bubble setup along the hyperboloid (3). Thus, in principle, Bucher's scenario seems to provide an attractive starting point for a brane-world cosmology.

3. Initial perturbations

Aside from solving the homogeneity and flatness problems, a serious candidate for a theory of initial conditions should explain the origin of the primordial cosmological perturbations. In brane collision scenarios these may be seeded by preexisting fluctuations in the shape of the colliding branes. These produce distortions of the "Big Bang" hypersurface as well as initial perturbations in the distribution of the energy density.

Quantum fluctuations on expanding vacuum bubbles have been studied in Refs. [11]. For a bubble which is centered at the origin of coordinates, the perturbed worldsheet $\tilde{X}^A$ is conveniently parametrized as

$$\tilde{X}^A(\xi^\mu) = X^A(\xi^\mu) + N^A(\xi^\mu)\chi(\xi^\mu) = (1 + \chi/R_0)X^A.$$

Here, N^A is the unit normal to the unperturbed worldsheet of the bubble, which has the internal geometry of a 4D de Sitter space of radius R_0, ξ^μ are a set of coordinates in this space and X^A stands for the unperturbed worldsheet (2). It is known [11] that the normal displacement χ behaves like a worldsheet scalar field with the tachyonic mass $m^2 = -4R_0^{-2}$, which obeys an equation of the form

$$-\Box\chi - 4R_0^{-2}\chi = 0. \tag{6}$$

Here, $\Box$ is the covariant d'Alembertian in a 4D de Sitter space. We must also consider the second bubble, whose center is at a distance $2d$ from the first. For perturbations which are symmetric with respect to the plane $W = d$, the collision will still take place on this plane, the second bubble being just a mirror image of the first. In what follows, we shall restrict our attention to such "Z_2- symmetric"perturbations, since it can be shown that the antisymmetric mode does not contribute to cosmological perturbations to linear order.

An important observation is that, by conservation of momentum, the fluid lines will be orthogonal to the perturbed surface of collision, which is therefore a co-moving, or velocity orthogonal surface[12]. It is known (see next section) that the *growing* mode of the curvature perturbation $\mathcal{R}_c$ on co-moving surfaces is a constant of motion on scales larger than the Hubble rate, and hence $\mathcal{R}_c$ will be a quantity of interest to us. In the present case, however, it will be equally important to consider the *decaying* mode, since its initial amplitude is comparable to that of the growing mode. Therefore in order to characterize the initial perturbation we will also need the density perturbation δ_c on the same initial surface.

3.1. Initial value of $\mathcal{R}_c$

Due to local shifts δt in the time of collision caused by fluctuations of the bubble shape, the Big Bang surface will no longer be the smooth hyperboloid (3). As explained in the preceeding paragraphs, we can restrict attention to perturbations which are symmetric with respect to the plane $W = d$. Then, to linear order in δt, the metric induced on the Big Bang surface is easily obtained from (4), and it is given by

$$^{(3)}ds^2 = \left(1 + 2\frac{\delta t}{t}\right) t^2 d\Omega_H^2.$$

Since we live in an almost flat universe, we shall be interested in lengthscales which are short compared to the curvature scale of the unperturbed surface. On these scales $d\Omega^H$ can be replaced with a flat metric. Then, the intrinsic curvature $^{(3)}R$ of the perturbed surface is easily found to be

$$^{(3)}R = \frac{4}{t^2}\Delta\mathcal{R}_c = \frac{4}{t^2}\Delta\frac{\delta t}{t},$$

where Δ is the co-moving Laplacian. The first equality is just the conventional definition of the curvature perturbation used by most authors [16]. In a flat FRW with arbitrary scale factor and in an arbitrary gauge this definition takes the form

$$^{(3)}R \equiv \frac{4}{a^2}\Delta\mathcal{R}, \tag{7}$$

where now $^{(3)}R$ is the perturbation of the intrinsic curvature scalar in the constant time surfaces of the corresponding gauge.

To proceed, we must find δt as a function of the normal displacement χ. Since all points on the unperturbed hyperboloid are equivalent, we may just consider the vicinity of the point $\vec{X} = 0$, where $t = T$. It should be noted that the field χ lives on the unperturbed bubble, which hits the collision plane at some value β_d of the unperturbed boost parameter determined by $d = R_0 \cosh \beta_d$. However, the actual collision does not take place at that time, but at the time when the perturbed bubble reaches $W = d$, at some value β_c of the unperturbed boost parameter given by $d = (R_0 + \chi)\cosh\beta_c$. Thus, the shift in the time of collision is given by

$$\begin{aligned} \delta t = \delta T &= \tilde{T} - T \\ &= (R_0 + \chi)\sinh\beta_c - R_0 \sinh\beta_d \approx -\frac{\chi}{\sinh\beta_d}. \end{aligned}$$

We conclude that the curvature perturbation on the initial co-moving surface is given by

$$\mathcal{R}_{c(i)} = \frac{\delta T}{T} = -\frac{\chi}{R_0 \sinh^2\beta_d}. \tag{8}$$

Let us now estimate the size of this effect. As mentioned above, χ satisfies the equation for a scalar field on the worldsheet de Sitter space, Eq. (6). The corresponding canonical field ϕ with dimensions of mass is related to χ by [11] $\phi = \sigma^{1/2}\chi$, where σ is the tension of the wall. As is well known, when a given mode of a nearly massless scalar field crosses the de Sitter horizon R_0^{-1}, it "freezes" with some amplitude $\phi_k \sim R_0^{-1}$. However, in our case this amplitude does not stay constant. Due to the tachyonic mass $m^2 = -4H^2$ in (6), a mode with wavenumber k grows proportionally to e^{β} as

$$\chi_k(\beta) \sim \sigma^{-1/2} R_0^{-1} e^{\beta - \beta_k}, \tag{9}$$

where β_k is the value of the time-like boost parameter [see Eq. (2)] at which the physical wavelength of the mode $k^{-1}R_0 e^{\beta}$ becomes larger than the inverse expansion rate on the bubble, R_0. Here we adopt the convention that the co-moving curvature scale corresponds to $k \sim 1$. It is easy to see that at the time of collision we have $\exp(\beta_d - \beta_k) \sim k^{-1}(d/R_0)$, and therefore,

$$\chi_k(\beta_c) \sim \sigma^{-1/2} \frac{d}{R_0^2} k^{-1}.$$

Using $\exp \beta_d \approx d/R_0$ in Eq. (8) we find that by order of magnitude,

$$\mathcal{R}_{c(i)}(k) \sim \left(\frac{1}{S_E}\right)^{1/2} \frac{R_0}{d} k^{-1}. \tag{10}$$

Here $S_E \sim \sigma R_0^4 \gg 1$ is the Euclidean action of the instanton describing the nucleation of the bubble. It can be shown that this perturbation is rather minute even on scales comparable to the curvature scale $k \sim 1$, where it falls short from the observed value by some 20 orders of magnitude [6].

3.2. Initial value of δ_c

To evolve the cosmological perturbations, two initial conditions are needed for each wavelength. In addition to the initial curvature $\mathcal{R}_c$ we will also need the density perturbation δ_c. By assumption, a fixed fraction of the energy of the collision goes into the brane, and therefore the density at the moment of collision is proportional to the Lorentz factor $\gamma = \cosh \tilde{\beta}$. Since the boost parameter is additive, the perturbed one will be given by $\tilde{\beta} = \beta_c + \dot{\chi}$, where

$$\dot{\chi} = \frac{1}{R_0} \frac{\partial \chi}{\partial \beta}$$

is the derivative of the perturbation with respect to the proper time $\tau = R_0 \beta$ measured by an observer on the bubble. Thus, the change in the boost parameter is given to linear order by $\delta\beta = \tilde{\beta} - \beta_d = \dot{\chi} - (\chi/R_0)\coth\beta_d$. With these relations, we obtain the density perturbation at the moment of collision as

$$\delta_{c(i)} = \tanh\beta_d \delta\beta = \dot{\chi}\tanh\beta_d - \frac{\chi}{R_0}. \tag{11}$$

As noted above, in the case of our interest, the r.m.s. fluctuation in χ grows exponentially fast with β. However, the combination which enters δ_c anihilates the leading term in this exponential dependence, and leaves only a contribution which decays with β.

To extract the spectral index, we need to look at the detailed form of the mode functions. For definiteness, we shall use the open chart on the de Sitter worldsheet of the bubble. The quantum state of perturbations on a nucleating bubble is uniquely determined by de Sitter invariance [13]. It is given by the so called Bunch-Davies vacuum (also known as Euclidean vacuum). The corresponding positive frequency modes in the open chart have been studied by [14, 15], and for a scalar field with mass $m^2 = -4R_0^2$ they are given by $\chi_k(\beta) \propto P_2^{ik}(\cosh\beta)/\sinh\beta$, where P_2^{ik} are the Legendre functions with the branch cut from $-\infty$ to 1 on the real axis. In the above expression we have ignored a contribution proportional to the decaying mode, which is accompanied by the factor $e^{-\pi k}$ and which is therefore irrelevant at wavelenths much shorter than the curvature scale. Expanding the Legendre function for large β, we have

$$\chi_k(\beta) \sim \frac{\sigma^{-1/2}R_0^{-1}}{\sinh\beta_k}\left[\sinh\beta + \frac{k^2+4}{6\sinh\beta} + O(e^{-3\beta})\right]. \tag{12}$$

Here, the normalization is fixed as in Eq.(9), the only difference being that we have been careful to keep more terms in the expansion because the leading one clearly does not contribute to (11). Substituting (12) in (11) and using $\exp\beta \approx d/R_0$ we have

$$\delta_{c(i)} \sim \left(\frac{1}{S_E}\right)^{1/2}\frac{R_0}{d}k. \tag{13}$$

For $k \sim 1$, this is of the same order as the curvature perturbation given in (10), where both are very small. However, (13) can be much larger on small scales. Unfortunately, this is due to a strong tilt in the power spectrum, corresponding to $n_s = 3$ in the standard notation.

4. Evolution of the perturbations

Eqs. (8) and (11) provide the initial conditions (position and momentum, as it were) for the evolution of cosmological perturbations in the FRW phase.

The evolution equations for the perturbations in the co-moving gauge are given by [6, 16]

$$\frac{d\mathcal{R}_c}{dN} = -c_s^2 \frac{\delta_c}{(1+w)}, \tag{14}$$

$$\frac{\mathcal{H}}{a^2}\frac{d}{dN}(a^2\mathcal{H}\delta_c) = k^2(1+w)\mathcal{R}_c, \tag{15}$$

Here, we have introduced the conformal Hubble rate $\mathcal{H} = a'/a$, where prime denotes a derivative with respect to the conformal time η, defined by $a(\eta)d\eta = dt$. In the above equation, $w = P/\rho$ is the parameter characterizing the equation of state of the fluid, $c_s^2 = \delta P/\delta\rho$ is the sound speed, and $dN = \mathcal{H}d\eta$.

To discuss the evolution of perturbations it is useful to formally integrate the second equation,

$$\begin{aligned} \delta_c &= \frac{1}{a^2\mathcal{H}}\int_{N_i}^{N}(a^2\mathcal{H})\frac{k^2}{\mathcal{H}^2}(1+w)\mathcal{R}_c dN \\ &+\frac{1}{a^2\mathcal{H}}(a^2\mathcal{H}\delta_c)_{(i)}. \end{aligned} \tag{16}$$

The function $a^2/\mathcal{H}$ is an increasing function, so provided that $\mathcal{R}_c$ stays approximately constant at late times (which will be a self-consistent assumption), the integral in the right hand side is dominated by the contribution from the neighborhood of the upper boundary of integration. If we consider a simple case in which $c_s^2 = w = $const., then $a = (\eta/\eta_0)^{2/1+3w}$ and $\mathcal{H} = (2/1+3w)\eta^{-1}$. Performing the integral in (16) we have

$$\delta_c \approx \frac{2+2w}{5+3w}\frac{k^2}{\mathcal{H}^2}\mathcal{R}_c + \frac{1}{a^2\mathcal{H}}(a^2\mathcal{H}\delta_c)_{(i)}, \tag{17}$$

where we have assumed $\mathcal{R}_c \approx const.$ at late times. Initially, the first term in the right hand side is suppressed with respect to the second at least by a factor of $(k^2/\mathcal{H}^2) \ll 1$. Substituting the dominant part into Eq.(14), we have

$$\mathcal{R}_{c(f)} \approx \mathcal{R}_{c(i)} - (a^2\mathcal{H}\delta)_{(i)}\int_{\eta_i}^{\infty}\frac{d\eta}{a^2}\frac{c_s^2}{1+w}. \tag{18}$$

Then, this integration is performed to obtain

$$\mathcal{R}_{c(f)} \approx \mathcal{R}_{c(i)} - \frac{2w}{3(1-w^2)}\delta_{c(i)}. \tag{19}$$

In obtaining (19) we have neglected the contribution from the first term in the right hand side of (17). It is easy to check that this term does not have any effect until the wavelength of the mode is comparable to the hubble radius $k \sim \mathcal{H}$. At that time, the second term in (17) is unimportant, and the amplitude of density perturbations at horizon crossing can be read off from the first term

$$\delta_c|_{hc} \approx \frac{2+2w}{5+3w}\frac{k^2}{\mathcal{H}^2}\mathcal{R}_{c(f)}. \tag{20}$$

Substituting (19) into the previous equation and using (10) and (13), we have

$$\delta_c|_{hc} \sim \mathcal{R}_{c(f)} \sim \frac{R_0}{d}S_E^{-1/2}k,$$

which of course displays the same strong spectral tilt as the initial density perturbation $\delta_{c(i)}$.

5. conclusions

In Bucher's scenario, two bubbles nucleate in the 5-dimensional Minkowski bulk, and their collision forms a brane where gravity is localized. This would solve the homogeneity and flatness problems, provided that the separation $2d$ of the bubble nucleation points is sufficiently large compared with the bubble radius R_0.

We have evaluated the initial spectrum of scalar density perturbations, assuming that they originate from quantum fluctuations on the bubbles. The initial value of the density contrast $\delta_{c(i)}$ turns out to have a very steep spectrum corresponding to $n_s = 3$, while the spectral index for the curvature perturbation $\mathcal{R}_{c(i)}$ is $n_s = -1$. The amplitude of both perturbations is comparable at wavelengths of the order of the curvature scale, and therefore $\delta_{c(i)}$ has a larger amplitude on scales relevant to present observations. Given the initial values, we have evolved the perturbations to find that the final value of $\mathcal{R}_c$ (at the time of horizon crossing) becomes comparable to the initial value of the density perturbation $\delta_{c(i)}$ [See Eq. (18)]. This final value of $\mathcal{R}_c$ also gives the order of magnitude of the density contrast at horizon crossing. Thus, we conclude that the relevant spectral index is $n_s = 3$, which is definitely too steep to explain observations. Hence, some other mechanism must be invoked in this model for the generation of cosmological perturbations. We should perhaps stress that here we have restricted attention to vacuum bubbles whose boundary is described by the Nambu action. It would be interesting to investigate if the nature of the spectrum of fluctuations improves if we consider, say, an induced Einstein term on the bubble wall.

Although here we have focussed on a particular realization of the brane Big Bang, we hope to have learned a few lessons which may be useful in more general cases [5, 4], whose study is likely to provide some interesting results in the near future.

acknowledgements

I am grateful to Martin Bucher for very useful early discussions on the subject of cosmological perturbations in bubble collisions, and for sharing his preliminary notes with me prior to publication. I would also like to thank the organizers of the RESCEU meeting for their generous hospitality. This work is partially supported by CICYT under grant AEN98-0431.

References

[1] M. Bucher, hep-th/0107148.
[2] L. Randall and R. Sundrum, Phys. Rev. Lett. **83**, 3370 (1999); Phys. Rev. Lett. **83**, 4690 (1999).
[3] G. Dvali, G. Gabadadze and M. Porrati, Phys. Lett. B485 (2000) 208.
[4] J. Khoury, B.A. Ovrut, P.J. Steinhardt, and N. Turok, hep-th/0103239; R. Kallosh, L. Kofman, and A. Linde, hep-th/0104073; J. Khoury, B.A. Ovrut, N. Seiberg, P.J. Steinhardt, and N. Turok, hep-th/0108187;
[5] U. Gen, A. Ishibashi, and T. Tanaka, hep-th/0110286.
[6] J. Garriga and T. Tanaka, hep-th/0112028.
[7] J. Blanco-Pillado and M. Bucher, hep-th/0111089.
[8] S.Coleman and F. De Luccia, Phys. Rev. **D21**, 3305 (1980).
[9] T. Shiromizu, K. Maeda and M. Sasaki, Phys. Rev. D **62**, 024012 (2000); P. Binetruy, C. Deffayet, U. Ellwanger and D. Langlois, Phys. Lett. B **477**, 285 (2000); E. E. Flanagan, S. H. Tye and I. Wasserman, Phys. Rev. D **62**, 024011 (2000); J. Garriga and T. Tanaka, Phys. Rev. Lett. **84**, 2778 (2000); J. Garriga and M. Sasaki, Phys. Rev. D62 (2000) 043523.
[10] P. Bowcock, C. Charmousis, R. Gregory, Class. Quant. Grav. 17 (2000) 4745.
[11] J. Garriga and A. Vilenkin, Phys. Rev. D **45**, 3469 (1992); J. Garriga, Phys. Rev. D49 (1994) 6327.
[12] D. Langlois, K. Maeda and D. Wands, gr-qc/0111013.
[13] B. Allen, Phys. Rev. D **32**, 3136 (1985); T. Vachaspati and A. Vilenkin, Phys. Rev. D **42**, 3401 (1990).
[14] M. Sasaki, T. Tanaka, and K. Yamamoto, Phys. Rev. D **51**, 2979 (1995); K. Yamamoto, M. Sasaki, and T. Tanaka, Astrophys. J. **455**, 412 (1995).
[15] M. Bucher, A.S. Goldhaber, and N. Turok, Nucl. Phys. B **43**, 173 (1995).
[16] For review of the standard cosmological perturbation see, e.g. H. Kodama and M. Sasaki, Prog. Theor. Phys. Suppl. **78** 1 (1984); V.F. Mukhanov, H.A. Feldmann and R.H. Brandenberger, Phys. Rep. **115**, 203 (1992).

Why Does Inflation Start at The Top of The Hill?

S. W. Hawking
DAMTP, Center for Mathematical Science, University of Cambridge, Wilberforce Road, Cambridge, CB3 0WA, UK

Inflation is now generally accepted as the best explanation, for the isotropy and flatness of the universe. Its prediction of a nearly scale free spectrum of fluctuations, has been spectacularly confirmed by recent observations of the microwave background. The appearance of Doppler peaks at small angular scales, shows the fluctuations have correlations beyond the particle horizon, of the hot big bang model. This rules out local processes, such as cosmic strings, because they would produce correlations only within the particle horizon. On the other hand, a period of exponential expansion in the early universe, would remove the particle horizons, and allow causal contact between regions that are widely separated now. The only alternative explanation of the Doppler peaks, would be to suppose there were long range correlations in the big bang, established during some pre big bang scenario, like the Ekpyrotic universe. I will leave it to others, to point out the flaws in the Ekpyrotic scenario. One problem is that one is chasing a moving target, because the theory seems to change every week. Nevertheless, I don't expect the Ekpyrotic universe will survive like the theory of inflation, which has been around twenty years.

There have been various proposals as to how inflation started, and ended. They all involve a scalar field, φ, with a potential, V.

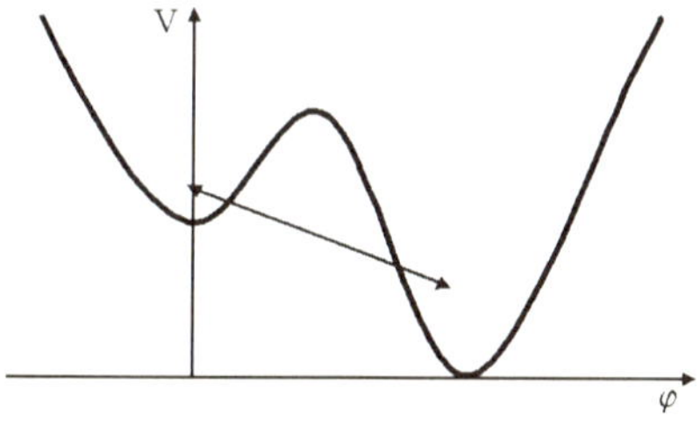

Fig. 1. Old Inflation

In the old inflation scenario, the universe was supposed to start with infinite temperature, at the big bang singularity. As the universe expanded and cooled, phi was supposed to be trapped in a local minimum of V. The energy momentum tensor of this constant phi field, would act like a cosmological constant, and cause the universe to expand exponentially. However, the local minimum, would be only a meta-stable false vacuum. It should decay to the true vacuum at the global minimum of V, through a Coleman De Lucia instanton.

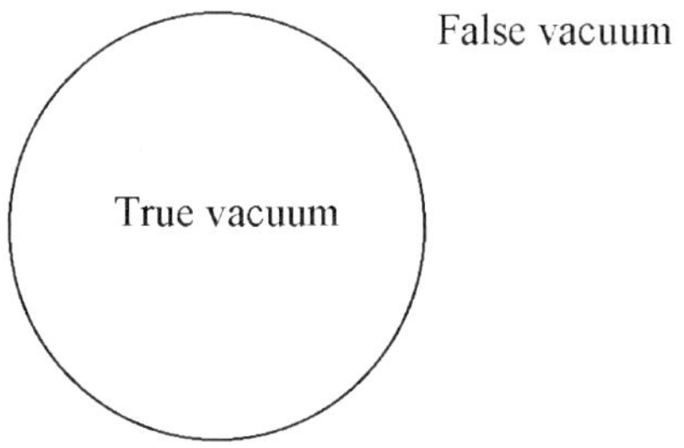

Fig. 2. Coleman-de Lucia Instanton

This consists of a bubble of the true vacuum, surrounded by the false vacuum. The trouble was, if the bubble formation rate was high, the false vacuum would decay before there had been enough inflation. On the other hand, if the bubble formation rate was low, the bubbles would appear so far apart, that the bubbles would never meet up and complete the decay of the false vacuum.

To overcome this difficulty, the new inflation scenario was proposed, I won't

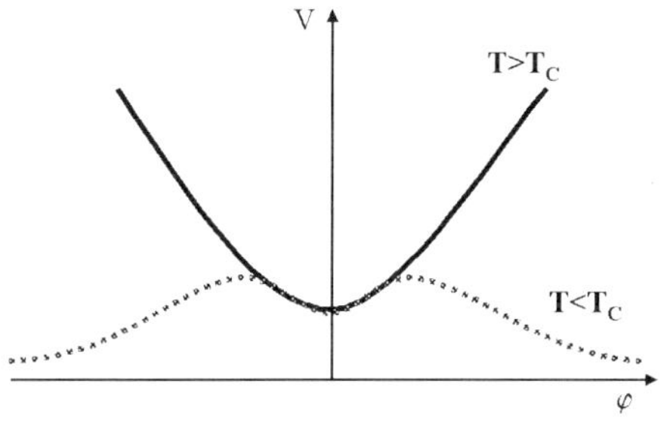

Fig. 3. New Inflation

say by whom, because that was a sore point, though I hope that is now a thing of the past. The idea was that the effective potential, V, would depend on temperature, and therefore on time in the early universe. It was still thought that the scalar field would be trapped in a local minimum. However, it was now supposed that the height of the barrier around the local minimum, would decrease, as the universe cooled. Eventually, the scalar field would quantum tunnel through the barrier, and would then run down the hill to the global minimum of the potential. If the hill was not too steep, the scalar field would roll down the hill slowly in the interior of the bubble. The interior would expand almost exponentially, by a large factor. Inflation would take place in a single bubble, and there was no need for different bubbles to join up, as in old inflation.

This was the new inflation proposal of Andrei Linde, which I heard when I saw him in Moscow in 1981. I realized there was something wrong.a bubble of the true vacuum, would have been bigger than the Hubble radius, which was not possible. Instead, ee'an Moss and I, showed that the scalar field would stay constant in space.

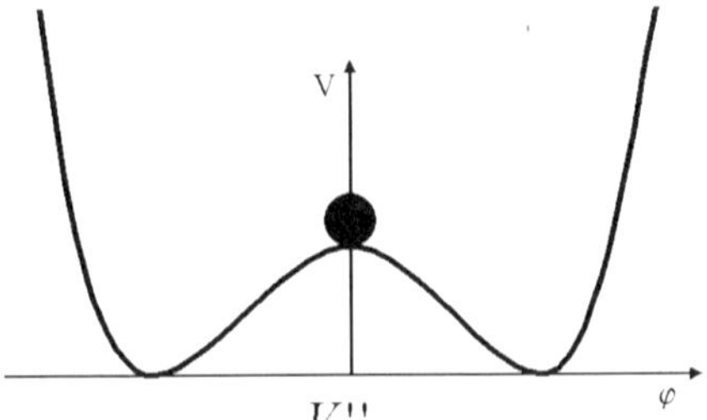

$$0 > \frac{V''}{H^2} > -4$$

Fig. 4. Hawking-Moss Instaton

It would jump to the top of the barrier between the false vacuum, and the true. The top is the barrier, is an equilibrium position, but it is an unstable equilibrium. The scalar field would not remain sitting at the top of the hill,, but would run down. Again, provided the hill was not too steep, there would be enough inflation to explain the isotropy and flatness of the universe.

It might seem surprising that the scalar field, should jump to the top of the barrier everywhere. The reason is that because of the horizon, de Sitter effectively has a finite spatial size. This means that a constant scalar field, is the tunneling configuration with least energy, if the second derivative of the potential, is fairly low.

New inflation fell out of favor for various reasons. One was that although inflation is supposed to explain the initial conditions of the hot big bang model, it needed a fairly isotropic and flat hot big bang to start it off. Another was that thermal fluctuations would have been large, and would have knocked the scalar field off the top of the hill. However, work I have done with Thomas Hertog at Cambridge, indicates that the scalar field can still start naturally at the top of the hill, even without an early hot big bang stage.

The usual approach to inflation, is to assume some initial condition for the universe, and evolve it forward in time. This could be described as the bottom up approach to cosmology. It is an essentially classical picture, because it assumes there is a single well defined metric for the universe. By contrast, Hertog and I adopted a quantum approach, based on the no boundary proposal. This says

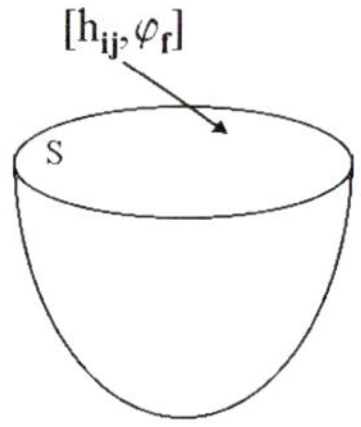

$$\Psi\left[h_{ij},\varphi_f,S\right]\approx \int_{h_{ij},\varphi_f} d[g]d[\varphi]e^{-S_E[g,\varphi]}$$

Fig. 5. No Boundary Proposal

that the amplitude for an observable like the three metric on S, is given by a path integral over all metrics whose only boundary is S. It leads to a top down view of the universe. The histories that contribute to the path integral, depend on the observable being measured. There will be an amplitude for empty flat space, but it is not of much significance. By the weak anthropic principle, we are interested only in quantum states of the universe, that are compatible with life.

The full conditions for the development of life, will be very complicated. It seems to take a long time to evolve intelligence. In fact, it is not clear we have achieved it yet. Thus a necessary condition for intelligent life, would be that the universe is expanding at something like the present rate. One could therefore ask for amplitudes for different conformal three geometries, on a surface with trace K, equal to the present Hubble constant. However, the heating phase after the end of inflation, would have been chaotic, and impossible to followed in detail. We therefore considered the quantum amplitudes, on an expanding surface in the inflationary period, and then evolved to the future using classical laws.

We considered a model in which the potential, had a local maximum at $\varphi = 0$. The second derivative at the maximum is assumed to be sufficiently low, that there is not a Coleman De Lucia solution that straddles the maximum, but

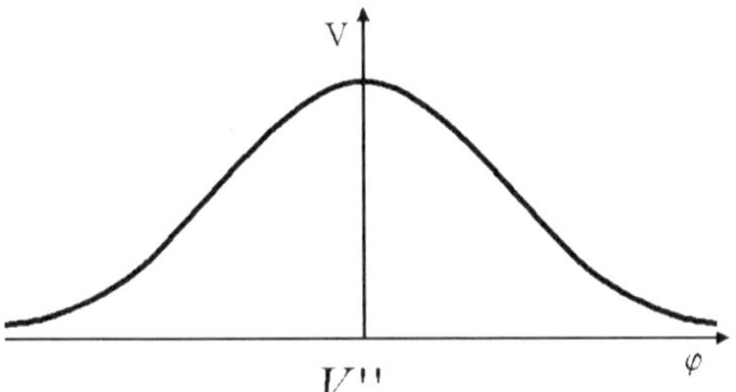

$$0 > \frac{V''}{H^2} > -4$$

Fig. 6.

only a constant Hawking Moss instanton, that sits on top of the hill. One now calculates the amplitude, or wave function of the universe, as a function of k, the second fundamental form, and φ on the surface. We first calculate the wave function for real Euclidean k, and then analytically continue to imaginary, or Lorentzian k. For small φ, and any Euclidean k, there will be a Euclidean O(4) symmetric solution of the field equations, with the given boundary conditions. This solution will be part of a deformed sphere, or Hawking Turok instanton. In it, the scalar field will run up hill from the non singular south pole, to the prescribed. For $k = 0$, the action of this Hawking Turok instanton, will be more

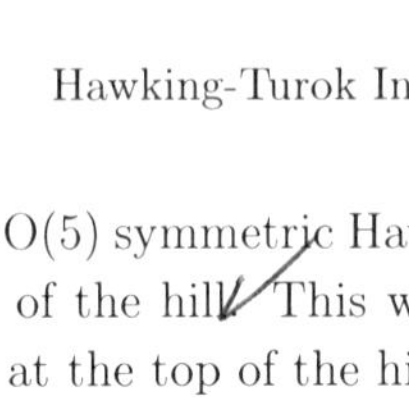

Fig. 7. Hawking-Turok Instanton

negative than the action of the O(5) symmetric Hawking Moss instanton, in which the scalar field sits at the top of the hill. This would seem to suggest that the universe is least likely to start at the top of the hill.

However, one is not interested in a Euclidean spacetime, but in a Lorentzian expanding universe. In this, Euclidean k will be pure imaginary. One therefore has to analytically continue to the imaginary k axis. In the O(5) Hawking Moss case, the real part of the action, is constant on the imaginary axis. However, a perturbation calculation we have done, indicates that for the O(4) case, the real

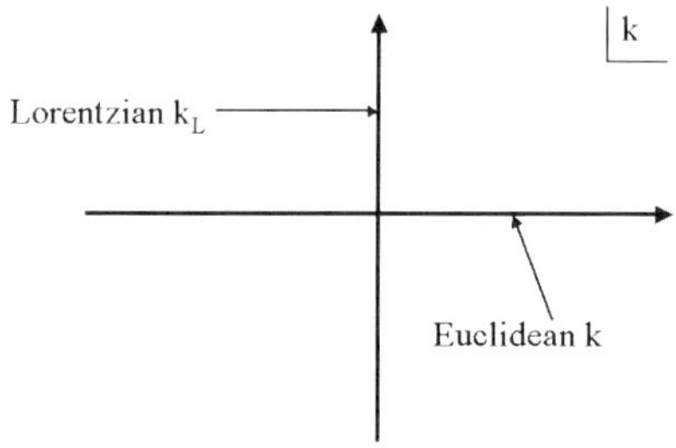

Fig. 8. Complex k-plane

part of the action, increases on the imaginary axis away from $k = 0$, and becomes more than the O(5) action. This indicates that in an inflationary universe, the scalar field is more likely to start at the top of the hill, and roll down, than to start lower down. The reason is that although being at the top of the hill, costs potential energy, it saves gradient energy, by having a scalar field that is constant in space and time. If the top of the hill is fairly flat, the gradient energy is dominant, and the universe starts with a constant scalar, at the top of the hill.

Hertog and I did this work, not to revive the new inflation scenario, but because we are interested in trace anomaly driven inflation. This was the first accelerated expansion model to be proposed, pre-dating even the use of the term, inflation. It is based on the large N approximation.

The standard model of particle physics, contains nearly a hundred fields. If as we suspect, the standard model is embedded in a super symmetric theory, the number of fields would be at least double, and maybe very much higher. Thus the large N approximation should hold in cosmology, even at the origin of the universe.

In the large N approximation, one performs the path integral over the matter fields in a given background, to obtain an effective action, which is a functional of the background metric.

$$S = -\frac{1}{16\pi G}\int d^4x\sqrt{g}(R + W[\mathrm{g}])$$
$$W[\mathrm{g}] = \int d[\phi]\exp(-S_m[\phi, \mathrm{g}])$$

One then argues that the effect of gravitational fluctuations are small, in comparison to the large number of matter fluctuations. Thus one can neglect graviton loops, and look for a stationary point of the combined gravitational action, and the effective action for the matter fields.

Einstein Equations

$$R_{ij} - \frac{1}{2} R g_{ij} = 8\pi G \langle T_{ij} \rangle$$

This is equivalent to solving the Einstein equations, with the source being the expectation value of the matter energy momentum tensor. Finally, one can calculate linearized fluctuations about this stationary point metric, and check they are small. This is confirmed observationally, by measurements of the cosmic microwave background, which indicate that the primordial metric fluctuations, were of the order of ten to the minus five.

The large N approximation was first applied to cosmology in the 70s, particularly by the Russians. One of the main motivations, was to obtain a model of the universe, without an initial singularity. Instead, Grishcuk and Zeldovitch proposed that the universe was in a de Sitter phase for an infinite time, before exiting to a decelerating expansion. This model was developed in more detail by Starobinski. He assumed there where a large number of conformally invariant matter fields, which would give the energy momentum tensor a trace anomaly, that was a known function of the local curvature.

Trace Anomaly

$$\langle T \rangle = aF - cG + d\nabla^2 R$$

$$\begin{aligned}
F &= C_{ijkl}C^{ijkl} = \text{Weyl squared} \\
G &= R_{ijkl}R^{ijkl} - 4R_{ij}R^{ij} + R^2 \propto \text{Euler density} \\
a &= \frac{1}{120(4\pi)^2}(N_S + 6N_F + 12N_V) \\
c &= \frac{1}{360(4\pi)^2}(N_S + 11N_F + 62N_V) \\
d &= \frac{1}{180(4\pi)^2}(N_S + 6N_F - 18N_V)
\end{aligned}$$

$$N_S = \text{number of real scalars}$$
$$N_F = \text{number of Dirac fermions}$$
$$N_V = \text{number of vectors}$$

In a de Sitter background, the trace free part of the energy momentum tensor would be zero, by symmetry. Thus the energy momentum tensor, would be proportional to the metric, and de Sitter space would be a stationary point of the combined action. However, in order to get the universe to exit the de Sitter phase, Starobinski had to assume there were also non conformally invariant matter fields, that added a non conformally invariant local term to the effective action.

I must admit I didn't take this Russian model very seriously at the time. This was before we realized the importance of exponential expansion or inflation, in solving the fine tuning problems of the Hot Big Bang, like horizons and space curvature. Also, why should the universe have expanded for an infinite time in a de Sitter phase, before becoming unstable, and exiting inflation. What was the clock, that told the instability to turn on. However, we can now recognize this initial de Sitter phase, as corresponding to the quantum creation of the universe from nothing, via an instanton which was the Euclidean four sphere.

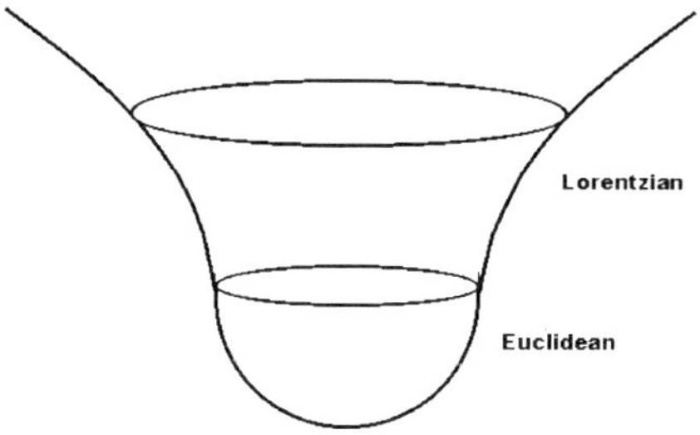

Fig. 9. Quantum Creation of the Universe

Moreover, the AdS/CFT correspondence, now provides us with a way of calculating the effective action of matter fields on backgrounds without symmetry. This was not available in the early days, so Starobinski had to neglect non local terms in the effective action. I shall therefore re-apraise the Starobinski model, in the light of modern knowledge. My talk will be based on joint work with Harvey re-al and Thomas hertog.

The AdS/CFT prescription for calculating the effective action on a background metric, is performed in Euclidean space, like all good quantum field theory calculations.

One takes the four dimensional metric, to be the boundary of a solution of the Einstein equations, with a negative cosmological constant, in five dimensions. One takes the action of this solution, adds counter terms that depend on the geometry of the boundary, and takes the limit that the AdS length scale, and the five dimensional Newton's constant go to zero.

One can now calculate the combined four dimensional gravitational, and matter field effective actions. The only O(4) stationary point metrics, are flat space, and the four sphere. The latter can be regarded as the Euclidean version

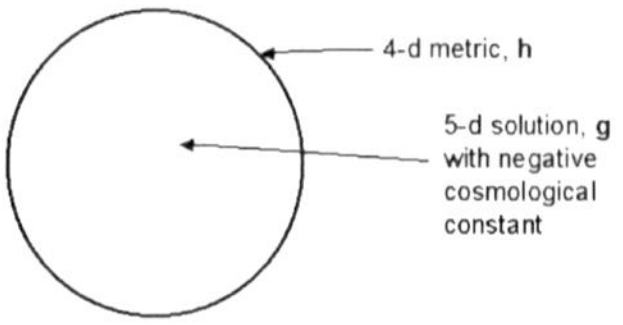

$$W[\mathbf{h}] = -\frac{1}{16\pi G_5}\int d^5x\sqrt{g}\left(R+\frac{12}{l^2}\right) + \text{surface counter terms}$$

Fig. 10. AdS/CFT

of a de Sitter space, where the cosmological constant is provided by the trace anomaly, of a large N conformal field theory. These two solutions are the final and initial stages, of an inflationary model. However, in order to get a solution that interpolates between the two, one has to add an R squared term to the gravitational action, as Starobinski discovered. This can be justified as the local counter term in the effective action, of non conformal invariant fields. With this addition, the expansion changes from exponential, to matter dominated, in a time scale that depends on the coefficient of the R squared term.

There are thus seven contributions to the combined gravitational and effective matter action:

$$S = -\frac{1}{16\pi G}\int d^4x\sqrt{h}R + W[\mathbf{h}]$$

where

$$\begin{aligned} W[\mathbf{h}] &= -\frac{1}{16\pi G_5}\int d^5x\sqrt{g}\left(R_5+\frac{12}{l^2}\right) - \frac{1}{8\pi G_5}\int d^4x\sqrt{h}K \\ &+\frac{3}{8\pi G_5 l}\int d^4x\sqrt{h} + \frac{1}{32\pi G_5}\int d^4x\sqrt{h}R \\ &-\frac{N}{32\pi^2}\left(\log\frac{l}{r}+\beta\right)\int d^4x\sqrt{h}\left(R_{ij}R^{ij}-\frac{1}{3}R^2\right) + \frac{-\alpha N}{192\pi^2}\int d^4x\sqrt{h}R^2 \end{aligned}$$

and

$$\frac{l^3}{G^5} = \frac{2N}{\pi}$$

The combined action is stationary under all perturbations, h_{ij}, of the metric of the boundary, if the boundary is a four sphere with radius, r, of order N to the half. This corresponds to a de Sitter solution, where the cosmological constant, is provided by the trace anomaly of large numbers of matter fields. For large N, the radius of the de Sitter space, will be large in Planck units. Thus gravitational fluctuations will be small, confirming the consistency of the large N approximation.

One can decompose the perturbations in the bulk and boundary, into harmonics under the isometry group, O(5), of the Euclidean de Sitter solution. The harmonics can be divided into scalar, transverse vector, and transverse trace less tensor. The vector harmonic perturbations are pure gauge, so they don't affect the action. The gauge freedom leads to closed loops of Fadeev Popov ghosts, but they can be neglected in the large N approximation.

The scalar part of the perturbed combined action, is

$$S = \frac{-3\alpha}{4NG^2} \int d^4x \sqrt{\gamma_{ij}}\varphi \left(\frac{1}{2\alpha} + \nabla^2\right) \left(\nabla^2 + 4\right) \varphi$$

where γ_{ij} and ∇ are the metric and connection on a unit four sphere. One can think of the non derivative terms in φ, as the potential, V.

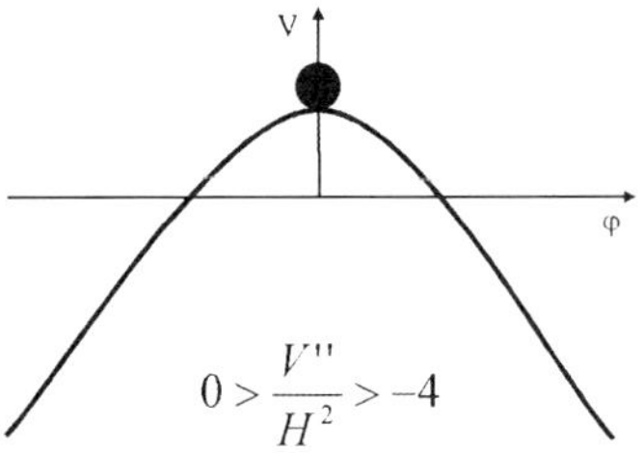

Fig. 11. Hawking-Moss Instanton

The unperturbed de Sitter solution at $\varphi = 0$, is at the top of the hill. If alpha is greater than 2, the top of the hill is flat enough, that the universe is most likely to start at the top. This justifies Starobinskis trace anomaly model, and explains why the universe starts in an unstable de Sitter state. The growth time of the instability, will be 12 α times the radius of the Instability of De Sitter four sphere:

$$\varphi \sim \exp\left(\frac{t}{12 \mid \alpha \mid r}\right)$$

Since α is a counter term, there is no reason it should not be quite large. If it is less than ten, there will be enough inflation to solve the flatness problem. A large alpha, will also give scalar fluctuations in the microwave background that are small.

One of the big objections to trace anomaly inflation, has been that it has ghosts, and classroom and quantum instabilities. Ghost poles in the propagator,

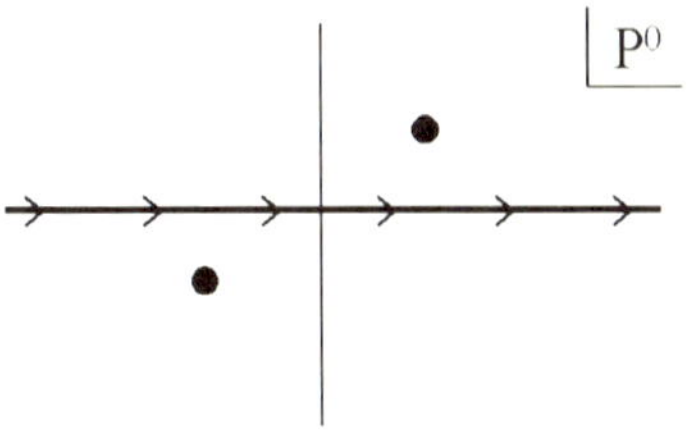

Fig. 12. Complex Poles

are normally taken to be a fatal flaw in a field theory. The reason is that they seem to indicate, that one could have asymptotic ingoing and outgoing states, which had negative norm. This would mean that the evolution from the initial state, to the final state, is not described by a unitary S matrix. Unitarity is usually taken to be an essential property, of any respectable field theory.

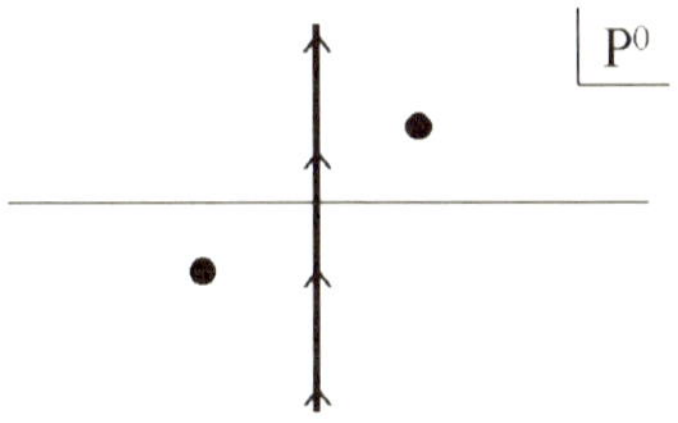

Fig. 13. Complex Poles

Hertog and I have shown however, that one can live with higher derivative ghosts. They just mean that one can not measure all the degrees of freedom, so one doesn't have unitarity. Defining the theory in the Euclidean regime, and then Wick rotating to the Lorentzian, implicitly imposes the final boundary condition, that the fields remain bounded. This removes the instabilities, like with the classical radiation reaction force.

What we can observe, are the fluctuations in the microwave background. These can be calculated in the large N theory, even though it has ghosts. The amplitude of the fluctuations, will be determined in the initial de Sitter stage. If the gravitational action, were all that contributed, the amplitude of the tensor fluctuations, would be of the order of one over horizon size. Since the horizon size will be roughly root N, the tensor fluctuations, would be one over root N. Starobinski visited DAMTP in Cambridge in 1980, and took the opportunity of being outside Soviet scientific Censorship, to publish a remarkable paper in Physics letters. In it, he showed how to calculate tensor fluctuations of inflationary models, before the term, inflation, had even been applied to the early universe. Starobinski concluded that the trace anomaly inflationary model, had to be abandoned, because it would require 10 to the 10 matter fields, to reduce the tensor fluctuations, to the observational limit.

Starobinski assumed that the amplitude of the tensor fluctuations, was not significantly changed by the coupling to the matter effective action. However, this assumption can now be examined, using AdS/CFT. It turns out that matter loops can greatly reduce the fluctuations, so that they can be compatible with the observations, with only a realistic number of matter fields. For example, one can have $N = 10$ to the 4, and $\beta = 10$ to the 3.

The large N trace anomaly inflationary model, requires tuning of the counter terms, alpha and beta, to fit the observations. But then, any other explanation of the microwave fluctuations, also involves tuning. No one can predict an amplitude of 10 to the minus 5, in a natural way. Maybe we have to resort to the anthropic principle.

To summarize, Hertog and I have shown that one doesn't need an initial hot stage, to explain why inflation began at a local maximum of the potential. If the curvature of the potential at the maximum is sufficiently low, a constant Hawking Moss type field at the maximum, will have lower energy than an inhomogeneous Coleman De Lucia type field. This makes it favorable for inflation to start at the top of the hill, in a de Sitter state. Quantum fluctuations will then disturb the scalar field, and cause it to run down the hill. This scenario is realized in the trace anomaly model, in which inflation is driven by the quantum effective action, of a large number of matter fields. The effective action can be calculated using ADS, CFT. In the Euclidean regime, one has a ball of five dimensional ADS, bounded by a four dimensional sphere, representing the de Sitter phase of our universe.

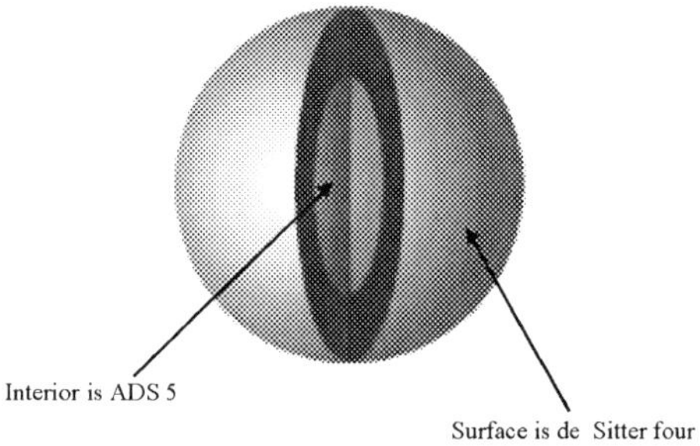

Fig. 14. Universe in a Nutshell

This could well be described as, the universe in a nutshell.

Measurements of The CMB: Current Status and Future Prospects

Lyman Page
Dept. of Physics, Princeton University, Princeton, NJ 08540-0708

Abstract

The study of the Cosmic Microwave Background (CMB) is a rapidly maturing field. In remarkable advances in the past two years, experiments have begun to probe the detailed structure of the CMB angular spectrum thereby providing insight into the contents of the universe. We briefly review the BOOMERanG and DASI results with an emphasis on what the data tell us with only minimal cosmological assumptions. After a short discussion of a grand future for CMB physics, key features of the *MAP* satellite, now in orbit, are described. We also discuss how we anticipate *MAP* will advance the field beyond what we know now.

1. Introduction

This is a very exciting time for those studying the CMB. The decades of work that have gone into developing sophisticated technologies and observational techniques are beginning to pay off. A number of teams now have measured features in the CMB angular spectrum, measured its frequency spectrum, and have deduced cosmological parameters that are in rough accord with what one obtains with other methods based on entirely different physics.

The CMB is a particularly potent probe of cosmology because its characteristics may be computed to high accuracy and it may be measured to high accuracy. As long as the systematic errors associated with the experimental technique are small, the detected signal has direct cosmological import. Potentially confounding emissions from our galaxy and sources between us and the surface of last scattering are small. To first order, such "foregrounds" may be ignored at many angular scales and frequencies[31]. To the extent the foregrounds do affect a result, they may be measured and subtracted: the CMB frequency spectrum is unique.

2. Current Status

The past two years have seen wonderful measurements of the CMB anisotropy. The first peak, the position and amplitude of which were known within the CMB community in the last millennium (e.g., [2, 1, 9, 19, 17]), has now been made manifestly apparent in high signal to noise maps by the BOOMERanG [25] and and MAXIMA teams [15]. The location of this peak is a measure of the angular diameter of the sound horizon when the photons first decoupled from the primordial baryons. The sound horizon gives a physical length scale in the early universe which, from our vantage, is affected primarily by the geometry of space. In a flat geometry, its angular diameter is $\approx 1/2°$; in a closed geometry, it will appear larger and in an open geometry smaller (e.g.,[10, 18, 3]). The location of the peak at $l \approx 215$, as shown in Figure 1, indicates that the geometry of the universe is parsimoniously described as flat.

This is a triumph for physical cosmology. The mechanisms leading to a peak have been known for over thirty years and detailed models of the shape predated all detections of the anisotropy. As the net matter density inferred from the distribution and velocities of galaxies is about 1/3 the critical density, there must be 2/3 the critical density in some new form of energy. This conclusion is independent, though similar, to that drawn from the supernovae results as discussed at this conference by Saul Perlmutter.

Many of the models that predicted the first peak also predict that there should be higher order peaks or "acoustic oscillations." The second peak, at $l \approx 400$, is particularly sensitive to the baryon density. More baryons lead to more inertia in the primordial plasma, in turn leading to relatively more pronounced compression peaks (1st and 3^{ed}) than rarefaction peaks (2nd and 4th). Though there were early indications from BOOMERanG that a second peak might be suppressed[5], a more complete analysis that included much more data[25] showed evidence of it. Independently, the DASI experiment saw the drop off from the first peak and also showed evidence for the second peak. The data from these experiments, along with calibration and beam errors, is shown in Figures 2 & 3.

The significance of the detection of the second peak depends on one's prior assumptions and analysis method[24, 12]. A combination of all the data by Wang, Tegmark, and Zaldarriaga[33] shows clear evidence for it. A non-believer might point to the fact that χ^2/ν for a fit to a flat line for the $l > 400$ data, is 23/11 for BOOMERanG (correlations not published) and is 8.66/4 for DASI (correlations included). In other words, random noise would produce the same fluctuations away from the $l > 400$ mean in 5% of the cases for BOOMERanG (from an analysis with correlations [6]) and in 7% of the cases for DASI. The third peak is

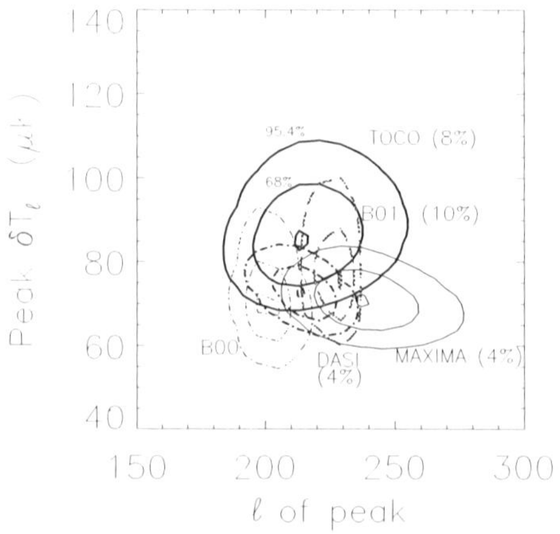

Fig. 1. The position and amplitude of the first peak following the Gaussian temperature method [19]. For TOCO[22] $l_{peak} = 216 \pm 14$ and $\delta T_l = 86 \pm 8$ μK. The preliminary BOOMERanG[5] data give $l_{peak} = 201$ and $\delta T_l = 70$ μK; the reanalyzed and expanded data set [25] gives $l_{peak} = 226$ and $\delta T_l = 77$ μK. (The North American Flight is omitted.) The BOOMERanG beam uncertainty is not accounted for; it will tend to broaden the distribution in l. MAXIMA[21] yields $l_{peak} = 238$ and $\delta T_l = 71$ μK and DASI [14] gives $l_{peak} = 213$ and $\delta T_l = 74$ μK. Calibration uncertainty has been taken into account though the correlations between bands have not. When treated consistently, TOCO, B01, MAXIMA, and DASI pick out values for the peak position and amplitude that are within 2σ of each other. For all experiments the angular spectra for the peak fits are normalized at $l = 10$ to DMR[19]; this is most important for DASI and less important for the other experiments. From Miller *et al.* 2001. [23]

on less firm footing than the second.

Extracting cosmological parameters from the CMB anisotropy started soon after its discovery[29]. The current CMB data give solutions for roughly three to four cosmological eigenparameters (e.g., [3, 13]) with similar results arising from just the BOOMERanG data or from all data combined. An analysis of all data [33] gives $\Omega_b h^2 = 0.02^{+0.06}_{-0.01}$, $\Omega_m h^2 = 0.1^{+0.07}_{-0.06}$, $\Omega_\Lambda = 0.6^{+0.32}_{-0.55}$, $n_s = 0.93^{+0.75}_{-0.16}$ (2σ) though the errors are correlated. It is worth keeping in mind that the "lambda adiabatic CDM" model is assumed for all of the parameter extraction.

3. Future Prospects

The future of CMB observations is bright. The recent successes have given us just a glimpse of what may be learned. The *MAP* satellite, which is discussed in greater detail below, is in orbit and taking data. The Planck satellite, with

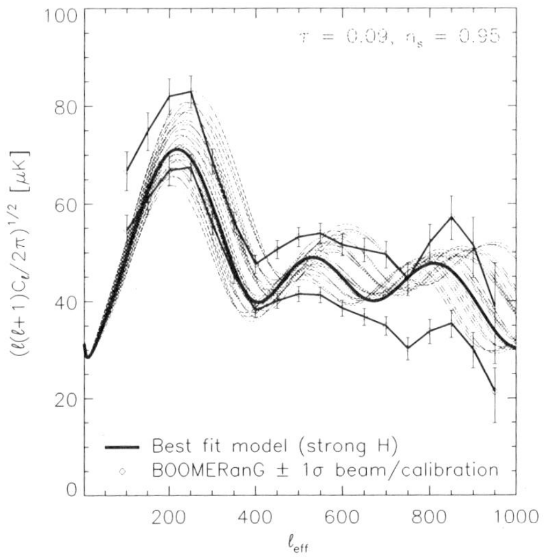

Fig. 2. The BOOMERanG data[25]. The top and bottom curves show the 1σ errors for the beam and calibration uncertainties added in quadrature. Assuming a flat prior ($\Omega = 1$), the best adiabatic model fit to the data is $(\Omega_b h^2, \Omega_c h^2, \Omega_\Lambda, n_s, \tau) = (0.021 \pm 0.003, 0.13 \pm 0.04, 0.57^{+0.12}_{-0.37}, 0.95^{+0.09}_{-0.08}, 0.09^{+0.13}_{-0.07})$, shown in black. The set of models shown in red are derived from the best fit, allowing the parameters to vary by $\pm 1\sigma$. The tilt n_s and depth to reionization τ are kept fixed because their combined action is similar to an overall scale factor. From M. Nolta.

higher resolution and greater sensitivity than *MAP*, is planned for a 2007 launch. The frontiers after the primary anisotropy is mapped out include measuring the polarization and, through the "secondary anisotropies," the onset of cosmic structure formation. A host of balloon and ground based experiments are planned for making the many observations for which space is not necessary and for laying the groundwork for future space missions.

If the currently most favored class of models holds, the polarization in the CMB, through it dependence on gravity waves, will give us information about fundamental fields in the $t < 10^{-20}$ s universe. It may possibly distinguish between such alternatives as inflation and a cyclic universe[30]. As the observable signal is very small, and may be masked by foreground effects, a great deal of work is left to do. Already though, researchers are discussing a dedicated satellite (CMBPOL) for measuring the polarization.

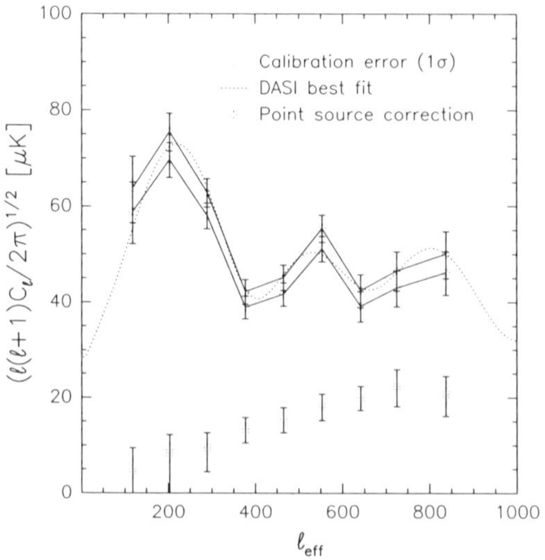

Fig. 3. The DASI data[27]. The upper and lower lines through the data points show the calibration error. The points on the bottom show the point source contribution to the CMB data. They add in quadrature and so are a smaller fraction of the CMB signal than a direct read of the plot would indicate. The dotted line shows their best model. From M. Nolta

4. *MAP*

The next large influx of CMB data is expected to come from the *MAP* satellite. The primary goal of *MAP* is to produce a high fidelity, polarization sensitive map of the full sky. *MAP* was successfully launched on June 30, 2001 at 3:46:46 from Kennedy Space Flight Center. The launch by Boeing was within 183 ms of the opening window! So far, the mission has proceeded as planned and the instrument performance is nominal.

Data from *MAP* will stand apart from previous data sets in a number of ways:

- *MAP* will produce all sky maps in five frequency bands with more than 45 times the sensitivity of *COBE/DMR* and 30 times the angular resolution.

- *MAP*'s maps, to a good approximation, will have the same relative temperature errors on large angular scales ($> 90°$) as on small ones ($< 1°$). To extract cosmological parameters, current experiments pin the C_l spectrum at low l to *COBE*.

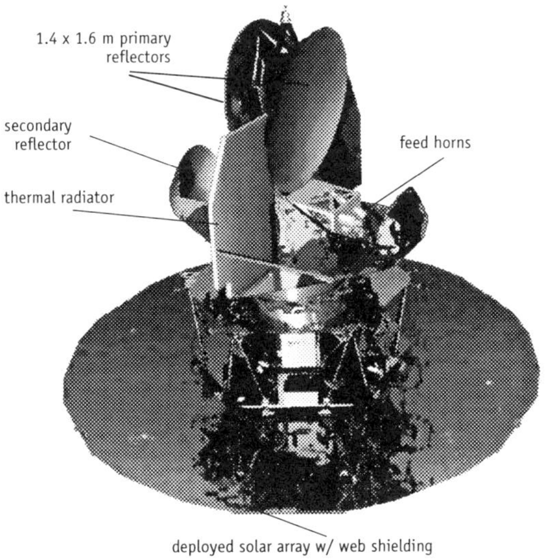

Fig. 4. Picture of *MAP* with the solar arrays deployed. The receivers are located directly beneath the primary reflectors and are passively cooled by the thermal radiators. The Earth, Moon, and Sun are beneath the solar arrays.

- Systematic uncertainties will be controlled at new levels. Calibration and/or beam uncertainty dominate all current measurements and there may be lower level systematic errors of which we are not aware. *MAP* will be calibrated on the CMB dipole to better than 1% and the beams will be mapped in flight multiple times. Because of the long time spent in a very stable orbit, numerous other consistency and systematic checks are possible.

- True maps will be produced with small pixel-to-pixel correlations. The fidelity of a map to the underlying true flux distribution depends on the level of filtering in going from the time stream to the spatial map. To a very good approximation, *MAP* senses radiation from only the direction in which it is pointing and the reconstruction from the differential data to a map constrains artifacts at the sub-μK level. Only the slightest filter is applied. Up to an unknown single additive constant, what you see in the map will be what's in the sky.

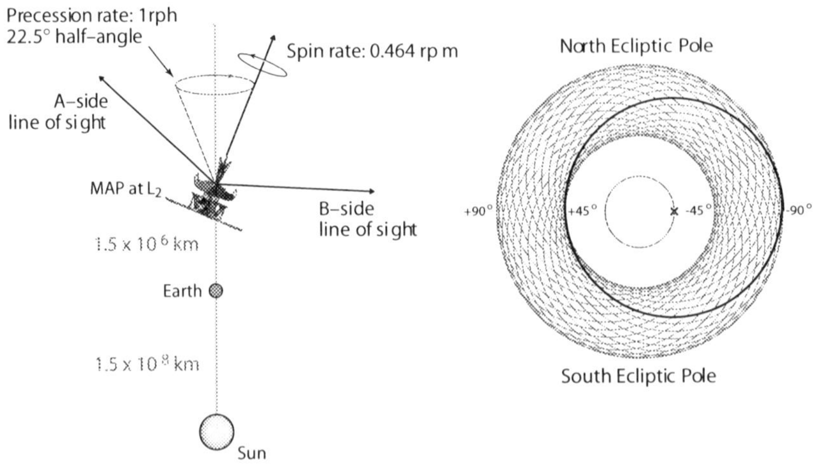

Fig. 5. *MAP*'s scan pattern from L2. The dark circle on the left hand drawing depicts the path covered by two beams for one rotation; the innermost circle is the path of the spin axis during one precession. The four modulation time scales are: 2.5 kHz switching between the two sides, 2.1 min spin period, 1 hr precession to cover $\approx 30\%$ of sky, and 6 months to cover the whole sky.

4.1. Essential components of the design

MAP is a differential microwave radiometer. Table 1 gives the approximate characteristics of the instrument though the flight performance may differ slightly. The components of the *MAP* mission—receivers, optics, scan strategy, thermal design, and attitude control—all work together. Without any one of them, the mission would not achieve its goals. One guiding philosophy, however, is that a differential measurement with a symmetric instrument is highly desirable as discussed, for example, by Dicke (1968). The reason is that in such a system the receivers outputs are, to first order, insensitive to changes in the satellite temperature or other radiative properties.

MAP uses a pair of composite back-to-back offset shaped Gregorian telescopes that focus celestial radiation onto ten pairs of back-to-back corrugated feeds. The primary optical axes of the two telescopes are separated by 141°. Separate feeds accept radiation in five frequency bands between 20 and 100 GHz. Two orthogonal polarizations are selected by an orthomode transducer at the base of each feed.

There are two keys to making large area maps with negligible pixel-to-pixel correlations: (a) stable receivers and (b) a highly interlocking scan strategy. One

Band	$f_{\rm center}$ (GHz)	Bandwidth (GHz)	# Signals	θ_E (deg)	θ_H (deg)
K	23	5.3	2	0.95	0.75
$\rm K_a$	33	7	2	0.7	0.6
Q	41	8.4	4	0.45	0.5
V	61	12	4	0.3	0.35
W	95	17	8	< 0.25	< 0.25

Table 1. *MAP* Frequencies, Noise Bandwidths, and Beam Sizes. The beams are elliptical and so both E and H plane Gaussian FWHM are given. There is one pair of feeds in both K and $\rm K_a$ bands. The two signals correspond to two polarizations. The sensitivity for each band will be ≈ 35 μK for each 3.2×10^{-5} sr pixel for two years.

measure of the stability of the receivers is its "$1/f$" knee. This is the frequency at which the power spectral density (PSD) of the receiver output is double that of the value at a high frequency. Laboratory tests showed that the $1/f$ knee is < 10 mHz and that the PSD from 20 mHz to 3 kHz is white. The stability of the test setup limited the low frequency performance; the receivers will be more stable in space.

The receiver stability is due to a number of factors. Foremost is the design. *MAP* uses NRAO HEMT amplifiers in a correlation configuration with 2.5 kHz phase switching. Thus, the detector output is proportional to the difference in polarized flux at 2.5 kHz from two celestial pixels separated by $\approx 141°$. Second is the thermal stability. The passive thermal design combined with a constant Sun angle and an L2 orbit 1.5×10^6 km from Earth results in a thermal drift rate measured in mK/hr, unattainable from any platform near Earth.

With the highly interlocking scan shown in Figure 5, the value in each pixel on the sky is referenced to that in hundreds of other pixels. As the cross-linking occurs on multiple time scales, there is a highly selective spatio-temporal filter that links the time ordered differential data to an all sky map. There are a number of proven algorithms for processing the data [34, 16] that are used by the *MAP* team.

4.2. Science from MAP

There is still a large gap between the models that are currently under consideration and the possible models that can describe the CMB. Despite the recent observational advances, the favored model of the late 1990's is still the most favored model[20] though our confidence in it has been considerably enhanced. Only with a map with negligible systematic artifacts can different models be tested. The following list is a sample of the anticipated science:

- CMB. *MAP* will measure the CMB angular spectrum to $l \approx 1000$ with a cosmic variance limit at $l \approx 500$, greatly extending *COBE*'s of $l \approx 20$. A simulation of the angular spectrum is shown in Figure 6. With the data one can test for Gaussanity, polarization, and the polarization-temperature cross-correlation. *MAP* can tell if there are significant isocurvature or tensor modes or if just six parameters describe the $\approx 10^6$ data points.

- Large scale correlations of other surveys with the CMB. Is there a SZ decrement in the supergalactic plane?[7] Are there correlations with point sources due to "dark energy" or quintessence?[8].

- ISM. Is there significant emission from spinning dust?[11], Does a 10^4K plasma describe all the free-free emission? Is dust naturally associated with free-free/H_α emission? What is the polarization of the ISM?

- Non-linear structures. *MAP* should detect the SZ effect in the largest ten or so clusters [28] and there should be a detectable lensing signal from a cross correlation with the Sloan Digital Sky Survey [26].

- Sources. *MAP* will give a full sky census of bright extragalactic sources. It will shed light on their emission properties between 30 and 100 GHz and will be sensitive to new classes of sources.

- Calibration. One of the largest systematic errors in many experiments is calibration. *MAP* will calibrate celestial sources to the Earth's velocity through the CMB dipole. This enables the precise comparison between multiple cosmic probes.

5. Acknowledgments

Thank you for a wonderful, stimulating, and fun conference. It was great. This work was supported by the NSF and NASA. *MAP* is a collaboration between NASA (Chuck Bennett [PI], Gary Hinshaw, Al Kogut, & Ed Wollack), Princeton (Chris Barnes, Norm Jarosik, Michele Limon, Lyman Page, David Spergel & David Wilkinson), Chicago (Steve Meyer), UCLA (Ned Wright), UBC (Mark Halpern), and Brown (Greg Tucker). *MAP*'s success is the result of a dedicated and incredibly hard working spacecraft team at NASA/GSFC led by the Project Manager Liz Citrin. More information may be found at http://map.gsfc.nasa.gov

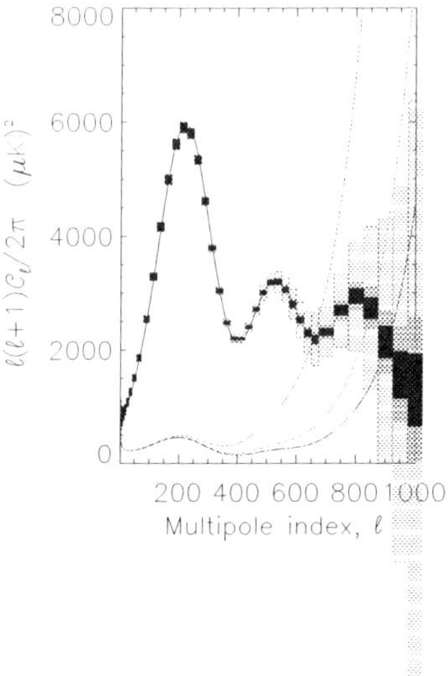

Fig. 6. An estimate of the 1σ errors for a concordance angular spectrum[32] based on Table 2. The smooth curves rising at high l indicate the error per l; the width of the box indicates the window in l over which the curve is averaged. The thickest error bars are for the 6 month maps, the next are for the 2 yr maps, and the narrowest are for the 4 yr maps. The assumed sky coverage is 80%.

References

[1] Bahcall, N., Ostriker, J. P., Perlmutter, S., & Steinhardt, P. J. 1999, Science, 284, 1481 (astro-ph/9906463)

[2] Bartlett, J. G., Blanchard, A., Douspis, M., & Le Dour, M. 1998, in Proc Evol of Large Scale Structure (Garching) (astro-ph/9810318)

[3] Bond, J. R., Crittenden, R., Davis, R. L., Efstathiou, G., & Steinhardt, P. J. 1994, PRL, 72, 1, 13

[4] J.R. Bond, G. Efstathiou & M. Tegmark, MNRAS, 291, L33-L41, 1997.

[5] de Bernardis, P. *et al.* 2000, Nature, 404, 6781, 955

[6] de Bernardis, P. *et al.* 2001, Ap. J., in press (astro-ph/0105296)

[7] Boughn, S, Ap.J, astro-ph/9807308

[8] Boughn, S & Crittenden, R. astro-ph/0111281

[9] Dodelson, S. & Knox, L. 2000, Phys. Rev Lett., 84, 3523 (astro-ph/909454)

[10] Doroshkevich, A. G., Zel'dovich, Ya. B. & Sunyaev, R. 1978, Sov. Astron. 22, 5

[11] B. T. Draine & A. Lazarian, Review for the Sloan Summit on Microwave Foregrounds, ed. A. de Oliveira-Costa and M. Tegmark astro-ph/9902356.

[12] Durrer, R., Novosyadlyj, B., Apunevych, S. astro-ph/0111594
[13] Efstathiou, G., astro-ph/0109151.
[14] Halverson,N. *et al.* 2001, Ap.J., in Press (astro-ph/0104489)
[15] Hanany, S. *et al.* 2000, Ap.J., 545, 5 (astro-ph/0005123)
[16] Hivon, E. *et al.* 2001, Submitted to Ap.J., (astro-ph/0105302)
[17] Hu, W., Resceu Symposium 1999, astro-ph/0002520.
[18] Kamionkowski, M. *et al.* 1994, astro-ph/94010003.
[19] Knox, L. & Page, L. 2000, PRL, 85, 1366 (astro-ph/0002162)
[20] Lange, A. *et al.* PRD, 63, 2001.
[21] Lee, A. T. *et al.* 2001, (astro-ph/0104459)
[22] Miller, A. *et al.* 1999, Ap.J., 524, L1
[23] Miller, A. *et al.* 2001, Accepted for publication, ApJS, astro-ph/
[24] Miller, C, Nichol, R., Genovese, C., & Wasserman, L., astro-ph/0112049.
[25] Netterfield C. B. *et al.* 2001, (astro-ph/0104460) Ap.J., in press
[26] H. Peiris & D. N. Spergel ApJ, 540, 605-613, 2000.
[27] Pryke, C. *et al.* 2001 (astro-ph/0104490)
[28] Refregier, A., Spergel, D. N., Herbig, T., ApJ 531, 31 (2000) astro-ph/9806349
[29] Smoot, G. F. *et al.* 1992, Ap.J., 396, L1
[30] Steinhardt, P, and Turok, N. hep-th/0111030
[31] Tegmark, M. & Efstathiou, G. P. 1996, MNRAS, 281, 1297-1314.
[32] Wang, L., *et al.* 2000, Ap.J, 530,17 astro-ph/9901388
[33] Wang, X., Tegmark, M. & Zaldarriaga, M. (astro-ph/0105091)
[34] Wright, E., Hinshaw, G. & Bennett, C. (astro-ph/9510102)

Cosmic Microwave Background Anisotropy with Late Time Entropy Production

Takeo MOROI and Tomo TAKAHASHI
Department of Physics, Tohoku University, Sendai 980-8578, Japan

Abstract

We discuss effects of cosmological moduli fields on the cosmic microwave background (CMB). If a modulus field ϕ once dominates the universe, the CMB we observe today is from the decay of ϕ and its anisotropy is affected by the primordial fluctuation in the amplitude of the modulus field. As a result, constraints on the inflaton potential from the CMB anisotropy can be relaxed. In addition, with the cosmological moduli fields, *correlated* mixture of adiabatic and isocurvature fluctuations may be generated, which results in enhanced CMB angular power spectrum at higher multipoles relative to that of lower ones. Such an enhancement can be an evidence of late time entropy production due to the cosmological moduli fields, and may be observed at on-going and future experiments.

1. Introduction

In superstring theory [1], it is well known that there are various flat directions parameterized by scalar fields. (Hereafter, we call these fields as "moduli" fields.) Since their potential is usually generated by effects of supersymmetry (SUSY) breaking, their masses are expected to be of the order of the gravitino mass. Although masses of the moduli fields can be as light as (or even lighter than) the electroweak scale, moduli fields do not affect collider experiments since their interactions are suppressed by inverse powers of the gravitational scale.

Cosmologically, however, they may cause serious problems [2]. If the mass of the moduli fields $m_\phi \lesssim O(10\ \text{TeV})$, reheating temperature of the universe becomes lower than ~ 1 MeV. With such a low reheating temperature, the success of the standard big-bang nucleosynthesis (BBN) is spoiled. For lighter moduli fields ($m_\phi \lesssim O(100\ \text{MeV})$), they survive until today and overclose the universe. One solution to these difficulties is to push up the mass of the moduli fields [3]. In particular, in Ref. [4], it was pointed out that the scenario with heavy moduli can naturally fit into the framework of the anomaly-mediated SUSY breaking [5]. Indeed the reheating temperature can be higher than ~ 1 MeV if $m_\phi \gtrsim O(10\ \text{TeV})$. In this case, the BBN occurs after the decay of the modulus field. Although the thermal history after the BBN is mostly the same as the standard one, cosmology before the modulus decay is completely different. Importantly, fluctuations of moduli fields affect the cosmic microwave background (CMB). In this talk, we consider this scenario and study its consequence in the CMB.

2. Effects of moduli fields on CMB

Now we discuss the CMB angular power spectrum from the scenario with the cosmological modulus field [6]. First, we emphasize that there are two independent sources of the CMB anisotropy, i.e., primordial metric perturbation $\Psi_{\rm i}$ induced by inflaton field fluctuation and amplitude fluctuation of the modulus field $\delta\phi_{\rm i}$ which is of order $H_{\rm inf}/2\pi$ where $H_{\rm inf}$ is the Hubble parameter during inflation. Therefore, the resultant CMB angular power spectrum C_l is given in the following form:

$$C_l = C_l^{(\rm adi)} + C_l^{(\delta\phi)}. \tag{1}$$

Here, $C_l^{(\rm adi)}$ is from the perturbation in the inflaton field, which is of order $\Psi_{\rm i}^2$. On the contrary, $C_l^{(\delta\phi)}$ is from the primordial fluctuation of the modulus amplitude, which is of order $\delta\phi_{\rm i}^2$. (Notice that there is no term which is of order $\Psi_{\rm i}\delta\phi_{\rm i}$ since two fluctuations are uncorrelated.) $C_l^{(\rm adi)}$ can be calculated by following the standard method. In calculating $C_l^{(\delta\phi)}$, we must specify the origin of CDM and baryon. Here, we assume that CDM is generated by the decay of ϕ. In this case, after the decay of the modulus field, there is no entropy between CDM and radiation. On the contrary, we consider two possibilities of generating baryon asymmetry: (i) the baryon asymmetry is (somehow) generated at the time of (or after) the decay of ϕ, or (ii) the Affleck-Dine (AD) mechanism [7] generates the baryon number.

Let us first consider the case (i). In this case, there is no entropy between baryon and radiation, so the cosmic fluctuations are same as the conventional adiabatic case once the modulus field decays. Thus, if we neglect the scale dependences of $\Psi_{\rm i}$ and $\delta\phi_{\rm i}$, $C_l^{(\delta\phi)}$ is proportional to $C_l^{(\rm adi)}$. In this case, the CMB angular power spectrum is the same as the usual adiabatic case if the normalization of the initial fluctuations are properly chosen. However, this fact has significant implications when we construct a model of inflation. First of all, since curvature perturbation induced by the modulus field $\Psi^{(\delta\phi)}$ is of order $H_{\rm inf}/\bar{\phi}_{\rm i}$ where $\bar{\phi}_{\rm i}$ is the initial amplitude of ϕ, large cosmic perturbation can be generated by lowering $\bar{\phi}_{\rm i}$ even if $H_{\rm inf}$ is small. Furthermore, usually, scale dependence of $\delta\phi_{\rm i}$ is milder than that of $\Psi_{\rm i}$. Thus, when $C_l^{(\rm adi)} \ll C_l^{(\delta\phi)}$ is realized, the resultant CMB angular power spectrum may be like that from the scale-invariant adiabatic perturbation even if $\Psi_{\rm i}$ has a strong scale dependence. These facts relax constraints on the potential of the inflaton field [6, 8].

Next we consider the case (ii). Here we assume that the initial value of the

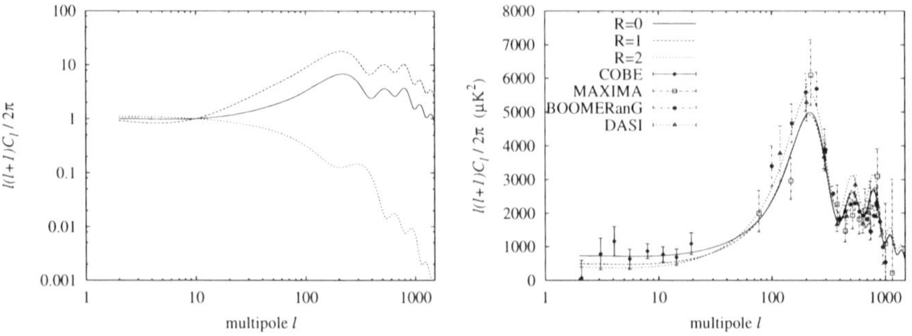

Fig. 1. *Left*: The CMB angular power spectrum $C_l^{(\delta\phi)}$ for the case with the correlated isocurvature perturbation in the baryonic sector (dashed line), as well as C_l for purely adiabatic (solid line) and purely baryonic isocurvature (dotted line) cases. We used the normalization $[l(l+1)C_l/2\pi]_{l=10} = 1$. *Right*: The CMB angular power spectrum C_l for $R=0$ (solid), $R=1$ (dashed), and $R=2$ (dotted) where $R \equiv S_{\rm i}/\Psi^{(\rm adi)}$. The overall normalization of C_l is determined to be best fitted to the the observational data [9]. In these figures, the cosmological parameters are taken such that $h = 0.65$, $\Omega_{\rm b}h^2 = 0.019$, $\Omega_{\rm m} = 0.4, \Omega_\Lambda = 0.6$, and scale-invariance is assumed both for $\Psi_{\rm i}$ and $S_{\rm i}$.

fluctuation in the AD field is negligibly small.* Even when there is no primordial fluctuation in the AD field, however, non-vanishing isocurvature fluctuation is generated in the baryonic sector if $\delta\phi_{\rm i} \neq 0$. Since, initially, there is no fluctuation in the baryon energy density, the entropy between baryon and ϕ exists; $S_{b\phi} \equiv \delta_b - \delta_\phi = S_{\rm i} \neq 0$, where $\delta_x \equiv \delta\rho_x/\rho_x$. This entropy is conserved until the modulus field decays, and it becomes the entropy between photon and baryon after the decay of ϕ. Therefore, in calculating $C_l^{(\delta\phi)}$ which is generated by the primordial fluctuation in the modulus amplitude, initial condition for the baryonic density fluctuation is different from the conventional adiabatic one, and is given by, in deep radiation dominated epoch [6],

$$\delta_b = S_{\rm i} + (3/4)\delta_\gamma = 4.5\Psi^{(\delta\phi)} + (3/4)\delta_\gamma. \tag{2}$$

Importantly, the curvature perturbation $\Psi^{(\delta\phi)}$ is induced by the initial entropy perturbation $S_{\rm i}$. In addition, notice that the first term in the right-handed side of the above equation does not exist in the usual adiabatic ones. Initial conditions for other perturbations are the same as the usual adiabatic case. The first term in Eq. (2) gives rise to the non-vanishing entropy in the baryonic sector and hence we call it as the "isocurvature" term. Here, it should be emphasized that such an isocurvature fluctuation is *correlated* with the contribution from $\Psi^{(\delta\phi)}$ which gives rise to the effect like the conventional adiabatic fluctuation. Thus, the effect

*This may happen when, for example, the effective mass of the AD field is comparable to $H_{\rm inf}$ during the inflation.

of this "isocurvature" fluctuation is completely different from the conventional *uncorrelated* isocurvature fluctuation.

In Fig. 1., we show the total CMB angular power spectrum with such a correlated isocurvature fluctuation in the baryonic density fluctuation. Since there are two sources of the cosmic perturbations, $\Psi_{\rm i}$ and $S_{\rm i}$ (or equivalently, $\delta\phi_{\rm i}$), we define $R \equiv S_{\rm i}/\Psi^{(\rm adi)}$, where $\Psi^{(\rm adi)}$ is the gravitational potential in the adiabatic mode at the radiation dominated epoch after the modulus decay. (For simplicity, we neglect the scale dependence of R.) Treating R as a free parameter, we plot the total C_l for several values of R in Fig. 1. As one can see, correlated isocurvature perturbation in the baryonic sector may result in an enhancement of C_l at higher multipoles relative to that at lower ones. Therefore, on-going and future experiments will give us interesting tests of the scenario with the cosmological moduli fields.

3. Conclusion

We have studied the effects of the cosmological moduli fields on the CMB anisotropy. In the scenario with the cosmological moduli fields, *correlated* isocurvature fluctuation may exist in the baryonic sector which results in enhanced CMB angular power spectrum at higher multipoles relative to that at lower ones. In addition, even in the case where there is no isocurvature perturbation, the cosmological modulus field may have important implication to the model-building of the inflation.

References

[1] See, for example, J. Polchinski, *String Theory* (1998, Cambridge Univ. Press).

[2] G.D. Coughlan et al., Phys. Lett. **B131** (1983) 59.

[3] T. Moroi, M. Yamaguchi and T. Yanagida, Phys. Lett. **B342** (1995) 105; M. Kawasaki, T. Moroi and T. Yanagida, Phys. Lett. **B370** (1996) 52.

[4] T. Moroi and L. Randall, Nucl. Phys. **B570** (2000) 455.

[5] L. Randall and R. Sundrum, Nucl. Phys. **B557** (1999) 79; G.F. Giudice, M.A. Luty, H. Murayama and R. Rattazzi, JHEP **9812** (1998) 027; J.A. Bagger, T. Moroi and E. Poppitz, JHEP **04** (2000) 009.

[6] T. Moroi and T. Takahashi, Phys. Lett. **B522** (2001) 215.

[7] I. Affleck and M. Dine, Nucl. Phys. **B249** (1985) 361.

[8] K. Enqvist and M.S. Sloth, hep-ph/0109214; D.H. Lyth and D. Wands, hep-ph/0110002.

[9] C. Bennett et al., Astrophys. J. Lett. **464** (1996) L1; C.B. Netterfield et al., astro-ph/0104460; A.T. Lee et al., Astrophys. J. Lett. **561** (2001) L1; N.W. Halverson et al., astro-ph/0104489.

Instanton-Induced Cosmological Constant

Jun'ichi Yokoyama
Department of Earth and Space Sciences, Graduate School of Science, Osaka University, Toyonaka 560-0043, Japan

Abstract

Assuming that the cosmological constant vanishes in the true ground state with the lowest possible energy density, it is shown that the observed small but finite vacuum-like energy density can be explained without introducing any small numbers in a theory with two or more degenerate perturbative vacua which are unstable due to quantum tunneling. An example is given making use of the topological vacua in non-Abelian gauge theories.

Recent progress in observational cosmology indicates that there are *two* cosmological constant problems. One is the older problem why the vacuum energy density, ρ_v, is vanishingly small compared with its natural magnitude $\rho_v = M_G^4$, where $M_G = M_{Pl}/\sqrt{8\pi} = 2.4 \times 10^{18}$ GeV is the reduced Planck scale [1]. The second, newer problem is why it is nevertheless finite with $\rho_v \simeq \rho_{cr0} \simeq 10^{-120} M_G^4$ as indicated by the $m - z$ relation of type Ia supernovae [2].

In this talk I present a posssible explanation of the small but finite cosmological constant without introducing any small numbers. That is, I attempt to solve the second problem under the assumption that the first one is solved in the true ground state by arguing that we have not fallen into that state. Other proposed solutions to the second problem, such as quintessence [4] (see also [5] for earlier work) or meV-scale false vacuum energy [6], are also based on such a hypothesis. My starting point is that the energy eigenvalue of the true ground state of a theory with two or more degenerate perturbative vacua, which cannot be transformed from one another without costing energy, is smaller than that of a quasi-ground state localized around one of these states in field space by an exponentially small amount. Here, by perturbative vacua I mean a state with the lowest energy density without taking possible tunneling effect to another perturbative vacuum into account. My hypothesis is that the cosmological constant vanishes *not* in these degenerate perturbative vacua but in the absolute ground state with quantum tunneling effects taken into account, because there are both classical and quantum contributions to the cosmological constant and what we observe is their sum.

For illustration let us first consider an abstract field theory model whose perturbative vacuum states are classified into two distinct categories labeled by $|+\rangle$ and $|-\rangle$ with $\langle +|-\rangle = 0$ at the lowest order. I also assume that, although the transition from $|+\rangle$ to $|-\rangle$ is classically forbidden, there is an instanton solution which describes quantum tunneling from $|+\rangle$ to $|-\rangle$ and vice versa. By nature the instanton is localized in space and (Euclidean) time with a finite Euclidean action S_0. Then the true ground state, $|S\rangle$, with this tunneling effect taken into account, is given by the symmetric superposition of $|+\rangle$ and $|-\rangle$, namely, $|S\rangle = (|+\rangle + |-\rangle)/\sqrt{2}$, where I have assumed that $|+\rangle$ and $|-\rangle$ are normalized.

Now we evolve $|S\rangle$ with Euclidean time T to calculate $\langle S|e^{-\mathcal{H}T}|S\rangle$ by summing up contributions of instantons and anti-instantons as

$$\begin{aligned}
\langle S|e^{-\mathcal{H}T}|S\rangle &= \frac{1}{2}\left(\langle +|e^{-\mathcal{H}T}|+\rangle + \langle -|e^{-\mathcal{H}T}|-\rangle + \langle +|e^{-\mathcal{H}T}|-\rangle + \langle -|e^{-\mathcal{H}T}|+\rangle\right) \\
&= e^{-\rho_0 VT}\left[\sum_{j=0}^{\infty}\frac{1}{(2j)!}\left(kVTe^{-S_0}\right)^{2j} + \sum_{j=0}^{\infty}\frac{1}{(2j+1)!}\left(kVTe^{-S_0}\right)^{2j+1}\right] \\
&= \exp\left(-\rho_0 VT + kVTe^{-S_0}\right), \qquad (1)
\end{aligned}$$

where $\mathcal{H}$ is the Hamiltonian, $k \equiv m^4$ is a positive constant and V represents spatial volume [7]. Here ρ_0 is the energy density of the perturbative vacua, $|+\rangle$ and $|-\rangle$, which are presumably translationally invariant. Thus the energy density of the true ground state $|S\rangle$ is given by $\rho_S = \rho_0 - m^4 e^{-S_0}$. It is this energy density that vanishes under our hypothesis. This in turn implies that we find a nonvanishing vacuum energy density $\rho_0 = m^4 e^{-S_0}$ in either of the perturbative vacuum states, $|+\rangle$ or $|-\rangle$.

Thus if our Universe is in one of the perturbative vacuum state because it is too young to be relaxed into the true ground state $|S\rangle$, we observe a nonvanishing vacuum energy density $\rho_0 = m^4 e^{-S_0}$ today. Since the tunneling rate per unit volume per unit time is given by $\Gamma \simeq m^4 e^{-2S_0}$ apart from a prefactor of order unity [8], the requirement that there should be no transition in the horizon volume in the cosmic age reads

$$\Gamma H_0^{-4} \simeq 9M_G^4/m^4 \lesssim 1, \qquad (2)$$

where $H_0^2 \cong \rho_0/(3M_G^2)$ is the current Hubble parameter squared. We therefore find that, if the parameters satisfy $m \gtrsim M_G$ and $S_0 = 120 \ln 10 + 4\ln(m/M_G)$, we can account for the observed small value of the cosmological constant without introducing any small numbers.

So far is a generic study to generate an exponentially small difference in energy density using a theory with degenerate perturbative vacua whose real ground state is given by their superposition. Next in order to see how this mechanism may be implemented in a more specific theory with this property, let us consider a famous example of an SU(N) ($N \geq 2$) gauge theory with the coupling constant g whose perturbative vacuum states are classified in terms of the winding number n and denoted by $|n\rangle$ [9, 10]. States with different winding numbers cannot be transformed from each other by a continuous gauge transformation [10] and there is an energy barrier between them. I take $N = 2$ for simplicity hereafter.

The true ground state in the presence of quantum tunneling, which is described by the instanton solution with the Euclidean action $S_0 = 8\pi^2/g^2$, is given by an infinite sum of these $|n\rangle$ states as

$$|\theta\rangle = \sum_{n=-\infty}^{\infty} e^{in\theta}|n\rangle, \tag{3}$$

where θ is a real parameter [9, 10]. One can easily find that this state is a real eigenstate of the Hamiltonian $\mathcal{H}$ as

$$\langle\theta'|\, e^{-\mathcal{H}T}\, |\theta\rangle = \exp\left(2KVTe^{-S_0}\cos\theta\right)\delta\left(\theta-\theta'\right), \tag{4}$$

where $KVTe^{-S_0}$ represents contribution of a single instanton or anti-instanton with K being a positive constant.

Then the above equality (4) shows that each θ-vacuum $|\theta\rangle$ has a different energy density than the perturbative vacuum $|n\rangle$ by

$$\Delta\rho = -2Ke^{-S_0}\cos\theta. \tag{5}$$

Apparently the $\theta = 0$ vacuum has the lowest energy (and no CP violation), but one cannot immediately conclude that this is the only vacuum state, because the other θ-vacua are also stable against gauge-invariant perturbations [9, 10]. Here let us take the CP-conserving $\theta = 0$ vacuum as the reference ground state of the full theory where the cosmological constant is normalized to vanish. For quantum gravitational approach to realize $\theta = 0$ vacuum as the ground state without introducing an axion, see [13]. With this normalization of Λ, the vacuum energy density in each perturbative vacua is found to be

$$\langle n|\rho_{\rm v}|n\rangle = 2Ke^{-S_0}. \tag{6}$$

In the case of pure gauge theory the prefactor K is divergent due to the contribution of arbitrary large instantons [14]. In order to obtain a physical cutoff scale let us introduce an SU(2) doublet scalar field Φ with a potential $V[\Phi] = \lambda(|\Phi|^2 - M^2/2)^2/2$, following 't Hooft [14]. For $M \neq 0$, the instanton solution should be replaced by the constrained-instanton solution [15] and we obtain a finite result for K, to find

$$\rho_v \cong M^4 \left(\frac{8\pi}{g^2}\right)^4 e^{-\frac{8\pi^2}{g^2}}, \qquad \Gamma \simeq M^4 \left(\frac{8\pi}{g^2}\right)^4 e^{-\frac{16\pi^2}{g^2}}. \tag{7}$$

Demanding that $\rho_v = 10^{-120} M_G^4$ and that the tunneling rate in the current horizon should be smaller than unity in the cosmic age, $\Gamma H_0^{-4} \lesssim 1$, we find

$$\frac{\pi}{\alpha} + 2\ln\alpha = 60\ln 10 + 2\ln\left(\frac{M}{M_G}\right), \qquad M \gtrsim \alpha M_G, \tag{8}$$

where $\alpha \equiv g^2/(4\pi)$ is the coupling strength at the energy scale $M/\sqrt{2}$. If the inequality (8) is marginally satisfied, we find $\alpha = 1/44.4$ and $M = 5 \times 10^{16}$GeV. If, on the other hand, we take $M = M_G$ so that the cutoff scale of instanton is identical to the presumed field-theory cutoff, we find $\alpha = 1/47$.

Thus if our Universe happened to be created in a state with some specific winding number and remains there up until now, we would observe a nonvanishing vacuum energy density (7). Although θ-vacuum is the real ground state of the theory, there is no *a priori* reason that the Universe is created in this state. In fact, in the possibly chaotic initial state of the early universe [16], there may well be a small domain where the scalar field has a nonvanishing expectation value with a constant SU(2) phase and the gauge field vanishes. If such a region is exponentially stretched by cosmological inflation [17] and our Universe is contained in it, the state of our Universe would be more like the perturbative vacuum $|n = 0\rangle$ than some θ-vacuum. Then it is not surprising that we observe a nonvanishing vacuum energy density (7) today. Since the Hubble paramter during the last 70 e-folds of inflation should be much smaller than the above-obtained lower bound on M [18], quantum fluctuations generated along the phase direction of the fields during inflation is much smaller than M, and it does not affect the realization of the state $|n = 0\rangle$. This also means that gravitational effects are negligibly small for the instanton configuration even during inflation. Furthermore thermal transition to a state with a different winding number is suppressed since the reheat temperature after inflation, T_R, is typically much smaller than M. In fact, to avoid overproduction of gravitinos, it should satisfy $T_R < 10^{12}$GeV [19]. Hence we have only to worry about quantum transition as we have already done.

In summary, we have pointed out that in a field theory with two (or more) degenerate perturbative vacua, the vacuum energy density of the true ground state is smaller than that in a perturbative vacua by an exponentially small amount if quantum tunneling between degenerate vacua is possible, and that this may be utilized to explain the observed small value of the cosmological constant without introducing any small quantities.

References

[1] S. Weinberg, Rev. Mod. Phys. **61**, 1 (1988).

[2] S. Perlimutter et al., Astrophys. J. **517**, 565 (1999); B.P. Schmidt et al., Astrophys. J. **507**, 46 (1998);A.G. Riess et al., Astron. J. **116**, 1009 (1998).

[3] A.D. Dolgov, in The very early universe, eds. G.W. Gibbons and S.T. Siklos (Cambridge, UK 1982).

[4] I. Zlatev, L. Wang and P.J. Steinhardt, Phys. Rev. Lett. **82**, 896 (1999); P.J. Steinhardt, L. Wang and I. Zlatev, Phys. Rev. D **59**, 123504 (1999)

[5] Y. Fujii, Phys. Rev. **D26**, 2580 (1982); L.H. Ford, Phys. Rev. **D35**, 2339 (1987); P.J. Peebles and B. Ratra, Astrophys. J. **325**, L17 (1988); Y. Fujii and T. Nishioka, Phys. Rev. **D42**, 361 (1990); K. Sato, N. Terasawa, and J. Yokoyama, in Proc. XXIVth Recontre de Moriond, "The Quest for the Fundamental Constants in Cosmology," eds. J. Audouze and J. Tran Thanh Van. (Editions Frontières, France 1990) 193; T. Nishioka and S. Wada, Int. J. Mod. Phys. **A8**, 3933 (1993).

[6] N. Arkani-Hamed, L.J. Hall, C. Kolda, H. Murayama, Phys. Rev. Lett. **85**, 4434 (2000); S.M. Barr and D. Seckel, hep-ph/0106239.

[7] S. Coleman, Aspects of symmetry, (Cambridge, UK 1985).

[8] A. Lapedes and E. Mottola, Nucl. Phys. B **203**, 58 (1982).

[9] C.G. Callan Jr, R.F. Dashen, and D.J. Gross, Phys. Lett. **63B**, 334 (1976).

[10] R. Jackiw and C. Rebbi, Phys. Rev. Lett. **37**, 172 (1976).

[11] A.M. Polyakov, Phys. Lett. **59B**, 82 (1975); A.A. Belavin, A.M. Polyakov, A.S. Schvartz, and Yu.S. Tyupkin, Phys. Lett. **59B**, 85 (1975).

[12] G. 't Hooft, Phys. Rev. Lett. **37**, 8 (1976).

[13] H.B. Nielsen and M. Ninomiya, Phys. Rev. Lett. **62**, 1429 (1989).

[14] G. 't Hooft, Phys. Rev. **D14**, 3432 (1976); **D18**, 2199 (1978) (E).

[15] I. Affleck, Nucl. Phys. B **191**, 429 (1981); M. Nielsen and N.K. Nielsen, Phys. Rev. **D61**, 105020 (2000).

[16] A.D. Linde, Phys. Lett. **129B**, 177(1983).

[17] For a review of inflation, see, e.g. A.D. Linde, Particle Physics and Inflationary Cosmology (Harwood, Chur, Switzerland, 1990); K.A. Olive, Phys. Rep. **190**,181(1990).

[18] V.A. Rubakov, M.V. Sazin, and A.V. Veryaskin, Phys. Lett. **115B**, 189(1982); L.F. Abbott and M.B. Wise, Nucl. Phys. B **244**, 541 (1984).

[19] M. Kawasaki and T. Moroi, Prog Theor. Phys. **93**, 879 (1995).

Cosmological Aspects of Affleck-Dine Mechanism and Q-ball Formation in the Gauge-Mediated SUSY Breaking Models

Shinta KASUYA
Research Center for the Early Universe, University of Tokyo, Bunkyo-ku, Tokyo 113-0033, Japan

Abstract

We consider the Affleck-Dine baryogenesis comprehensively in the minimal supersymmetric standard model with gauge-mediated supersymmetry breaking. Considering the high temperature effects, we see that the Affleck-Dine field is naturally deformed into the form of the Q ball. In the natural scenario where the initial amplitude of the field and the A-terms are both determined by the nonrenormalizable superpotential, we obtain only very a narrow allowed region in the parameter space in order to explain the baryon number and the dark matter of the universe. Therefore, the Affleck-Dine baryogenesis is successful, although difficult.

1. Introduction

The Affleck-Dine (AD) mechanism [1] is the most promising scenario for explaining the baryon number of the universe in the gauge-mediated SUSY breaking models. It is based on the dynamics of a scalar field ϕ (flat directions) carrying baryon number, which is regarded as the AD field. During inflation, the expectation value of the AD field develops at a very large value. After inflation the inflaton field oscillates about the minimum of the effective potential, and dominates the energy density of the universe like matter, while the AD field stays at the large field value. It starts the oscillation, or more precisely, rotation in its effective potential when $H \sim m_{\phi,eff}$, where H and $m_{\phi,eff} \equiv |V''(\phi)|$ are the Hubble parameter and the curvature of the potential of the AD field. Once it rotates, the baryon number will be created as $n_B \sim \omega\phi^2$, where $\omega = \dot{\theta}$ is the velocity of the phase of the AD field. The AD field feels spatial instabilities [2] during its rotation, and these instabilities grow very large and the field deforms into clumpy objects: Q balls, a kind of nontopological soliton, whose stability is guaranteed by the existence of some charge Q [3]. The baryons should be explained by the evaporated charges from produced Q balls. We will see whether cosmological

Q-ball model can be successful in the gauge-mediation mechanism.

2. Flat directions as the Affleck-Dine field

The effective potential of the flat direction is lifted by SUSY breaking effects. In the gauge-mediated SUSY breaking scenario, It reads as

$$\begin{aligned} V(\Phi) \;=\;& M_F^4 \log\left(1+\frac{|\Phi|^2}{M_S^2}\right) + m_{3/2}^2\left[1+K\log\left(\frac{|\Phi|^2}{M^2}\right)\right]|\Phi|^2 \\ &+\lambda^2\frac{|\Phi|^{2n-2}}{M^{2n-6}} + \left(\lambda A_\lambda \frac{\Phi^n}{M^{n-3}} + h.c.\right) \\ &-c_H H^2|\Phi|^2 + \left(\lambda a_H H\frac{\Phi^n}{M^{n-3}} + h.c.\right) \\ &+c_T^{(1)}T^2|\Phi|^2 + c_T^{(2)}T^4\log\frac{|\Phi|^2}{T^2}. \end{aligned} \qquad (1)$$

In the first line, $M_S \sim M_F^2/m_\phi$ is a messenger scale. The second term comes from the gravity mediation effects, since the gravity effects always exist. The K is the coefficient of the the one-loop corrections [4], and is usually negative. However it may be positive in some cases, and Q balls are not formed until the amplitude of the field becomes as small as it stays in the logarithmic potential. We call it *delayed* case. $M = 2.4 \times 10^{18}$ GeV is the reduced Planck mass. The second line represents for the nonrenormalizable terms. The third line shows terms which depends on the Hubble parameter H, during inflation and inflaton oscillation stage which starts just after inflation. c_H is a positive constant and a_H is a complex constant with a different phase from A_λ in order for the AD mechanism to work. The last line is the finite temperature potential. The first and second term denote direct and indirect coupling of the flat direction to thermal bath [5, 6], respectively.

3. Charge evaporation at high temperature

Since the Q-ball formation takes place nonadiabatically, (almost) all the charges are absorbed into produced Q balls [7, 8]. However, at finite temperature, the minimum of the free energy of the AD field is achieved for the situation that all the charges exist in the form of free particles in thermal plasma [9]. This situation is realized through charge evaporation from Q-ball surface [9, 10], which becomes the baryons in the universe.

The rate of charge transfer from inside the Q ball to its outside is determined by the diffusion rate at high temperature and the evaporation rate at low temperature [11]. We finally obtain the total evaporated charge as

$$\Delta Q \sim 4.6 \times 10^{17} \left(\frac{m_\phi}{\mathrm{TeV}}\right)^{-2/3} \left(\frac{M_F}{10^6 \mathrm{GeV}}\right)^{-1/3} \left(\frac{Q}{10^{24}}\right)^{1/12}. \tag{2}$$

4. Q-ball formation

Q-balls are produced while the AD field is rotating. We obtained the relation between the typical charge of the Q balls and the initial amplitude of the AD field [12] as

$$Q = \beta \left(\frac{\phi_0}{m(T)}\right)^4, \tag{3}$$

where $\beta \simeq 6 \times 10^{-4}$, and $m(T) = M_F$ for $T < M_F$, while $m(T) = T$ for $T > M_F$.

5. Consistent cosmological Q-ball scenarios

Now we can see whether there is any consistent scenario to explain the amounts of baryons and dark matter of the universe naturally, which means that the initial conditions and the A-terms are provided by the non-renormalizable superpotential. The details can be seen in Ref.[12], so we will summarize the results here. There are only three successful cases. The first one is when the potential is dominated by the thermal logarithmic term, which is shown in Fig. 1.. Here, the successful scenario is achieved if the parameters are as follows: $M_F \sim 5 \times 10^2$ GeV, $T_{RH} \sim 5 \times 10^4$ GeV, and $Q \sim 2 \times 10^{24}$.

The second case is the delayed case with the Q-balls formed when the thermal logarithmic term is dominated. In this case, the allowed parameter sets are $m_{3/2} \sim$ GeV, $M_F \sim 5 \times 10^2$ GeV, $T_{RH} \sim 10^7$ GeV, and $Q \sim 2 \times 10^{24}$.

The last case is the delayed case with the Q-balls formed when the zero-temperature logarithmic term is dominated. In this case, the allowed parameter sets are $m_{3/2} \sim 0.1$ GeV, $M_F \sim 5 \times 10^4$ GeV, $T_{RH} \sim 5$ GeV, and $Q \sim 10^{20}$.

6. Conclusion

In the natural scenario where the initial amplitude of the field and the A-terms are both determined by the nonrenormalizable superpotential, we need very low reheating temperature and very low M_F, which represents for the expectation value of the F-term of the messenger sector, for all the successful scenarios. Therefore, the model is constrained rather severely in order to explain the baryons and the dark matter of the universe in these Q-ball scenarios.

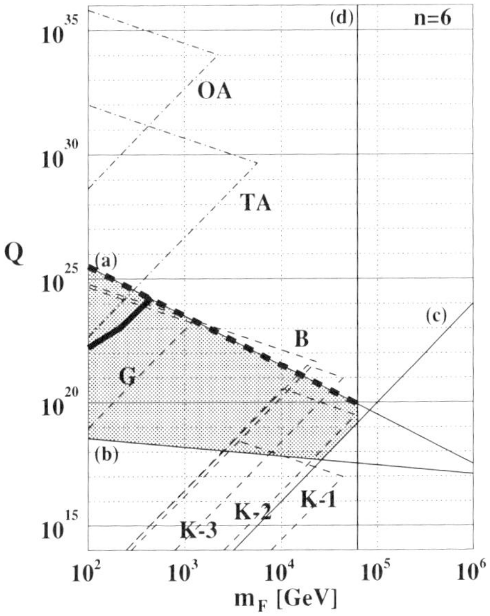

Fig. 1. Summary of constraints on the parameter space (Q, M_F) with $n = 6$ and $m_\phi = 100$ GeV in the generic logarithmic potential where the thermal terms are dominated when the Q-ball formation occurs. Lines (a), (b), and (c) denote the B-DM, survival, and stability conditions, respectively. Line (d) shows the conditions $T \gtrsim M_F$.

References

[1] I. Affleck and M. Dine, Nucl. Phys. **B249**, 361 (1985).
[2] A. Kusenko and M. Shaposhnikov, Phys. Lett. **B418**, 46 (1998).
[3] S. Coleman, Nucl. Phys. **B262**, 263 (1985).
[4] K. Enqvist and J. McDonald, Phys. Lett. **B425**, 309 (1998).
[5] A. Anisimov and M. Dine, hep-ph/0008058, 2000.
[6] M. Fujii, K. Hamaguchi, and T. Yanagida, Phys. Rev. **63**, 123513 (2001).
[7] S. Kasuya and M. Kawasaki, Phys. Rev. **D61**, 041301 (2000).
[8] S. Kasuya and M. Kawasaki, Phys. Rev. **D62**, 023512 (2000).
[9] M. Laine and M. Shaposhnikov, Nucl. Phys. **B532**, 376 (1998).
[10] K. Enqvist and J. McDonald, Nucl. Phys. **B538**, 321 (1999).
[11] S. Kasuya and M. Kawasaki, Phys. Rev. Lett. **85**, 2677 (2000).
[12] S. Kasuya and M. Kawasaki, Phys. Rev. **D64**, 123515 (2001).

Global Structure of the Brane Universe

Hideki ISHIHARA and Izumi TANAKA
Department of Physics, Graduate School of Science, Osaka City University, Sumiyoshi, Osaka 558-8585, Japan

Abstract

Global structure of the Friedmann-Robertson-Walker (FRW)-brane universe is studied. In contrast to the spacelike initial singularity of the usual FRW universe, the initial singularity of the FRW-brane universe is point-like from the viewpoint of causality including gravitational waves propagating in the bulk. Existence of null singularities ('seam singuralities') is also shown in the flat and open FRW-brane universe models.

1. Introduction

Inspired by Horava-Witten theory[1], new attempts have been made on schemes of the Z_2-symmetric brane universe [2] where matter fields are confined to a hypersurface embedded in a higher-dimensional spacetime, while only gravitational fields propagate through all of spacetime. Randall and Sundrum[3] have proposed a static vacuum brane solution, which is embedded in the five-dimensional anti-de Sitter bulk spacetime. Subsequently, many authors investigate on cosmological brane models which contains matter on the brane[4].

If the universe is a sub-spacetime embedded in a higher dimensional spacetime, there may be great changes of basic properties of the cosmological model. For example, it is pointed out that the causal structure of the closed FRW-brane universe is altered owing to gravitational waves propagating in the bulk[5]. In this article, we study the global structure of the FRW-brane universe and point out new properties of the singularities of the FRW-brane universe models.

2. Expansion of universe

We consider a brane universe embedded in the five-dimensional anti-de Sitter spacetime with a negative cosmological constant $-4/l^2$. The five-dimensional anti-de Sitter spacetime can be decomposed locally as $\mathcal{N}^2_{(k)} \times \mathcal{M}^3_{(k)}$ where $\mathcal{N}^2_{(k)}$ is a two-dimensional static spacetime and $\mathcal{M}^3_{(k)}$ is a maximally symmetric 3-

dimensional space characterized by a constant curvatures. The index k takes one of three values, $+1, 0, -1$, corresponding to the positive, zero and negative constant curvature of $\mathcal{M}^3_{(k)}$, respectively. The metric has a form

$$ds_5^2 = -f_{(k)} dt_{(k)}{}^2 + f_{(k)}^{-1} dr_{(k)}{}^2 + r_{(k)}{}^2 d\Omega^2_{(k)}, \tag{1}$$

where

$$f_{(k)} = \left\{ \begin{array}{c} 1 + r_+{}^2/l^2, \\ r_0{}^2/l^2, \\ r_-{}^2/l^2 - 1, \end{array} \right. \quad d\Omega^2_{(k)} = \left\{ \begin{array}{ll} d\chi_+{}^2 + \sin^2\chi_+ \{d\theta^2 + \sin^2\theta \; d\phi^2\} & \text{for } k = +1, \\ d\chi_0{}^2 + \chi_0{}^2 \{d\theta^2 + \sin^2\theta \; d\phi^2\} & \text{for } k = 0, \\ d\chi_-{}^2 + \sinh^2\chi_- \{d\theta^2 + \sin^2\theta \; d\phi^2\} & \text{for } k = -1, \end{array} \right. \tag{2}$$

respectively.

Let us consider a embedding of a 4-dimensional hypersurface (three-brane), Σ, with the structure of $\mathcal{N}^1 \times \mathcal{M}^3_{(k)}$, where $\mathcal{N}^1$ is a trajectory of Σ in $\mathcal{N}^2_{(k)}$. We assume the position of $\mathcal{N}^1$ in $\mathcal{N}^2_{(k)}$ is given by

$$r_{(k)} = a(\tau), \quad t_{(k)} = T(\tau), \tag{3}$$

where a and T are functions of a parameter τ. When τ is normalized by $-d\tau^2 = -f_{(k)} dT^2 + f_{(k)}^{-1} da^2$, the induced metric on Σ has a form of the Robertson-Walker metric $ds^2 = -d\tau^2 + a^2(\tau) d\Omega^2_{(k)}$, where $k = +1, 0, -1$ corresponds to the model with closed, flat and open spatial section, respectively.

If matter is confined on Σ, the surface energy-momentum tensor of the brane, S_{ab}, has a form of $S_{ab} = -\sigma\gamma_{ab} + T_{ab}$, $(a, b = 0, 1, 2, 3)$, where γ_{ab} is the induced metric on Σ, σ is a constant which characterizes the brane tension, and T_{ab} is the energy-momentum tensor of matter on Σ. Since self-gravitating brane in the five-dimensional bulk should satisfy the metric junction condition[6] with Z_2 symmetry, the extrinsic curvature of Σ, K_{ab}, satisfies $K_{ab} = -\kappa^2 \frac{1}{2}(S_{ab} - \frac{1}{3} S \gamma_{ab})$, where κ^2 is the five-dimensional gravitational constant.

We assume, for simplicity, the energy-momentum tensor of matter takes the form of perfect fluid $T_{ab} = (\rho + p) u_a u_b + p\gamma_{ab}$, where u^a is a 4-velocity field of the comoving observers on Σ. From the junction condition we obtain the modified Friedmann equation[4]

$$\left(\frac{1}{a} \frac{da}{d\tau} \right)^2 = -\frac{k}{a^2} + \frac{8\pi}{3} \frac{\kappa^4 \sigma}{6} \rho + \frac{\kappa^4}{36} \rho^2, \tag{4}$$

and the conservation equation $\frac{d}{d\tau}(\rho a^3) + p \frac{d}{d\tau}(a^3) = 0$, where fine tuning $1/l^2 = \kappa^4 \sigma^2/36$ has been done so as to vanish the 4-dimensional effective cosmological constant on Σ.

3. Embedding

We can rewrite (4) as

$$\left(\frac{da}{dT}\right)^2 = f_{(k)}^2 \left[1 - f_{(k)} \left\{l^{-2}a^2 + \alpha^2 a^{2-q} + \beta^2 a^{2-2q}\right\}^{-1}\right], \tag{5}$$

where the equation of state is assumed as $p = (\frac{1}{3}q - 1)\rho, (q = const.)$, and α and β are constants. Equation $r_{(k)} = a(T) = a(t_{(k)})$ determines the embedding of Σ. Now, we introduce a new coordinate system,

$$\begin{aligned} y_*^0 &:= t_+,\ y_*^1 := r_* \cos\chi_+,\ y_*^2 := r_* \sin\chi_+ \cos\theta, \\ y_*^3 &:= r_* \sin\chi_+ \sin\theta\cos\phi,\ y_*^4 := r_* \sin\chi_+ \sin\theta \sin\phi, \end{aligned} \tag{6}$$

that makes the metric has a conformally flat form in the temporal-radial subspace for convenience to discuss the causal structure. In this coordinate system, radial null geodesics are represented by straight lines at $\pm 45°$. Since this coordinate covers the whole bulk spacetime we use it for $k = 0, \pm 1$ cases.

Embedded hypersurfaces, Σ, in y_*^0, y_*^1, y_*^2 spacetime are shown in Fig.1. for $q = 4$ case, where y_*^3, y_*^4 are suppressed by setting $\theta = 0, \pi$ since the directions y_*^2, y_*^3, y_*^4 are isotropic.

It is clear that the brane in each case is smooth everywhere except the vertex point, i^-, located on $y_*^a = 0 (a = 0, 1, 2, 3, 4)$. In the case $k = +1$, Σ is timelike everywhere except i^-, and isotropic in $y_*^1, y_*^2, y_*^3, y_*^4$ directions. On the other hand, in the case $k = 0$, a null line

$$y_*^0 + y_*^1 = 0, \quad y_*^2 = y_*^3 = y_*^4 = 0 \tag{7}$$

lies on Σ. Since the induced metric on Σ degenerates there, then this line is a singular subspace on the brane universe. It seems that one sheet of brane is curled up and seamed two edges of the brane together along the null line (7). We call the line, therefore, 'seam singularity'. Similarly, in the case $k = -1$, a null cone

$$-(y_*^0)^2 + (y_*^2)^2 + (y_*^3)^2 + (y_*^4)^2 = 0, \quad y_*^1 = 0 \tag{8}$$

lies on Σ. In the open case in Fig.1., two seams represent the cross section of the null cone (8) and the hyperplane $y_*^3 = y_*^4 = 0$. In this case, the seam singularity is not a line but a three-dimensional surface which divides Σ into two parts.

We construct the total spacetime as follows; cut the five-dimensional anti-de Sitter spacetime along Σ, keep the side surrounded by Σ, make a copy of the part of five-dimensional anti-de Sitter spacetime with the boundary Σ, join these two with identifying the both boundaries Σ. The brane and bulk begin

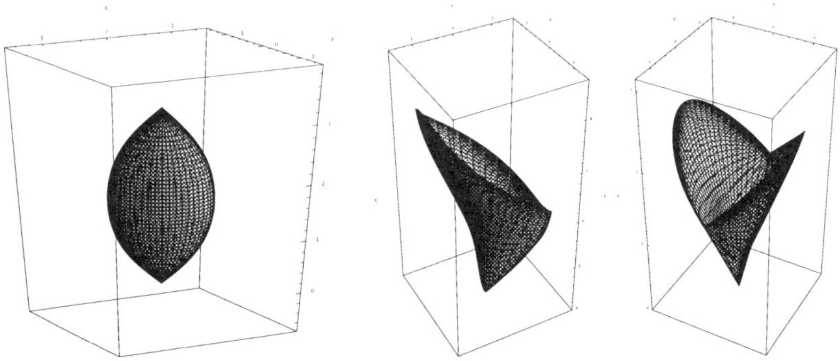

Fig. 1. Conformal diagram of closed, flat, and open FRW-brane universes.

simultaneously at the initial big bang singularity, i^-. Since the initial singularity i^- is point like, there is no particle horizon in the FRW-brane models.

We can consider a tangent space about an arbitrary point on the seam singularity because Σ is smooth there. Since the seam singularity is null subspace, the tangent space of Σ at a point on the seam singularity is spanned by a null vector and spacelike vectors. Any causal curve confined on Σ which starts from the point on the seam singularity is a null geodesic, which is a generator of the seam singularity itself. Therefore, there is no causal curve on Σ which connects a regular point on Σ and the seam singularity. It means that the seam singularity is invisible by use of physical phenomena confined on the brane. However, there exists a causal curve penetrating the bulk which connects a point on Σ and the seam singularity. It means that the seam singularity is visible by gravitational waves propagating in the bulk.

References

[1] P. Horava and E. Witten, Nucl. Phys. **B460**, 506 (1996).

[2] A. Lukas, B.A. Ovrut, K.S. Stelle, and D.Waldram, Phys. Rev. **D59**, 086001 (1999); Nucl. Phys. **B552**, 246 (1999).

[3] L.Randall, R.Sundrum, Phys. Rev. Lett. **83**, 4690 (1999).

[4] P.Binétruy, C.Deffayet, D.Langlois, Nucl. Phys. **B 565**, 269 (2000); C. Csáki, M. Graesser, C. Kolda, J. Terning, Phys. Lett. **B462**, 34 (1999); J.M. Cline, C.Grojean, G.Servant, Phys. Rev. Lett. **83**, 4245 (1999); D. Ida, JHEP **0009**, 014 (2000).

[5] H. Ishihara, Phys. Rev. Lett. **86**, 381 (2000); H. Ishihara, GR-QC 0107085.

[6] W. Israel, Nuovo Cimento **B44**, 1 (1966).

Bulk Gravitational Field and Cosmological Perturbations on the Brane

Kazuya KOYAMA
Graduate School of Human and Environment Studies, Kyoto University, Kyoto 606-8501, Japan
Jiro SODA
Department of Fundamental Sciences, FIHS, Kyoto University, Kyoto 606-8501, Japan

Abstract

We investigate the effect of the bulk gravitational field on the cosmological perturbations on a brane embedded in the 5D Anti-de Sitter (AdS) spacetime. We consider the effects of the bulk perturbations which do not contribute to the metric perturbations but do contribute to the matter perturbations. At low energies, they have imaginary mass $m^2 = -(2/3)\boldsymbol{k}^2$ in the bulk where $\boldsymbol{k}$ is the 3D comoving wave number of the perturbations. They diverge at the horizon of the AdS spacetime. The induced density perturbations behave as sound waves with sound velocity $1/\sqrt{3}$ in the low energy universe. At large scales, they are homogeneous perturbations that depend only on time and decay like radiation. They can be identified as the perturbations of the dark radiation. They produce isocurvature perturbations in the matter dominated era. Their effects can be observed as the shifts of the location and the height of the acoustic peak in the CMB spectrum.

1. Introduction

Recent developments of the particle physics revive the old idea that we are living in 4D brane in higher dimensional spacetime. Since Randall and Sundrum proposed fascinating model for the brane world, many works have been done about the consistency of the model with observations [1]. In their model, our 3-brane universe is located in the 5D Anti-de Sitter (AdS) spacetime. The essence of the model is that the spacetime is effectively compactified with curvature scale l of the AdS spacetime. Thus even the gravity can propagate in the whole higher dimensional spacetime, the 4D Newtonian gravity is reproduced at the scales larger than l on the brane.

After their work, the cosmological consequences of the model are actively investigated. It has been shown that the spatially homogeneous and isotropic universe can also be embedded in this model. In order to study further consistency of the model with the observations, it is needed to study the behaviour of the cosmological perturbations [2].

The purpose of this paper is to clarify the difference between the behaviour of the perturbations in the brane world model and the one in the conventional 4D model. For this purpose, it is desirable to obtain the effective 4D Einstein equations on the brane. In Ref.[3], we derived the effective 4D Einstein equations for scalar cosmological perturbations by solving the perturbations in the bulk. Then the behaviour of the corrections induced by the bulk gravitational field to the conventional 4D Einstein equation are analysed. Following that paper, we clarify the effect of the bulk gravitational fields on the evolution of the cosmological perturbations on the brane.

2. Calculations of the effective Einstein equation

We concentrate our attention on the scalar perturbations. The induced metric on the brane is taken as

$$ds_{brane} = -(1+2\Phi_0)dt^2 + e^{2\alpha_0(t)}(1-2\Psi_0)\delta_{ij}dx^i dx^j. \tag{1}$$

The effective Einstein equations are obtained in two ways. We first use the equations on the brane obtained from the power series expansion of the 5D Einstein equations. From the equations on the brane, we obtained the effective Einstein equations. It should be mentioned that we cannot derive the equation that corresponds to $(i \neq j)$ component of the 4D Einstein equations which gives the relation between the metric perturbations Φ_0 and Ψ_0. Two types of the corrections are found. One is written by the gradient of the metric perturbations. Another is independent of the metric perturbations and induces the density perturbations which behave as the sound waves with the sound velocity $1/\sqrt{3}$ at low energies. At large scales, they are homogeneous perturbations that depend only on time and decay like radiation. We identified them as the perturbations of the dark radiation.

Then we derive the effective Einstein equations again in another way by solving the perturbations in the bulk and imposing the junction conditions. The solution of the bulk perturbations h can be written as

$$h = e^{-2\alpha(y,t)} \int \frac{d^3\boldsymbol{k}}{(2\pi)^3} \int dm\ E(m,\boldsymbol{k}) Z_2(mle^{-\alpha(y,t)}) e^{-i\omega\tau(t,y)} e^{i\boldsymbol{kx}}, \tag{2}$$

where Z_2 is defined as the combination of the Hankel function of the first kind and the second kind;

$$Z_2(mz) = H_2^{(1)}(mz) + a(m)H_2^{(2)}(mz), \tag{3}$$

and $\omega^2 = m^2 + \boldsymbol{k}^2$. So far $E(m, \boldsymbol{k})$ and $a(m)$ are arbitrary coefficients. We should impose the junction conditions on the perturbations (2) on the brane. The point is that the junction conditions are nothing but the effective 4D Einstein equations. We obtain the equation which gives the relation between Φ_0 and Ψ_0. We can identify the perturbations in the bulk which induce the perturbations of the dark radiation. These perturbations do not contribute to the metric perturbations but do contribute to the matter perturbations on the brane. At low energies, they have imaginary mass $m^2 = -(2/3)\boldsymbol{k}^2$ in the bulk and diverge at the horizon of the AdS spacetime, where $\boldsymbol{k}$ is the 3D comoving wave number of the perturbations. The existence of them in the bulk depends on the boundary condition of the perturbations. We should impose two boundary conditions to completely determine the perturbations in the bulk. One is given by the equation of the state of the matter on the brane. The other choice of the boundary condition at the horizon determines the existence of the perturbations of the dark radiation. We will investigate the behaviour of the corrections according to the choice of the boundary conditions.

3. Modifications of the evolution

If we take the boundary condition that the perturbations do not diverge at the horizon of the AdS spacetime, the perturbations of the dark radiation do not appear. The other corrections are suppressed by $\boldsymbol{k}le^{-\alpha_0}$. Thus they correspond to the correction terms written by the gradient of the metric perturbations. Corrections also arise in the relation between Φ_0 and Ψ_0. The corrections become large at the scales smaller than the curvature scales of the AdS spacetime ($\boldsymbol{k}le^{-\alpha_0} \gg 1$) and in the high energy universe with the energy density larger than the tension of the brane ($\dot{\alpha}_0 l \gg 1$). Particularly, at high energies $\dot{\alpha}_0 l \gg 1$, the potential perturbation Φ_0 becomes dominant over the curvature perturbation Ψ_0. We discuss the evolution of the adiabatic perturbations including these corrections. The interesting point is that at sufficiently large scales ($\boldsymbol{k}le^{-\alpha_0} \to 0$) the Bardeen parameter is constant even at high energies. Then the potential perturbation Φ_0 are always constant if the barotropic parameter of the matter w is constant. At the scales below l, the correction becomes large and the evolution equation for the perturbations becomes non-local.

If we choose appropriate boundary conditions in the bulk, the perturbations of the dark radiation arise. The perturbations of the dark radiation induce the isocurvature perturbations in the dust dominated universe. The key feature of them is that they can play a role even in the low energy universe at scales larger than l where the former corrections are suppressed. At large scales the solution for the metric perturbation is given by

$$\Phi_0 = \frac{3(1+w)}{5+3w}\zeta_* + \frac{1}{3(1+3w)}\left(\frac{\rho_r}{\rho}\right)C_*, \tag{4}$$

for $w = p/\rho = const.$ where ρ_r is the energy density of the radiation, ζ_* is constant and C_* determines the amplitude of the perturbations of the dark radiation. At sub-horizon scales, they act as an extra force on the acoustic oscillations of the density perturbation. In the matter dominated era, the location and the height of the acoustic peak are shifted due to the extra force (see Fig.1). These shifts can be directly observed by CMB anisotropies. Recently, many works have been done about the test of the correlation between adiabatic and isocurvature perturbations using CMB spectrum. The detailed analysis of the CMB spectrum will reveal the existence of the dark sound waves.

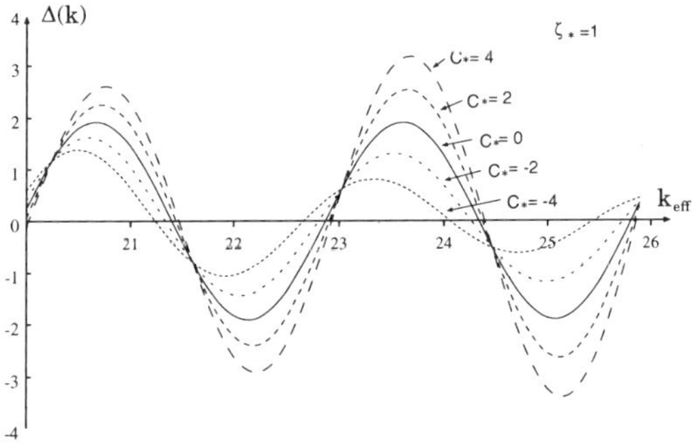

Fig. 1. The density fluctuations $\Delta(\boldsymbol{k})$ at $\rho_r/\rho_m = 0.1$ with $C_* = 0,\ 2\zeta_*,\ 4\zeta_*,\ -2\zeta_*,\ -4\zeta_*$ where we take $\zeta_* = 1$. Here ρ_m is the energy density of the matter. The horizontal coordinate is the value $\boldsymbol{k}_{eff} = \dot{\alpha}_0^{-1}e^{-\alpha_0}\boldsymbol{k}$ at $\rho_r/\rho_m = 1$. The perturbation with $\boldsymbol{k}_{eff} = 1$ crosses the horizon at $\rho_r/\rho_m = 1$. The initial conditions are set at $\rho_r/\rho_m = 100$.

References

[1] L. Randall and R. Sundrum, Phys. Rev. Lett. 83 (1999) 4690.
[2] K. Koyama, J. Soda, Phys. Rev. D62 (2001) 123502.
[3] K. Koyama, J. Soda, Phys. Rev. D65 (2002) 023514.

Brane World, Mass Hierarchy and the Cosmological Constant

Shoichi ICHINOSE
Laboratory of Physics, School of Food and Nutritional Sciences, University of Shizuoka, Yada 52-1, Shizuoka 422-8526, Japan

Abstract

The brane world based on the 6D gravitational model is examined. It is regarded as a higher dimensional version of the 5D model by Randall and Sundrum . The obtained analytic solution is checked by the numerical method. The mass hierarchy is examined. Especially the *geometrical see-saw* mass relation, between the Planck mass, the cosmological constant, and the neutrino mass, is suggested. Comparison with the 5D model is made.

1. Introduction

The higher dimensional approach is a natural way to analyze the 4D physics in the geometrical standpoint The history traces back to the work by Kaluza and Klein. Stimulated by the recent development of the string and D-brane theories, a new type compactification mechanism was invented by Randall and Sundrum[1, 2]. The domain wall configuration in 5D space-time, which is a kink solution in the extra dimension, is exploited. The D-brane inspired model has provided us with new possibilities for the extension of the standard model, with or without the supersymmetry. It has some advantages in the hierarchy problem and the chiral problem. We present a 6D soliton solution, and show that it provides a new dimensional reduction mechanism[3, 4].

2. Six Dimensional Model and Brane World Solution

We consider the 6D gravitational theory with the 6D Higgs potential.

$$S[G_{AB},\Phi] = \int d^6X\sqrt{-G}(-\frac{1}{2}M^4\hat{R} - G^{AB}\partial_A\Phi^*\partial_B\Phi - V(\Phi^*,\Phi)) \quad , \qquad (1)$$

where $V(\Phi^*,\Phi) = \frac{\lambda}{4}(|\Phi|^2 - {v_0}^2)^2 + \Lambda$, $(X^A) \equiv (x^\mu,\rho,\varphi), \mu = 0,1,2,3$. x^μ's are regarded as our world coordinates, whereas $(X^4, X^5) = (\rho,\varphi)$ the extra ones. ρ and φ are taken as in Fig.1. $M(>0)$ is the 6D Planck mass and is regarded as the *fundamental scale* of this dimensional reduction scenario. We take the line

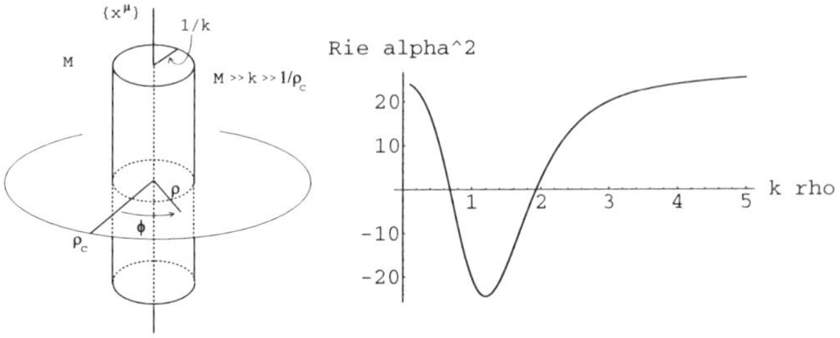

Fig. 1. [LEFT] The pole configuration. [RIGHT] (6D) Riemann scalar curvature $\hat{R}/\alpha^2$ in the 6th order approximation. The horizontal axis is $k\rho$.

element: $ds^2 = \mathrm{e}^{-2\sigma(\rho)}\eta_{\mu\nu}dx^\mu dx^\nu + d\rho^2 + \rho^2\mathrm{e}^{-2\omega(\rho)}d\varphi^2$ $(0 \leq \rho < \infty \quad , \quad 0 \leq \varphi < 2\pi)$, where $\eta_{\mu\nu} = \mathrm{diag}(-1,1,1,1)$. In this choice, the 4D Poincaré invariance is preserved. Two "warp" factors $\mathrm{e}^{-2\sigma(\rho)}$ and $\mathrm{e}^{-2\omega(\rho)}$ appear. The complex scalar field Φ is periodic with repect to φ. Taking a simple case : $\Phi_m(\rho,\varphi) = P(\rho)\mathrm{e}^{im\varphi}$, $(m = 0, \pm1, \pm2, \cdots)$, the Einstein equation reduces to $3\sigma'' - \frac{\sigma'}{\rho} + \sigma'\omega' + \omega'' - (\omega')^2 + 2\frac{\omega'}{\rho} = 2M^{-4}P'^2$; $-16\sigma'^2 + 4\frac{\sigma'}{\rho} - 4\sigma'\omega' + 4\sigma'' = 2M^{-4}V$; $4\sigma'' - 10\sigma'^2 = M^{-4}(P'^2 - m^2\frac{\mathrm{e}^{2\omega}}{\rho^2}P^2 + V)$.

The boundary condition, at $\rho = \infty$(infrared region), for $P(\rho)$ is taken as : $\rho \to \infty$, $P(\rho) \to +v_0$. As for σ and ω, we assume (from the "experience" in 5D Randall-Sundrum model[1, 2, 5, 6]) $\sigma' \to a(\mathrm{const}), \omega' \to b(\mathrm{const})$ as $\rho \to \infty$. Then, from the Einstein equations, we can deduce $m = 0$ and $\sigma' \to \alpha$, $\omega' \to \alpha$, $\alpha = +\sqrt{\frac{-\Lambda}{10}}M^{-2}$ as $\rho \to \infty$. Note that the present asymptotic requirement demands the *isotropic property* around the $\rho = 0$ axis $(m = 0)$, that is, the *pole* configuration. See Fig.1. In the above result we must have the condition: $\Lambda < 0$ (Anti de Sitter). We can also fix the boundary condition at $\rho = +0$(ultra-violet region) based on the power-behavior assumption and $\hat{R}$ regularity: As $\rho \to +0$, the three functions σ', ω' and P goes like $\sigma' \to s\rho^a$, $\omega' \to w\rho^b$, $P \to x\rho^c$ where s, w and x are some constants $(s \neq 0, w \neq 0)$, and $a = b = 1$, $c = 0$, $s = \frac{\lambda}{16}(x^2 - v_0{}^2)^2 + \frac{\Lambda}{4}$.

Let us take the following form for $\sigma'(\rho), \omega'(\rho)$ and $P(\rho)$ as a solution.

$$\sigma'(\rho) = \alpha\sum_{n=0}^{\infty}\frac{c_{2n+1}}{(2n+1)!}\{\tanh(k\rho)\}^{2n+1} ,$$

$$\omega'(\rho) = \alpha\sum_{n=0}^{\infty}\frac{d_{2n+1}}{(2n+1)!}\{\tanh(k\rho)\}^{2n+1} , \quad P(\rho) = v_0\sum_{n=0}^{\infty}\frac{e_{2n}}{(2n)!}\{\tanh(k\rho)\}^{2n} . \quad (2)$$

A *new mass scale* $k(> 0)$ is introduced here and $1/k$ is the "thickness" of the pole. The parameter k, with M and ρ_c(defined later), plays a central role in this dimensional reduction scenario. The distortion of 6D space-time by the existence of the pole should be sufficiently small so that the quantum effect of 6D gravity can be ignored and the present *classical* analysis is valid. This requires the condition[1]: $k \ll M$.

The infrared boundary conditions require the coefficient-constants of (2) to have the following constraints

$$1=\sum_{n=0}^{\infty}\frac{c_{2n+1}}{(2n+1)!}\quad,\quad 1=\sum_{n=0}^{\infty}\frac{d_{2n+1}}{(2n+1)!}\quad,\quad 1=\sum_{n=0}^{\infty}\frac{e_{2n}}{(2n)!}\quad. \tag{3}$$

We solve these constraints. All coefficients are fixed except one *free* parameter $e_0(=P(0)/v_0)$. The first two orders are concretely given as

$$\begin{array}{ll} c_1=\frac{M^{-4}}{4k\alpha}\{\frac{\lambda v_0{}^4(1-e_0{}^2)^2}{4}+\Lambda\}, & c_3=\frac{3\lambda^2 v_0{}^6}{32k^3\alpha}M^{-4}e_0{}^2(1-e_0{}^2)^2+c_1(2+\frac{5\alpha}{k}c_1), \\ d_1=-\frac{2c_1}{3}, e_0 : \text{free parameter}, & d_3=-\frac{4c_1}{3}(1+\frac{5\alpha c_1}{k}), e_2=-\frac{\lambda v_0{}^2}{4k^2}e_0(1-e_0{}^2). \end{array} \tag{4}$$

The general terms $(c_{2n+1}, d_{2n+1}, e_{2n}), n \geq 2$ are obtained in [3]. All coefficients are expressed by four parameters λ, v_0, Λ and e_0. The four ones have three constraints (3) from the boundary condition at the infrared infinity. Hence the present solution is *one-parameter family* solution. For an input value $e_0=-0.8$, the solution is concretely obtained as in Fig.1[RIGHT] and in Fig.2[LEFT]. Furthermore they are checked by the numerical method (Runge-Kutta) in Fig.2[RIGHT] where no assumption is made about the form of the solution. From the above solution (4), we can easily estimate the behavior of the vacuum parameters (Λ, v_0, λ) near the 4D world limit (thin pole limit) as : $-\Lambda \sim M^4k^2$, $v_0 \sim M^2$, $\lambda \sim M^{-4}k^2$, as $k \to \infty$. As for the k-dependence, this result is the *same* as the 5D model of Randall and Sundrum[1, 5].

3. Physical constants and See-Saw relation

Let us consider the case that the 4D geometry is slightly fluctuating around the Minkowski (flat) space: $ds^2 = \mathrm{e}^{-2\sigma(\rho)}g_{\mu\nu}dx^\mu dx^\nu + d\rho^2 + \rho^2\mathrm{e}^{-2\omega(\rho)}d\varphi^2$, $g_{\mu\nu} = \eta_{\mu\nu} + h_{\mu\nu}$. The gravitational part of 6D action (1) reduces to 4D action as

$$\int d^6X\sqrt{-G}M^4\hat{R} \sim M^4\int_0^{2\pi}d\varphi\int_0^{\rho_c}d\rho\,\rho\mathrm{e}^{-2\sigma(\rho)-\omega(\rho)}\int d^4x\sqrt{-g}R+\cdots\quad, \tag{5}$$

where the *infrared regularization* parameter ρ_c is introduced. ρ_c specifies the *size* of the extra 2D space. Using the asymptotic forms, $\sigma \sim \alpha\rho$, $\omega \sim \alpha\rho$ as $\rho \to \infty$ and $\alpha = \sqrt{\frac{-\Lambda}{10}}M^{-2} \sim k$ as $k \to \infty$, we can evaluate M_{pl} as : ${M_{pl}}^2 \sim M^4 2\pi\int_0^{\rho_c}d\rho\,\rho\,\mathrm{e}^{-3\alpha\rho} \sim \frac{M^4}{\alpha^2} \sim \frac{M^4}{k^2}$, where we have used the *4D reduction condition*: $k\rho_c \gg 1$. (This result is *different* from 5D model[1, 5]: ${M_{pl}}^2 \sim \frac{M^3}{k}$.) Writing the

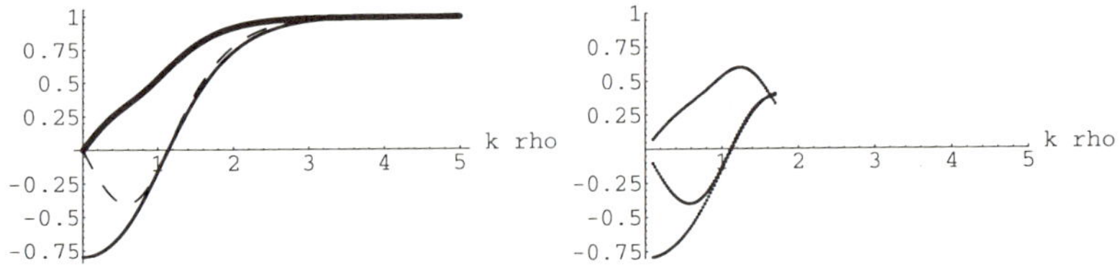

Fig. 2. Horizontal axis: $k\rho$. [LEFT] The analytic results of ω'/α(bold line), σ'/α(dashed line) and P/v_0(normal line). $e_0 = -0.8$. The graphs are depicted in the 6-th order approximation. [RIGHT] The numerical results for ω'/α(top), σ'/α(middle) and P/v_0(down). They are obtained by Runge-Kutta method. The initial point is $k\rho = 0.1$. The initial values are borrowed from the analytical results.

above result as $M_{pl}/M \sim M/k$, we notice this mass relation is the *geometrical see-saw* relation corresponding to the 2 by 2 matrix : $(0,\ M\ //\ M,\ M_{pl})$. This provides the geometrical approach to the see-saw mechanism which is usually explained by the diagonalization of the (neutrino) mass matrix. (See a textbook[7].) Similarly the 4D cosmological constant Λ_{4d} is evaluated as $\Lambda_{4d} \sim \frac{\Lambda}{\alpha^2} \sim -M^4 < 0$, $k\rho_c \gg 1$. Using the value $M_{pl} \sim 10^{19}$GeV , the "rescaled" cosmological parameter $\tilde{\Lambda}_{4d} \equiv \Lambda_{4d}/{M_{pl}}^2$ has the relation: $\sqrt{-\tilde{\Lambda}_{4d}} \sim k \sim M^2 \times 10^{-19}$ GeV . The unit of mass is GeV here and in the following. The observed value of $\sqrt{|\tilde{\Lambda}_{4d}|}$ is roughly 10^{-41}[8].

Some typical cases are 1) ($k = 10^{-41}, M = 10^{-11}$), 2) ($k = 10^{-13}, M = 10^3$) 3) ($k = 10, M = 10^{10}$) 4) ($k = 10^4, M = 10^{11.5}$) and 5) ($k = 10^{19}, M = 10^{19}$). Cases 3) and 4) are moderate cases which are acceptable except for the cosmological constant. At present any choice of (k, M) looks to have some trouble if we take into account the cosmological constant. We consider the observed cosmological constant value should be explained by some unknown mechanism. As in the Callan and Harvey's paper[9], we can have the 4D *massless chiral* fermion bound to the wall by introducing 6D *Dirac* fermion ψ into (1). If we *regulate* the extra axis by the finite range $0 \leq y \leq \rho_c$, the 4D fermion is expected to have a small mass $m_f \sim k\mathrm{e}^{-k\rho_c}$. If we take case 4) and regard the 4D fermion as a neutrino ($m_\nu \sim 10^{-11} - 10^{-9}$GeV), we obtain $\rho_c = (3.45 - 2.99) \times 10^{-3}\text{GeV}^{-1}$. When the quarks or other leptons ($m_q, m_l \sim 10^{-3} - 10^2$GeV) are taken as the 4D fermion, we obtain $\rho_c = (1.61 - 0.461) \times 10^{-3}\text{GeV}^{-1}$. It is quite a fascinating idea to *identify the chiral fermion zero mode bound to the pole with the neutrinos, quarks or other leptons.*

4. Discussion and conclusion

We add some numerical fact about the see-saw relation. In the case 1), the value of M ($\sim 10^{-11}$GeV) is the order of the neutrino mass. This choice looks ridiculous because the space-time behaves as six dimensional at the cosmological scale. The choice is, however, attractive in that it gives the right value of the cosmological constant. If this numerical fact is not accidental and has meaning, it says the cosmological size is related to the neutrino mass when it is "see-sawed" with the Planck mass. These three fundamental scales could be geometrically related. We hope the results of the present hierarchy model will lead to develop further rich possibilities in the cosmology.

References

[1] L.Randall and R.Sundrum, Phys.Rev.Lett. **83**(1999)3370,hep-ph/9905221

[2] L.Randall and R.Sundrum, Phys.Rev.Lett. **83**(1999)4690,hep-th/9906064

[3] S.Ichinose,Univ.of Shizuoka preprint, US-00-11, hep-th/0012255,"Pole Solution in Six Dimensions and Mass Hierarchy"

[4] S. Ichinose, Phys.Rev. **D64**(2001)025012 [Erratum ibid.**D65**(2001)029901], hep-th/0103211

[5] S.Ichinose,Class.Quant.Grav.**18**(2001)421,hep-th/0003275

[6] S.Ichinose,Class.Quant.Grav.**18**(2001)5239,hep-th/0107254

[7] R.N.Mohapatra and P.B.Pal,"Massive Neutrinos in Physics and Astrophysics", World Scientific, Singapore,1991

[8] Perlmutter et al.,Astro.Phys.Journ.**517**(1999)567.

[9] C.G.Callan and J.A.Harvey,Nucl.Phys.**B250**(1985)427

Primordial Black Hole Formation in Natural Double Inflation in Supergravity

Masahide YAMAGUCHI
Research Center for the Early Universe (RESCEU), School of Science, The University of Tokyo, 7-3-1 Hongo, Bunkyo-ku, Tokyo 113-0033, Japan

Abstract

We propose a natural double inflation model in supergravity. In this model, chaotic inflation first takes place by virtue of the Nambu-Goldstone-like shift symmetry. During chaotic inflation, an initial value of second inflation (new inflation) is set, which is adequately far from the local maximum of the potential due to the small linear term in the Kähler potential. Then, primordial fluctuations within the present horizon scale may be produced during both inflations. Primordial fluctuations responsible for anisotropies of the cosmic microwave background radiation and the large scale structure are produced during chaotic inflation, while fluctuations on smaller scales are produced during new inflation. Because of the peculiar nature of new inflation, they can become as large as 10^{-1}-10^{-2}, which may lead to the formation of primordial black holes.

1. Introduction

Primordial black holes (PBHs) are very interesting. Because massive compact halo objects (MACHOs) are observed through gravitational microlensing effects, which are a possible candidate of dark matter. Furthermore, PBHs evaporating now may be a source of antiproton flux observed by the BESS experiment or responsible for short gamma ray bursts (GRBs). PBHs may be realized in the context of double inflation. Though many double inflations have been considered, they are often discussed in a simple toy model with two massive scalar fields. In this talk, we propose a natural double inflation model in SUGRA, where chaotic inflation takes place first of all, during which an initial value of new inflation is dynamically set due to the supergravity effects. The energy scale of new inflation becomes of the same order as the initial value of new inflation so that produced density fluctuations may become as large as the order of unity due to the peculiar nature of new inflation, which straightforwardly may lead to PBHs formation.

2. Double inflation and primordial black hole formation

2.1. Model of double inflation

In this section we propose a double inflation model in supergravity. We introduce an inflaton chiral superfield $\Phi(x,\theta)$ and assume that the model, especially, Kähler potential $K(\Phi,\Phi^*,\ldots)$ is a function of $\Phi+\Phi^*$, which enables the imaginary part of the scalar component of the superfield Φ to take a value larger than the gravitational scale, and leads to chaotic inflation. Such a functional dependence of K can be attributed to the Nambu-Goldstone-like symmetry. We also introduce a spurion superfield Ξ describing the breaking of the shift symmetry and extend the shift symmetry as follows,

$$\begin{aligned} \Phi &\rightarrow \Phi + i\,CM_G, \\ \Xi &\rightarrow \left(\frac{\Phi}{\Phi+i\,CM_G}\right)^2 \Xi, \end{aligned} \tag{1}$$

where C is a dimensionless real constant. Below, the reduced Planck scale M_G is set to be unity. Under this shift symmetry, the combination $\Xi\Phi^2$ is invariant. Inserting the vacuum value into the spurion field, $\langle\Xi\rangle=\lambda$, softly breaks the above shift symmetry. Here, the parameter λ is fixed with a value much smaller than unity representing the magnitude of breaking of the shift symmetry (1).

We further assume that in addition to the shift symmetry, the superpotential is invariant under the $U(1)_R$ symmetry. The above Kähler potential is invariant only if the R-charge of Φ is zero. Then, we are compelled to introduce another supermultiplet $X(x,\theta)$ with its R-charge equal to two, which allows the linear term X in the superpotential. Furthermore we introduce the Z_2 symmetry and a spurion field Π with odd charge under the Z_2 symmetry and zero R-charge. The vacuum value $\langle\Pi\rangle=v$ softly breaks the Z_2 symmetry and suppress the linear term of X. After inserting vacuum values of spurion fields Ξ and Π, the superpotential neglecting higher order terms is given by

$$W \simeq vX - \lambda X\Phi^2 \tag{2}$$

$$= vX(1-g\Phi^2), \tag{3}$$

with $g\equiv\lambda/v$. Here we set both constants v and λ (g) to be real for simplicity. In the same way, the Kähler potential neglecting a constant term and higher order terms is given by

$$K = v_2(\Phi+\Phi^*) + \frac{1}{2}(\Phi+\Phi^*)^2 + XX^*. \tag{4}$$

Here $v_2 \sim v$ is a real constant representing the breaking effect of the Z_2 symmetry. Here and hereafter, we use the same characters for scalar with those for corresponding supermultiplets.

2.2. Density fluctuations and PBHs

Decomposing the scalar field Φ into real and imaginary components,

$$\Phi = \frac{1}{\sqrt{2}}(\varphi + i\chi), \tag{5}$$

the Lagrangian density $L(\varphi, \chi, X)$ is given by

$$L(\varphi, \chi, X) = \frac{1}{2}\partial_\mu\varphi\partial^\mu\varphi + \frac{1}{2}\partial_\mu\chi\partial^\mu\chi + \partial_\mu X\partial^\mu X^* - V(\varphi, \chi, X) \tag{6}$$

with the potential $V(\varphi, \chi, X)$ given by

$$\begin{aligned} V(\varphi, \chi, X) \;=\;& v^2 e^{-\frac{v_2^2}{2}} \exp\left\{\left(\varphi + \frac{v_2}{\sqrt{2}}\right)^2 + |X|^2\right\} \\ &\times\left\{ \left[1 - g(\varphi^2 - \chi^2) + \frac{1}{4}\, g^2(\varphi^2 + \chi^2)^2\right](1 - |X|^2 + |X|^4) \right. \\ &+ |X|^2 \left[\, 2g^2(\varphi^2 + \chi^2) \right. \\ &\quad -(v_2 + \sqrt{2}\varphi)\left\{\sqrt{2}\, g\varphi \left[\, 2 - g(\varphi^2 - \chi^2)\,\right] - 2\sqrt{2}\, g^2\varphi\chi^2\right\} \\ &\quad \left.\left. +(v_2 + \sqrt{2}\varphi)^2 \left\{1 - g(\varphi^2 - \chi^2) + \frac{1}{4}\, g^2\, (\varphi^2 + \chi^2)^2\right\} \,\right] \right\} \end{aligned} \tag{7}$$

While φ and X rapidly goes down to $\mathcal{O}(1)$ because of the exponential factor, χ can take a value much larger than unity. Then the scalar potential is approximated as

$$V \simeq \lambda^2 \left(\frac{\chi^4}{4} + 2\chi^2|X|^2\right), \tag{8}$$

with $\lambda = gv$, which causes chaotic inflation. During chaotic inflation, the potential minimum for φ, φ_{min}, is given by

$$\varphi_{min} \simeq -v_2/\sqrt{2}. \tag{9}$$

Thus the initial value of the inflaton φ of second inflation (new inflation) is set dynamically during chaotic inflation.

Once new inflation begins, χ rapidly goes to zero because the effective mass squared becomes $m_\chi^2 \simeq 6gH^2 \geq 6H^2$. Then, for $\chi \simeq 0$ and $X \ll 1$, the potential is given by

$$V(\varphi, \chi \simeq 0, X \ll 1) \sim v^2 \left\{1 - (g-1)\tilde{\varphi}^2 + 2(g-1)^2\tilde{\varphi}^2|X|^2 + \cdots\right\}, \tag{10}$$

where $\tilde{\varphi} = \varphi - \varphi_{max}$ and $\varphi_{max} \equiv v_2/[\sqrt{2}(g-1)]$. Thus if $g \geq 1$ ($\lambda \geq v$), new inflation takes place and φ rolls down slowly toward the vacuum expectation value

$\eta = \sqrt{2/g}$. Before new inflation starts, φ stays at φ_{min}, which is different from φ_{max}. Then, the initial value of $\tilde{\varphi}$, $\tilde{\varphi}_i$, for new inflation is given by

$$\tilde{\varphi}_i = -\frac{v_2}{\sqrt{2}}\frac{g}{g-1}. \tag{11}$$

Next let us discuss density fluctuations produced during new inflation. Though both $\tilde{\varphi}$ and X are effectively massless fields, only $\tilde{\varphi}$ contributes to growing adiabatic fluctuations. Then, with the fact that $X \ll 1$, the amplitude of curvature perturbation Φ_A on the comoving horizon scale at $\tilde{\varphi} = \tilde{\varphi}_N$ is given by

$$\Phi_A(N) \simeq \frac{f}{2\sqrt{3}\pi}\frac{v}{2(g-1)\tilde{\varphi}_N}. \tag{12}$$

Since $\tilde{\varphi}_i = \mathcal{O}(v)$, the amplitude of curvature perturbation Φ_A becomes as large as of the order of unity, which may produce PBHs. In fact, for example, taking the parameters as

$$\begin{aligned} \lambda &\sim 1.3 \times 10^{-6}, \\ v &\sim 1.1 \times 10^{-6}, \\ v_2 &\sim 0.93 \times 10^{-6}, \end{aligned} \tag{13}$$

MACHO PBHs are realized in our model of double inflation.

3. Discussion and conclusions

In this talk we have proposed a natural double inflation model in SUGRA. By virtue of the shift symmetry, chaotic inflation can take place, during which the initial value of new inflation is set. The initial value of new inflation is adequately far from the local maximum of the potential so that primordial fluctuations within the present horizon scale are attributed to both inflations. Due to the peculiar nature of new inflation, fluctuations on smaller scale are as large as of the order of unity, which may lead to PBHs formation. We find that if we take reasonable values of parameters, PBHs are produced in our double inflation model.

References

[1] M. Yamaguchi Phys. Rev. D64 (2001) 063502.
[2] M. Yamaguchi Phys. Rev. D64 (2001) 063503.
[3] See also references in the above two references.

The Origin of Inflation, Quantum Effects and AdS/CFT in Cosmology

Thomas Hertog
Centre for Mathematical Sciences, University of Cambridge
Wilberforce Road, Cambridge CB3 0WA, United Kingdom

Abstract

We discuss some aspects of trace anomaly driven inflation. We show that according to no boundary initial conditions, the inflationary phase is most likely to occur by semiclassical tunneling via a four sphere instanton. We explain how defining the theory in the Euclidean regime naturally removes the tensor instabilities associated with the higher derivative ghosts. We comment on the effect of the back-reaction of the matter fields on the primordial perturbations.

1. Introduction

The inflationary universe scenario provides an appealing explanation for the riddles of why the observed universe is so flat and isotropic, and it neatly provides a causal mechanism for generating the primordial inhomogeneities which seed the large-scale structure we observe today. However, the origin of inflation is still unclear. The lack of compelling theoretical evidence casts doubt on the claim that the potential energy of a scalar field could be the origin of an inflationary period in the early universe. Moreover, whether inflation actually occurs within a given theory is known to depend strongly on the initial conditions.

An appealing way to circumvent the no go theorems is to include quantum corrections to the classical supergravity equations, such as arise in string theory. In this talk, which is based on work with Stephen Hawking, we argue in favour of Starobinsky's scenario, in which inflation is driven by the trace anomaly of a large number of conformally coupled matter fields [1]. Starobinsky's model has a sound motivation in particle physics, since the Standard Model of particle physics contains nearly a hundred fields and this amount is at least doubled if the Standard Model is embedded in a supersymmetric theory. The SM fields might be expected to be effectively conformally invariant because their masses are negligible compared to the spacetime curvature during inflation.

Starobinsky showed that if one also adds an R^2 term to the action, which arises naturally in the renormalisation of the CFT, then the trace anomaly supports a finite de Sitter phase of expansion which decays into a matter dominated Friedman-Lemaitre Robertson-Walker universe. However, the de Sitter phase is unstable both to the future and to the past, so a priori it is not clear how the universe could have entered the de Sitter phase. In the next section, we show that this problem can be overcome by an appeal to quantum cosmology, which allows the de Sitter phase of the universe to be created by semiclassical tunnelling from nothing.

2. Inflation and Initial Conditions

In order for the universe to have an interesting dynamics which allows structure and complexity to develop, it must be out of equilibrium. All currently known cosmological scenarios for the early universe rely on an initial instability which induces the dynamical evolution. Our present understanding of M-theory does not provide any insight in how the initial conditions should be described, but it does suggest the initial quantum state of the universe is not unique. Usually, therefore, the required initial conditions are imposed by hand. The Hartle-Hawking no boundary proposal, however, gives a definite 'ansatz' for the wave function $\Psi[\Sigma_f, \mathbf{h}, \phi]$ of the universe. This yields a probability measure on the space of initial conditions for cosmology and in principle removes the initial singularity in the hot big bang model. Whether or not the no boundary proposal has anything to do with our actual universe should be decided on the basis of its predictions for observations. If according to the fundamental laws as we know them, universes such as the one we observe are likely to emerge from no boundary initial conditions, then this could be interpreted as evidence that our universe is in the no boundary state.

We now turn to the problem of initial conditions for trace anomaly inflation. Invoking weak anthropic arguments, we consider the quantum no boundary amplitude for an inflationary universe, specified by the values of the Hubble parameter and the matter fields on a spacelike three surface. The presence of a tachyon in the scalar fluctuation spectrum around the four sphere instanton solution in trace anomaly inflation indicates that the situation is similar to the problem of initial conditions in scalar field inflation with a double well potential, allowing the existence of a Hawking–Moss instanton and a family of Hawking–Turok instantons [2]. Hence we can simply look for complex solutions to the Euclidean field equations for gravity coupled to a scalar field with potential $V(\phi) = (1 - a\phi^2)^2$, bounded only by a homogeneous three sphere Σ_f with extrinsic curvature $k_f = 3\dot{a}/a$ and field value $\phi_f = \epsilon$.

Within the semiclassical approximation, there are two contributions to this

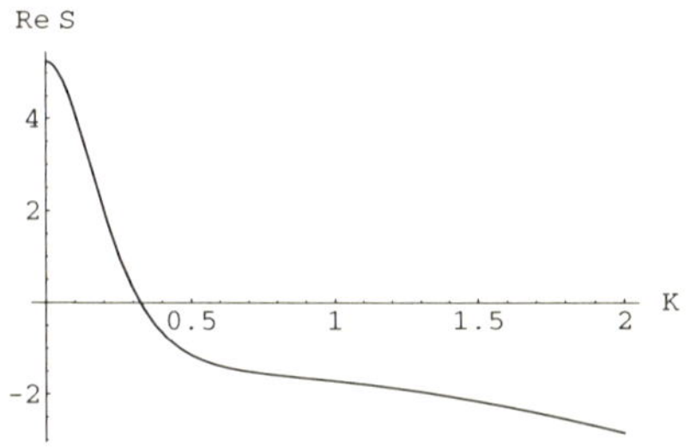

Fig. 1.

amplitude. The first comes from universes that are created via a Hawking–Moss instanton with $\phi_S = 0$ and which inflate for a long time before a quantum fluctuation causes the field to roll down to its final value ϵ. The second contribution comes from neighbouring Hawking–Turok instantons, in which the field starts with some value ϕ_S at the South Pole and slowly rolls up the potential to ϵ on Σ_f. To find the dominant contribution to the path integral, we compute the real part of $S \equiv S_{HM}[k_f, \epsilon] - S_{HT}[k_f, \epsilon]$, the difference between the Euclidean actions of both geometries.

Figure 1 shows $\Re[S]$ as a function* of $K = \frac{k_f}{\sqrt{9-k_f^2}}$, where we have set $a = 1$ in V . We find that the dominant contribution to the no boundary path integral for a large, Lorentzian inflating universe comes from spacetimes which are created by semiclassical tunnelling via a Hawking–Moss instanton [4]. The reason is that although being at the top of the hill costs potential energy, it saves gradient energy by having a scalar field that is constant in space and time.

3. Primordial Perturbations and Instabilities

In order to test trace anomaly driven, we must compare its predictions for the fluctuations in the cosmic microwave background with observation. The first calculations of perturbations in trace anomaly inflation, performed by Starobinsky and Vilenkin, were incomplete and complicated by the fact that tensor perturbations destroy the conformal flatness of a FLRW background, making the effective action for matter fields hard to calculate. However, since the Starobinsky scenario is based on the large N approximation, we now have a very attractive way of calculating the effective action of matter fields for a particular theory, namely $\mathcal{N} = 4$ $U(N)$ super Yang-Mills theory, using the AdS/CFT correspondence. The

*The dependence on ϵ enters only in the prefactor, at least for small ϵ.

fact that we are using $\mathcal{N} = 4$ super Yang-Mills theory is probably not significant and we expect our results to be valid for any theory that is approximately massless during the de Sitter phase.

Using AdS/CFT, we have computed the scalar and tensor metric propagators for trace anomaly driven inflation, taking full account of the back-reaction of matter fields [2]. Using the results of the previous section, we compute the propagators directly from the Euclidean no boundary path integral, by perturbatively expanding around the four sphere instanton. The Euclidean two-point correlators are then analytically continued to Lorentzian signature, where they describe the quantum mechanical vacuum fluctuations according to no boundary initial conditions.

We have included in our action higher derivative counterterms, which arise naturally in the renormalisation of the Yang-Mills theory. The square of the Ricci scalar adjusts the coefficient of the $\nabla^2 R$ term in the trace anomaly, which governs the instability of the de Sitter phase. The square of the Weyl tensor is known to give rise to spin-2 ghosts in a pure gravity theory. When gravity is coupled to the CFT, these ghosts appear in the graviton propagator as a pair of complex conjugate poles in the complex momentum plane [2] and they usually lead to instabilities. However, by defining the theory in the Euclidean regime and then Wick rotating to the Lorentzian, one automatically obtains an integration contour that goes around the complex poles. Hence the no boundary path integral implicitly imposes the final boundary condition that the fields remain bounded, hereby removing the instabilities [3] which have been regarded as the major objection against trace anomaly inflation.

Provided there are sufficiently many matter fields, the de Sitter radius is much larger than the Planck scale, so primordial inhomogeneities are naturally suppressed. Fluctuations are further suppressed, especially on small scales, by the non-local part of the matter effective action. To quantify the latter effect, however, one must perform a detailed computation of the perturbation spectrum through the Starobinsky instability at the end of inflation. At present it is an open question whether the suppression by the matter loops is sufficient to account for the near homogeneity of the microwave sky, without fine-tuning of the coefficients of the higher derivative terms in the action.

The suppression by the matter loops also means that any small-scale modification of four dimensional Einstein gravity will probably be unobservable in the CMB, since matter fields would dominate the graviton propagator at the scales at which such modifications might be expected to become important. To illustrate this point we have studied a Randall-Sundrum brane world analogue of trace anomaly inflation, in which our universe is regarded as a domain wall in five dimensional AdS space with a large N conformal field theory living on the wall [2]. We have shown that the effects of the matter fields dominate the RS corrections, rendering them unobservable. This result is probably not restricted to trace anomaly driven inflation, since it is simply a consequence of the presence of a large number of matter fields.

References

[1] A.A. Starobinsky, Phys. Lett. **91B**, 99 (1980).
[2] S.W. Hawking, T. Hertog, H.S. Reall, Phys. Rev. **D63**, 083504 (2001).
[3] S.W. Hawking and T. Hertog, [hep-th/0107088].
[4] T. Hertog, Ph.D dissertation, University of Cambridge (2001).

Galactic Distributions and Type Ia Supernova Data in the Accelerating Universe with a Local Void

Kenji TOMITA
Yukawa Institute for Theoretical Physics, Kyoto University, Kyoto 606-8502, Japan

Abstract

First it is shown that the cosmological models with a local void are well consistent with high-redshift supernova data. Next I treat the two-point correlations in the inner void region and the outer homogeneous region, and show that their inhomogeneity is cosnsistent with the behavior of the observed correlations in the 2dF and SDSS galactic redshift surveys.

1. Introduction

At present we have important cosmological observations such as the high-redshift supernovae, the anisotropy of the cosmic microwave background radiation and the two-point correlations of galaxies and clusters. Using these observations, many workers have tried to determine the model parameters for homogeneous models. Data of type Ia supernovae were brought by the two groups (HSST[1, 2] and SCP[3]) and the model parameters were derived using the magnitude-redshift relation. The best-fit values in the flat case are $\Omega_0 \simeq 0.3,\ \lambda_0 \simeq 0.7$.

However, Hudson et al.[4] and Willick[5] discovered independently the existence of the bulk flow on scale of about 150 Mpc, which is incompatible with the assumption of the homogeneity. In this sense there is an essential problem about "homogeneity".

After their works, I treated cosmological models with a local void on scales of about 200 Mpc. The motivation was to explain the above-mentioned bulk flow. It is assumed in them that we are in the inner low-density region with larger Hubble constant, and it was shown that the bulk flow is reproduced for an off-center observer using the difference of outer and inner Hubble velocities. Moreover, by the role of the local void, the acceleration was also produced near the present epoch without Λ. This is because the void itself plays the role of positive Λ. First I consider the magnitude-redshift relation in the supernova data, and next large-scale structures in the galactic distribution. The spatial distributions of luminous matter for $z < 0.2$ have been shown as the result of 2dF and SDSS galactic redshift surveys, which suggest various large-scale structures in the region with $z \sim 0.1$ corresponding to the boundary of the local void. However the existence of the boundary is not so clear from the apparent distributions themselves. In order to clarify it, it is necessary to have a quantitative analysis to the distributions. Here I treat the two-point correlations and find their inhomogeneity from comparison of the theoretical correlations with the observed ones.

2. Supernova data in cosmological models with a local void

The inhomogeneous models we consider consist of the inner (low-density) region V^{I} with larger Hubble constant and the outer (high-density) region V^{II} with smaller Hubble constant which are separated by a single shell[6]. It is assumed that the mass in it compensates the mass deficiency in V^{I}. So V^{I} and the shell are regarded as a local void and the wall, respectively. The Hubble constants, the density parameters and the cosmological constants in the two regions are expressed as $(H_0^{I}, \Omega_0^{I}, \lambda_0^{I})$ and $(H_0^{II}, \Omega_0^{II}, \lambda_0^{II})$.

The luminosity distance from sources to off-center observers is complicated in general, but it is given by Eq.(5) in Ref.[7] in a simpler way for a central (or approximately centeral) observer. In Ref.[7] the magnitude-redshift relation in these inhomogeneous models was derived using this luminosity distance in comparison with that in homogeneous models with nonzero cosmological constant, and it was shown that it can explain similarly the supernova data for $z \leq 1$.

In Ref.[8] the confidence contours in the outer parameters $(\Omega_0^{II}, \lambda_0^{II})$ diagram were derived for various ratio of Hubble constants $R \equiv H_0^{II}/H_0^{I}$ using two data sets of HSST and SCP, and it was shown that the Einstein-de Sitter model is included in the 68.3% region for $R = 0.82$. Fig. 1 represents the diagram for the magnitude-redshift relation in various models, relative to the empty model. It is found that the data for $z \leq 1$ are satisfied by the models with a local void (without cosmological constant) as well as homogeneous models with non-zero cosmological constant. In this diagram there is a point which stands for the new data for $z = 1.7$, given by Riess et al.[9] in April. For this data, our void model with outer Einstein-de Sitter region is naturally consistent, while homogeneous model with non-zero cosmological constant may not explain the data. If more data with z larger than 1.5 are provided, we shall be able to distinguish these two types of models more clearly.

3. The redshift dependence of two-point correlations of galaxies

Next let us treat the statistical approach to galactic distribution. One of the best statistical quantities is the 2-point correlation of galaxies. Recently

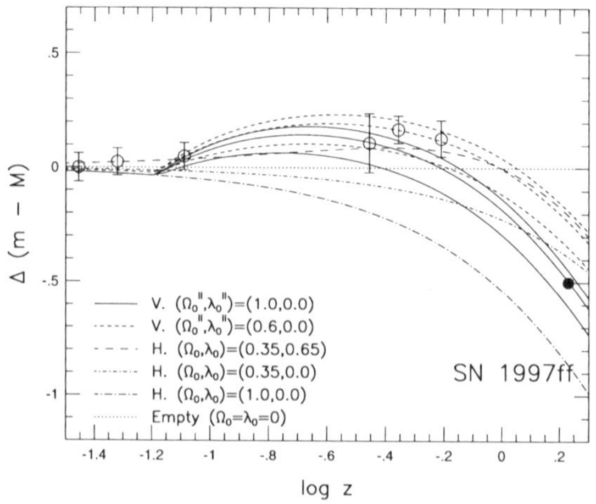

Fig. 1. The $\Delta(m-M)$ - $\log z$ diagram for models with a local v oid (V) and homogeneous models (H). Here $\Delta(m-M) \equiv (m-M)$ in each model $-$ $(m-M)$ in the empty model ($\Omega_0 = \lambda_0 = 0$). In the two V models, we have upper, middle and lower curves corresponding to $R\ (\equiv H_0^{\rm II}/H_0^{\rm I}) = 0.80, 0.82$ and 0.87, respectively. For comparison the SNIa binned data ($z < 1$) and the new data for SN 1997ff ($z = 1.7$) are also depicted in the diagram.

the observations of 2dF and SDSS gave us the important data in the region of $z < 0.2$[10, 11]. The data in both surveys show the inhomogeneity of the correlation length r_0, depending on samples. That is, correlation lengths depend not only on the luminosities but also on the redshifts.

For the change in the correlation length, we have the theories on the luminosity dependence, which are derived in theories of hierachical galaxy formation (for example, Benson et al.'s paper [12]). They show that generally r_0 increases with luminosity for the high-luminosity range, but it is nearly constant for the low-luminosity range of $M - 5\log h \geq -21$, where M is the b-band absolute magnitude and h is for $H_0 = 100h$. On the other hand, observations show r_0 changes by a factor of about 1.5, and the 2-point correlation ξ changes by a factor of about 2.5 in this range. So this factor represents purely the redshift dependence, or the spatial inhomogeneity in the observed region, that is, around $z = 0.07$ or 200 Mpc.

Now let us consider the 2-point correlations in our model with a local void. Here we adopt the case of outer Einstein-de Sitter model and take the parameter set $R = 0.82, \Omega_0^{\rm II} = 1.0$, and $\Omega_0^{\rm I} = 0.3$. Then if we consider the linear density perurbations ($\Delta_{\rm II}$ and $\Delta_{\rm I}$) in the outer and inner regions, the ratio ($\Delta_{\rm II}/\Delta_{\rm I}$) of their amplitudes changes from 1.0 to 1.6 at the interval from the initial time to the present time. So, the present ratio of ξ is estimated to be about $1.6^2 = 2.6$. This value is just equal to the above-mentioned observed ratio of 2-point correlations.

To take into account the effect of nonlinear evolution, I have recently performed N-body simulations ($N \sim 10^6$) for the gravitational evolution of dark matter particles in the region of $z \leq 0.1$ in which a low-density region on scale of 200Mpc is set, and derived the ratio of 2-point correlations in the outer and inner regions. The obtained values for the ratio is about 2.5 which is equal to the value in the linear analysis. This numerical analysis also can explain the spatial change in 2-point correlations in the outer and inner regions. The details will be shown in a forthcoming paper.

4. Concluding remarks

Cosmological models with a local void on scales of 200 Mpc have a strong point in that they can explain the above important observations, but they have also a weak point in that the realization probability is very small. Within our horizon size there may be only a few voids on such a large scale, if we assume a simple power spectrum in the standard dark matter scenario.

The author thanks Prof. J. Ostriker for a kind comment.

References

[1] A. G. Riess et al., Astron. J. 116 (1998) 1009.
[2] A. G. Riess et al., Astron. J. 118 (2000) 2668.
[3] S. Perlmutter et al., Astrophys. J. 517 (1999) 565.
[4] M. J. Hudson et al., Astrophys. J. 512 (1999) L79.
[5] J. A. Willick, Astrophys. J. 522 (1999) 647.
[6] K. Tomita, Astrophys. J. 529 (2000) 26, 38.
[7] K. Tomita, Mon. Not. R. Astron. Soc. 326 (2001) 287.
[8] K. Tomita, Prog. Theor. Phys. 106 (2001) 929, astro-ph/0104141.
[9] A. G. Riess et al., Astrophys. J. 560 (2001) 106.
[10] P. Norberg et al., Mon. Not. R. Astron. Soc. 328 (2001) 69.
[11] I. Zehavi et al., astro-ph/0106476.
[12] A. J. Benson et al., astro-ph/0103092.

Axion Telescope Experiment

M. Minowa
Department of Physics and Research Center for the Early Universe (RESCEU), School of Science, The University of Tokyo, 7-3-1 Hongo, Bunkyo-ku, Tokyo 113-0033, Japan

Abstract

We have constructed a long superconducting magnet called the axion telescope to search for axions which could have been produced in the solar core or other variaous celestial objects by exploiting their conversion to x-rays in the magnetic field. The axion telescope is equipped with PIN-photodiode x-ray detectors at the end and an altazimuth mount to track the object.

We put a limit on axion-photon coupling constant by the search for the solar axions. The limit is more stringent than the limt inferred from the solar age consideration and also more stringent than the recent helioseismological bound. We also put limits on the axion flux coming from the various celestial objects for the first time by the ground-based direct search.

1. Introduction

The axion is a light neutral pseudoscalar particle yet to be discovered. It was introduced to solve the strong *CP* problem[1, 2].

The expected property of the axion is characterized mostly by its mass, m_a. If $10^{-5} < m_a < 10^{-3}\,\mathrm{eV}$, the axion can be copiously produced in the early universe so that they can close the universe. Another astrophysically interesting mass region is at around one to a few eV. Such axions can be allowed for some hardonic axion models by other astrophysical or cosmological constraints. In this region, the axion would be produced in the solar core through the Primakoff effect. It has an energy spectrum very similar to that of black body radiation photons with an average energy of about 4 keV[3, 4]. It can be converted back to an x-ray in a strong magnetic field in the laboratory by the inverse process. The principle of the production and detection is illustrated in Fig. 1..

The conversion rate is given by:

$$P_{a\to\gamma} = \frac{g_{a\gamma\gamma}^2}{4}\left|\int_0^L Be^{iqz}dz\right|^2 \tag{1}$$

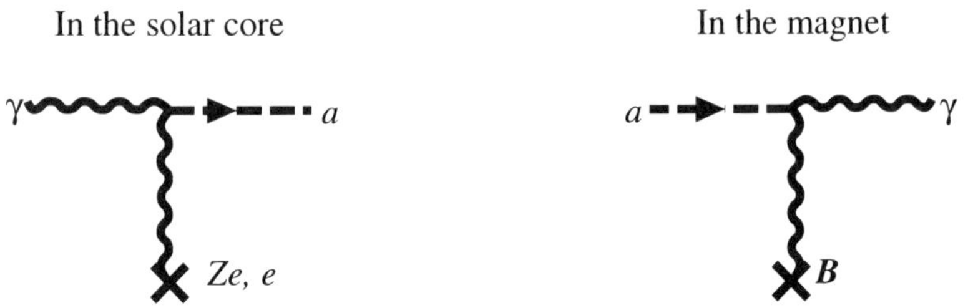

Fig. 1. The solar axions produced via the Primakoff process in the solar core are, then, reconverted into x rays via the reverse process.

where $g_{a\gamma\gamma}$ is the axion-photon coupling constant, z is the coordinate along the incident solar axion, B is the strength of the magnetic field, L is the length along the z-axis, and $q = |(m_\gamma^2 - m_a^2)/2E|$ is the momentum transfer by the virtual photon. Here, m_γ is the effective mass of the photon which is of course zero in vacuum. For heavier axions, momentum transfer, q, becomes not negligible, thus the sensitivity of the detector is lost. Coherence can, however, be restored by filling the conversion region with buffer gas since a photon of x-ray region has an effective mass in a medium. For light gas, such as hydrogen or helium, it is simply written as:

$$m_\gamma = \sqrt{\frac{4\pi\alpha N_e}{m_e}}. \tag{2}$$

We search for such x-rays coming from the direction of the sun with a newly developed instrument called the axion telescope.

2. Axion Helioscope

A schematic view of the helioscope is shown in Fig. 2.. It consists of three parts, a tracking system, superconducting coils and x-ray detectors. The tracking system is an altazimuth mount and drives a 3-m long vacuum cylinder of the helioscope to track the sun. Its trackable altitude ranges from $-28°$ to $+28°$ and almost any azimuthal direction is trackable. This view range corresponds to about 50% of duty cycle for the sun measurement in Tokyo. The other half of a day we measure the background. This helioscope mount is driven by two AC servo motors controlled by a personal computer (PC) through CAMAC bus. The PC regularly monitors two precision rotary encoders through CAMAC bus and forms a feedback control loop. The directional origin of the helioscope was measured using a theodolite. The azimuthal origin is determined from the observed direction

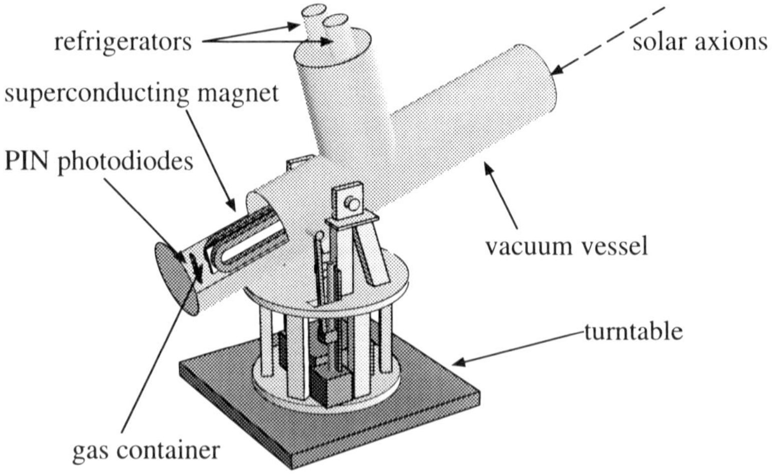

Fig. 2. The schematic view of the axion helioscope.

of Polaris and the absolute altitude is determined from a spirit level.

The superconducting coils and the x-ray detector are mounted in the cylinder. The aperture volume between the coils is $2300(\mathrm{L}) \times 92(\mathrm{H}) \times 20(\mathrm{W})\ \mathrm{mm}^3$ and the magnetic field strength in the aperture is 4 T. The coils are kept at 5–6 K during operation. They are directly cooled with two Gifford-McMahon refrigerators and no cryogen is needed. The magnet has also a persistent current switch. After excitation, it is switched into persistent current mode and the current leads are taken away. The details of this cryogen-free superconducting magnet are described elsewhere[5].

Sixteen PIN photodiodes, Hamamatsu Photonics S3590-06, are used as the x-ray detectors(Fig. 3.). The chip size is $11 \times 11 \times 0.5\,\mathrm{mm}^3$. Each chip is mounted on a Kapton film bonded to an Invar plate with cryogenic compatible adhesive. The x-ray detectors are mounted in a radiation shielding box made of oxygen-free high conductivity copper (OFHC Cu) which is operated at about 60 K. The copper shield is surrounded by a lead shield which is at room temperature.

The output from each photodiode is fed to a charge sensitive preamplifier whose first-stage FET is at the cryogenic stage near the photodiode chip and the preamprefier outputs are digitized using flash analog-to-digital convertors (FADC's), REPIC RPC-081's. We performed numerical pulse shaping to the raw waveform using the Wiener filter. The energy of an x ray is given by the peak height of a wave after shaping. Each detector was calibrated by 6-keV manganese

Fig. 3. 16 PIN photodiodes and head parts of th preamplifiers.

x ray from a ^{55}Fe source which can be exposed to the detector and hidden behind the shield during the measurement. The energy resolutions for 6-keV photon was 0.8–1 keV (FWHM).

3. Solar Axion Search

In the first stage of the solar axion search experiment, no buffer gas was used in the magnetic field region. The measurement consists of two parts, the solar observation and the background measurement. While the sun is in the trackable range, the helioscope is directed to the sun. For the rest of the time, the background spectrum was measured. The U.S. Naval Observatory Vector Astronomy Subroutines (NOVAS) [7] was used to calculate the sun position.

From the absence of the axion signal, an upper limit to the axion-photon coupling was set to be $g_{a\gamma\gamma} < 6.0 \times 10^{-10}\mathrm{GeV}^{-1}$ (95% CL) for $m_a < 0.03\,\mathrm{eV}$. The result is published in [6]. The upper limit is shown in Fig. 4. as a function of m_a. The limits from the second stage experiment are plotted together. There are also some other bounds plotted in the same figure. The SOLAX [8] is a solar axion experiment which exploits the coherent conversion on the crystalline planes in a germanium detector. The limit $g_{a\gamma\gamma} < 2.3 \times 10^{-9}\mathrm{GeV}^{-1}$ is the solar limit inferred from the solar age consideration. The limit $g_{a\gamma\gamma} < 1 \times 10^{-9}\mathrm{GeV}^{-1}$ is a new solar limit recently reported[9]. Above this line, standard solar models with energy losses by solar axions cannot fit to the helioseismological sound-speed profile. Our upper limit is the first to go lower than these two solar limits in this mass region.

In the second stage of the search experiment, the convergion region was filled with the buffer gas to retain the coherence of the convergion of the heavier mass axions.

The new device introduced this time is the gas container. We adopted cold helium gas as the dispersion-matching medium. Light gas is preferred since it minimizes x-ray absorption by the gas. Helium has another virtue that it remains at gas state at 5 K, the same temperature as the coil. At 5 K, the gas pressure corresponding to our ultimate goal, $m_a = 2.6\,\mathrm{eV}$, reaches only 0.13 MPa. The container body is made of four stainless steel square pipes welded to each other. The entire container body is wrapped with 5N high purity aluminium sheet to achieve high uniformity of temperature. At the end of the container, gas is separated from vacuum with an x-ray window which is transparent to x ray above 2 keV and can hold gas up to 0.3 MPa at liquid helium temperature.

The measurement with buffer gas was performed for ten photon mass settings to scan up to 0.26 eV.

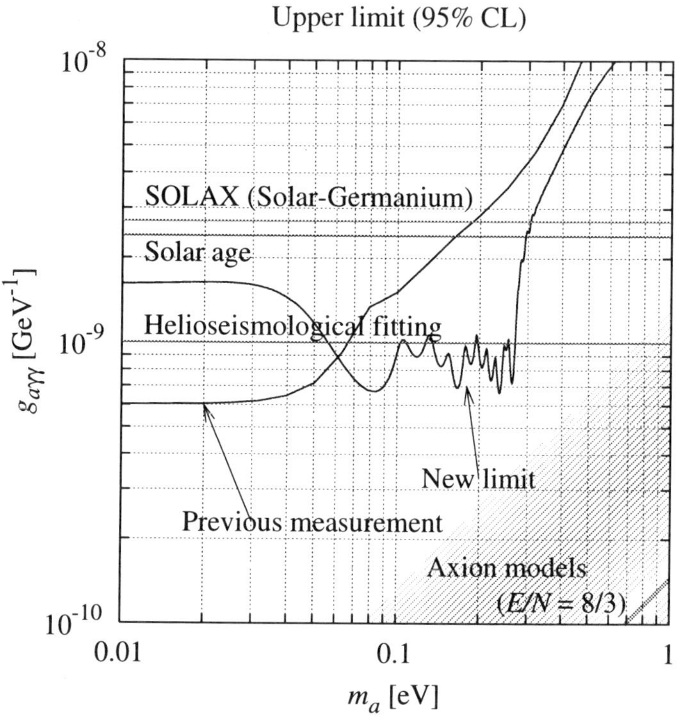

Fig. 4. The exclusion plot on $g_{a\gamma\gamma}$ to m_a at 95% confidence level is plotted where some other bounds are plotted together. The limit from the first stage experiment without buffer gas is shown as "Previous measurement". "New limit " corresponds to the second stage experiment with buffer gas. Dashed lines are the limit by SOLAX experiment, the limit inferred from the solar age consideration, and the recent helioseismological bound. The hatched area is the preferred axion models. The thick line corresponds to the case when a simple GUT is assumed.

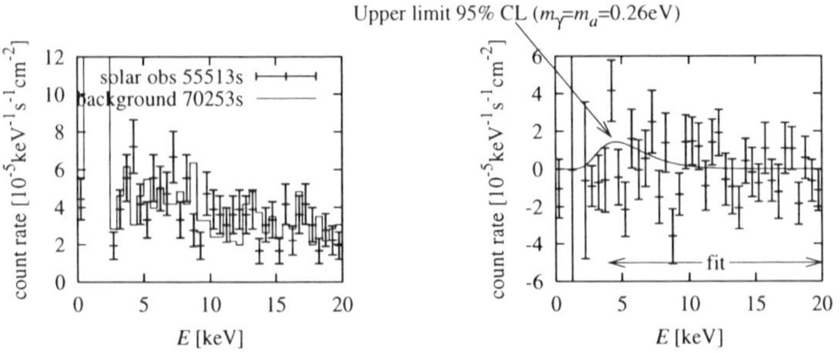

Fig. 5. The left figure shows the energy spectrum of the solar observation (error bars) and the background spectrum (solid line) when the gas density was tuned to $m_\gamma = 0.26\,\text{eV}$. The right figure shows the net energy spectrum of the left where the background is subtracted from the solar observation.

In Fig. 5., one of the energy spectra of the solar observation is shown together with the background spectrum. We searched for expected axion signals which scales with $g_{a\gamma\gamma}^4$ for various m_a in these spectra. The smooth curve in the figure represents an example for the expected axion signal for $m_a = m_\gamma = 0.26\,\text{eV}$ and $g_{a\gamma\gamma} = 7 \times 10^{-10}\text{GeV}^{-1}$.

In the analysis, the ten measurements with the ten different gas densities are combined together by using the total χ^2. The energy region 4–20 keV was used for fitting. Since no significant excess was seen for any m_a, we set upper limit on $g_{a\gamma\gamma}$ at 95% confidence level using the Bayesian method. In Fig. 4., upper limits on $g_{a\gamma\gamma}$ are plotted as a function of m_a. The result is published in [10].

Again, our limits are better than the limit inferred from the solar age consideration. They are also better than the helioseismological limit except in a small fraction of the mass region.

4. Other Axion Emitting Stellar Objects

Our helioscope is not necessarily only for the sun. In principle, it can be used to search for the axions from other stellar objects.

The arguments on the solar axion can be applied to other stars and axion luminosity of them can be calculated[11]. Compact objects like white dwarfs, neutron stars and black holes can also be the axion sources in two aspects. One is the thermal emissions from the objects themselves and another is the thermal emissions from their accretion disks. Axion luminosity of the thermal emission from the objects is calculated[12, 13] although axion luminosity from the accretions are not calculated.

Similar to neutrinos, the supernova explosion liberates its energy in the form of the axions [14, 15, 16]. The gamma ray bursts, which are the most violent explosions in the universe, are also the candidate. However, it is rather hard to observe these two objects by our apparatus because we do not know where to see in advance of the occurrence.

Except for the solar axion search, no ground-based celestial search experiments have been reported by now. There exist some observational constraints on $g_{a\gamma\gamma}$ extracted from the analyses of satellite x-ray observations[17, 14, 15, 16]. These limits are very stringent but valid only for a massless or a very light pseudoscalar, since the long conversion field requires strict coherency.

We show here the preliminary results of the celestial axion search experiments done by directing the axion helioscope toward the possible sources. During the search, no buffer gas was used in the magnetic field region. Our search consists of two methods. The one is the celestial scan. Fig.6. shows the region of the scan in the galactic coordinate.

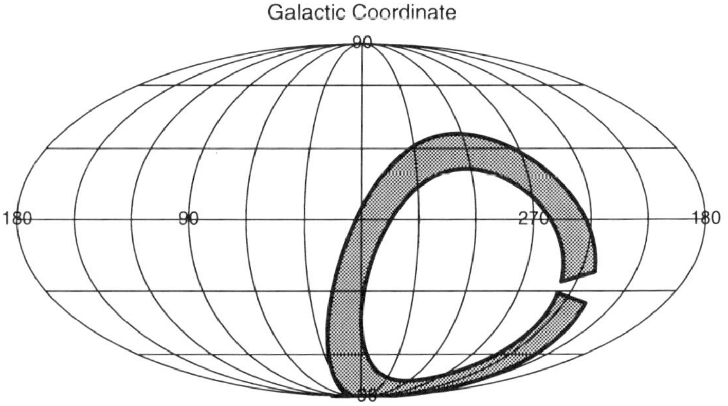

Fig. 6. The constrained region in the galactic coordinate. The shaded area is the constrained region. This region occupies about 10 % of the celestial sphere.

We scanned 10 % of the celestial sphere by the apparatus, and found no apparent signals. By assuming that the axion energy distribution is flat in 4–20 keV, the 95 % confidence limit on the differential flux is $d\Phi_a/dE_a < 3.3 \times 10^{16}$ keV^{-1}s^{-1}cm^{-2} for 4–20 keV, provided the axion mass of $m_a < 0.03$ eV and the axion-to-photon coupling of $g_{a\gamma\gamma} = 10^{-10}$ GeV^{-1}.

The other is the tracking. We examined the galactic center, Scorpius X-1,

Vela X-1 and Crab nebula by tracking these objects. The results are 10^2–10^3 more stringent than the scan result.

These results are the first reliable constraints for the axion emission from celestial objects other than the sun. Since the distances to the sources are very long, these limits are higher than the theoretically predicted values. However, these results give reliable informations for the celestial objects whose theories have much uncertainty, and can be strong constraints for exotic objects which might be proposed in the future.

5. Conclusion

We have developed the axion helioscope. The solar axion with mass up to 0.26eV has been searched for. But no evidence for the axion was seen. New upper limit on $g_{a\gamma\gamma}$ which ranges 6.0–$9.6 \times 10^{-10}\mathrm{GeV}^{-1}$ was set for $m_a < 0.26\,\mathrm{eV}$, which is far more stringent than the solar existence limit and is also more stringent than the recent helioseismological bound. This experiment is currently the only one existing experiment which has enough sensitivity to detect such solar axion that do not violate the solar model itself.

We also put limits on the axion flux coming from the various celestial objects for the first time by the ground-based direct search.

Acknowledgments

This research is supported by the Grant-in-Aid for COE research by the Japanese Ministry of Education, Science, Sports and Culture, by the Grant-in-Aid for scientific research by the Japan Society for the Promotion of Science, and also by the Matsuo Foundation. This work is done in collaboration with Y. Inoue, T. Namba, S. Moriyama, Y. Takasu, T. Horiuchi, and A. Yamamoto.

References

[1] R.D. Peccei, H.R. Quinn, Phys. Rev. Lett. 38 (1977) 1440; Phys. Rev. D 16 (1977) 1791; S. Weinberg, Phys. Rev. Lett. 40 (1978) 223; F. Wilczek, Phys. Rev. Lett. 40 (1978) 279.

[2] J.E. Kim, Phys. Rep. 150 (1978) 1; M.S. Turner, Phys. Rep. 198 (1990) 67; L.J Rosenberg, K. van Bi bber, Phys. Rep. 325 (2000) 1; G.G. Raffelt, Phys. Rep. 333-334 (2000) 593.

[3] P. Sikivie, Phys. Rev. Lett. 51 (1983) 1415.

[4] K. van Bibber *et al.*, Phys. Rev. D 39 (1989) 2089.

[5] Y. Sato, K. Makishima, S. Mizumaki, A. Yamamoto, S. Moriyama, Y. Inoue, M. Minowa, T. Namba, Y. Takasu, Development of A Cryogen-free Superconducting Dipole Magnet, *the 15th International Conference on Magnet Technology* (MT-15, Beijing, October 20–24, 1997), and KEK-PREPRINT-97-202.

[6] S. Moriyama, M. Minowa, Y. Inoue, T. Namba, Y. Takasu, and A. Yamamoto, Phys. Lett. B 434 (1998) 147.

[7] G. H. Kaplan *et al.*, Astronomical Journal 97 (1989) 1197;
URL: `http://aa.usno.navy.mil/AA/software/novas/`.

[8] A.O. Gattone, *et al.*, Nucl. Phys. B(Proc. Suppl.) 70 (1999) 59.

[9] H. Schlattl, A. Weiss, and G. Raffelt, Astropart. Phys. 10 (1999) 353.

[10] Y. Inoue, T. Namba, S. Moriyama, M. Minowa, Y. Takasu, T. Horiuchi, and A. Yamamoto, astro-ph/0012338.

[11] E. D. Carlson, Phys. Lett. B 344 (1995) 245.

[12] N. Iwamoto, Phys. Rev. D 53 (1984) 1198.

[13] D. E. Morris, Phys. Rev. D 34 (1986) 843.

[14] J. W. Brockway, E. D. Carlson, and G. G. Raffelt, Phys. Lett. B 383 (1996) 439.

[15] E. Massó, hep-ph/9704056.

[16] J. A. Grifols, E. Massó, and R. Toldrà, Phys. Rev. Lett. 77 (1996) 2372.

[17] E. D. Carlson, and W. D. Garretson, 336 (1994) 431.

Resolving the X-ray Spectral Paradox: XMM-Newton Spectra of Faint Sources in the Lockman Hole

Richard GRIFFITHS, Takamitsu MIYAJI and Andrew PTAK
Dept. of Physics, Carnegie Mellon University, 5000 Forbes Ave., Pittsburgh, Pennsylvania 15213, USA

Abstract

We report on the X-ray spectra of the faint sources detected by using the XMM-Newton Observatory in a 100ks exposure on the Lockman Hole, a region previously studied exceptionally well using ROSAT and other observatories. When the X-ray spectra of optically identified and unidentified X-ray sources are stacked together by optical type, we find that the summed spectra are best fit by (i) a spectral photon index of 1.9 for Type I AGN, (ii) a spectral index of 1.2 for Type II AGN, and (iii) a harder index of 1.1 for (non-ROSAT) unidentified sources selected in the 5 – 8 keV band. We conclude that the composition of the X-ray background between 2 and 10 keV is largely comprised of the latter two types of source.

Comparison between a CXO (1 Ms) and XMM (160ks) exposure in the HDF-N indicates agreement in spectral results, but also shows some confusion in the XMM sources at flux levels below about 5×10^{-15} ergs cm^{-2} s^{-1}. Although such deep exposures with CXO will be very limited in number, XMM exposures at such depths will be more common and the latter will eventually be used to study the cosmic variance in the deep survey sources and to study the evolution in the luminosity function of the multiple classes of galaxies and AGN which contribute to the faint source counts.

1. Introduction

CXO and XMM-Newton have enabled us to make rapid progress in deciphering the long standing problem of the origin of the X-ray background. A key factor in this problem has been the mismatch between the spectra of AGN and that of the XRB, a problem known as the 'spectral paradox', and we focus in this paper on the resolution of this paradox in broad terms. The XRB spectrum, approximated as a power law, has an effective photon index of 1.4 between 2 and 20 keV, and an energy density which reaches a peak near 30 keV. (see the 1992 review by Fabian & Barcons [1]. One of the major goals of the current generation of X-ray observatories has been to identify those source populations which are primarily responsible for the bulk of the overall energy density.

In this 2 – 10 keV energy range the XRB has now been resolved into point sources at the $\sim 80 - 90\%$ level by using CXO (e.g. [3], [4], [5], [6] and refs. therein) and by using XMM-Newton [7]. These observations have effectively been made at a typical photon energy of about 3 kev, since most of the detected photons are at the soft end of the specrum. The fraction of the XRB resolved by the CXO at 8 - 10 kev is about 20% in a few hundred ks.

XMM-Newton has great effectiveness in terms of collecting area for measurement of deep survey source spectra. In the Lockman Hole deep survey area, the optical identifications have largely been completed as a result of the deep ROSAT surveys ([2], [8]), and we are thus able to obtain X-ray spectra of the various classes of source contributing to the X-ray background. We comment here on the summed spectra of the broad source classes so as to be able to extrapolate to energies above 10 kev where the bulk of the XRB energy resides.

2. Observations and Results

2.1. The Lockman Hole

The Lockman Hole is a region in which the Galactic hydrogen column density N_H is only 6×10^{19} cm^{-2}, and it has therefore been the subject of intense study at soft X-ray energies using the ROSAT observatory [2]. As many as 1000 sources deg^{-2} were detected in this field (0.5 – 2 keV) and a considerable amount of follow-up optical spectroscopy has been performed using the Keck and other telescopes [8].

The XMM-Newton observatory was pointed at the Lockman Hole in Apr./May 2000, resulting in a total of ~100 ks. of useful data. A total of ~ 150 sources was detected, as reported in Hasinger et al. 2000 [7]. Of these sources, about 50 were previously detected with ROSAT, so that about 100 are new detections with XMM-Newton. The resulting log(N) - log(S) analysis and hardness ratios of the sources are also presented in ref. [7].

2.2. Spectra of Optically Identified Sources

In Table 1 and Figure 1, we show the summed x-ray spectra of sources for which the redshift has been measured via optical spectroscopy (from ref.[8]), together with the spectra of the unidentified sources. In summary, the best-fit power-law spectral indices for these summed spectra are 1.9 for the Type I objects, 1.2 for the Type II's, and 1.1 for unidentified sources (hard-selected), respectively.

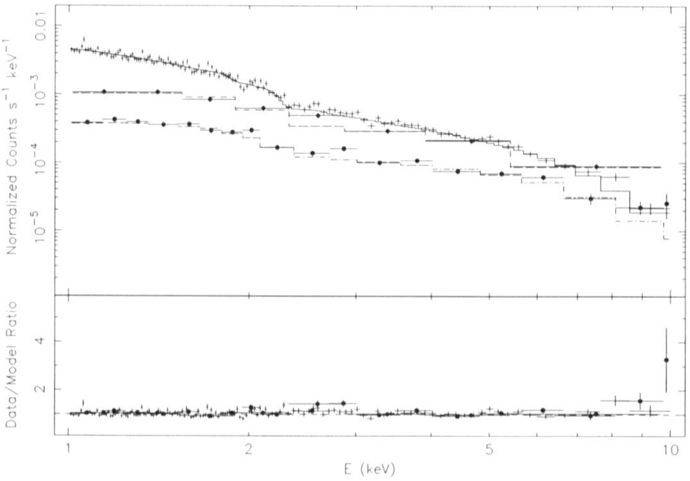

Fig. 1. Spectra Summed by AGN Type
top AGN Type I, *middle* AGN Type II, *bottom* unidentified sources, hard-selected

Table 1. X-ray Spectral Indices

Fit	N	Nh	Gamma	F(2-10)	Chisq/dof
		10^{21} cm^{-2}			
AGN I	42	0.5 (< 1.1)	1.87 (1.80-1.94)	2.32e-14	632/606
AGN I*	27	0.4 (< 1.0)	1.85 (1.78-1.92)	3.55e-14	600/562
AGN II	5	2.6 (< 5.4)	1.19 (0.93-1.45)	2.68e-15	71/71
No ROSAT ID	99	0.0 (< 1.4)	1.49 (1.39-1.68)	3.58e-15	711/635
No ROSAT ID*	40	1.8 (< 3.7)	1.11 (0.93-1.31)	7.06e-15	284/309

Notes to Table 1:
* = selected to have a > 90% confidence detection in the 5-8 keV bandpass.
F(2-10) is the mean flux/galaxy in cgs units. All AGN II galaxies have significant detection in the hard band.
N = number of galaxy spectra summed

AGN I = sources in Lehmann et al. [8] with class = a-c

AGN II = sources in Lehmann et al. [8] with class = d

3. Conclusions

- From a deep XMM-Newton observation of the Lockman Hole:

  (i) AGN of type I have an average photon spectral index of 1.9 (2 – 10 keV), as found by many previous experiments

  (ii) AGN of type II have an average index of 1.2 (2 - 10 keV)

  (iii) unidentified sources have a spectral index of 1.4 (2 –10 keV) and 1.1 (hard selected 5 – 8 keV).

- For sources identified with galaxies, the centroid source position as measured using XMM may not be sufficient to distinguish between an active nucleus and an off-nuclear transient or ultraluminous object of stellar or intermediate mass.

- Extrapolating to higher energies, it is evident that AGNs of Type II and the unidentified sources make up the bulk of the energy in the XRB.

- XMM deep surveys start to become confused at flux levels of 5×10^{-15} ergs cm^{-2} s^{-1} (2 - 10 kev), reached in exposure times of $\sim$ 150 ks.

- The multiple deep surveys with XMM will allow the study of cosmic variance and evolution of the source populations.

4. Acknowledgments

These observations were made using the XMM-Newton X-ray observatory, built, launched and operated by the European Space Agency with NASA participation. These results rely heavily on the analysis performed and reported by Guenther Hasinger and colleagues [7].

References

[1] A. C. Fabian and X. Barcons, Ann. Rev. Astron. Astroph. 30 (1992) 429.
[2] G. Hasinger et al., Astron. & Astroph. 329 (1998) 482.
[3] R. F. Mushotzky, L. L. Cowie, A. J. Barger, K. A. Arnaud, Nature 404 (2000) 459.
[4] G. P. Garmire et al., Ap. J. submitted (2002)
[5] A.E. Hornschemeier et al., Ap. J. 554 (2001) 742.
[6] W. N. Brandt et al., Astron. J. 122 (2001) 1.
[7] G. Hasinger et al., Astron. & Astroph. 365 (2001) L45
[8] Lehmann et al., Astron. & Astroph. 371 (2001) 833.

Observational Approach to Molecular Cloud Formation

Satoshi YAMAMOTO
Research Center for the Early Universe (RESCEU) and Department of Physics, School of Science, The University of Tokyo, 7-3-1 Hongo, Bunkyo-ku, Tokyo 113-0033, Japan

Abstract

We have constructed the 1.2 m submillimeter-wave telescope on a top of Mount Fuji for the exclusive use of large-scale observations of the Galactic molecular cloud with the CI lines. The telescope has been operated successfully during 4 observing seasons in a remote way from the University of Tokyo since 1998. With this telescope, we have already observed more than 40 square degrees of the sky with the CI 492 GHz line. The distribution of the CI emission is found to be different from those of the ^{13}CO or $C^{18}O$ emission in some clouds. These differences are discussed in relation to formation processes of molecular clouds.

1. Introduction

Star formation is an important elementary process in evolution of galaxies, and hence, it has extensively been studied by observations in various wavelengths from radio to X-ray and by theoretical considerations and simulations. As a result, a dense core in a molecular cloud is now established as a birthplace of a new star, and evolution of a protostar to a main sequence star has been explored in detail. On the other hand, formation processes of a parent molecular cloud are poorly understood. Since the initial mass function of newly born stars is suggested to depend on physical properties of molecular clouds, it is of crucial importance to understand formation processes of molecular clouds that would ultimately determine these properties.

In general, interstellar clouds can be divided into two large categories, ionized clouds and neutral clouds. The neutral clouds can further be classified into two. One is a rather diffuse atomic cloud where the interstellar UV radiation well penetrates, and the other is a dense molecular cloud where the interstellar UV radiation is almost shielded. Distributions of diffuse clouds have been explored most extensively with the HI 21 cm line observations [1], whereas distributions of molecular clouds have been surveyed with the rotational transitions of the CO molecule [2].

In order to understand the formation processes of molecular clouds from diffuse clouds, we have to identify a phase transition layer between the atomic gas and molecular gas. For this purpose, we are focusing on the neutral carbon atom, C. In a diffuse cloud, the major form of the gas-phase carbon is C^+, because the ionization potential of the C atom (11.3 eV) is lower than that of H (13.6 eV). As the cloud density increases and the UV radiation is shielded, the major form of the gas-phase carbon changes from C^+ to C, and then C to CO. Therefore, the neutral carbon atom traces a spatially and evolutionally intermediate region between the atomic and molecular gases. By comparing the distribution of the neutral carbon atom with those of CO and H, we would be able to understand observationally how molecular clouds are formed in the Galaxy.

The neutral carbon atom has two submillimeter-wave transitions among three fine structure levels in the ^3P ground electronic state. One is the $^3P_1 -^3 P_0$ transition at 492 GHz, and the other is the $^3P_2 -^3 P_1$ transition at 809 GHz. The 492 GHz line was first detected by Phillips et al. in 1980 [3]. Since then, detailed distributions of this line toward a few representative objects were studied with the large aperture submillimeter-wave telescopes such as the CSO 10 m antenna and the JCMT 15 m antenna. On the other hand, the *FIRAS* experiment of the *COBE* satellite demonstrated with its 7° beam that the CI emission is widespread over the Galactic plane [4]. This result clearly shows an importance of large-scale observations of the CI line. Plume and his collaborators developed the focal reducer system for the CSO 10 m antenna, and surveyed the CI distribution toward the S140 molecular cloud with the $2' - 3'$ beam [5]. This was a great step toward the large-scale observation of the CI line, but the observed area was still limited in comparison with those of CO and its isotopic species.

In contrast to the 492 GHz line, the observation of the 809 GHz line was very few. The 809 GHz line was first detected by Jaffe et al. [6]. Since then, no mapping observation has been reported for this line except for ours [7]. This is because of difficulty in realizing a sensitive receiver system at this frequency. However the observations of the 809 GHz line would be important for detailed understanding of the physical conditions of the CI emitting region.

Under these circumstances, we planned to construct a small submillimeter-wave telescope for the exclusive use of the CI observations in order to delineate the large-scale CI distributions in various molecular clouds. This is an observational approach to the molecular formation from a chemical point of view.

2. Mount Fuji Submillimeter-wave Telescope

The CI observations are not easy from the ground-based telescopes, because the submillimeter-wave from space is heavily absorbed by the water vapor contained in the earth atmosphere. Therefore the high-altitude and dry site is essential for the CI observations. We surveyed the possible observing sites on the basis of the meteorological data, and found that the Mount Fuji (el. 3775 m) is the unique place for the CI observation in Japan. In fact, we installed the 220 GHz tipping radiometer on a top of Mount Fuji in collaboration with Meteorological Agency of Japan, and continuously evaluated the atmospheric conditions during two years. As a result, it was found that about 45 % of the total time from November to March can be used for the CI observation [8]. Encouraged by this result, we decided to construct a submillimeter-wave telescope at Mount Fuji.

Figure 1 shows a schematic drawing of the telescope. A diameter of the main reflector is 1.2 m, which provides us with the beam size of 2.′2 at 492 GHz. The surface accuracy is better than 10 μm. An SIS mixer receiver for the 492 GHz and 809 GHz bands is installed at the Nasmyth focus. The Nb-based PCTJ junctions fabricated at Nobeyama Radio Observatory were employed as the mixers. The receiver temperature is 120 K (DSB) at 492 GHz and 580 K (DSB) at 809 GHz [9]. With this heterodyne mixer, the submillimeter-wave signal from space is down-converted to the intermediate frequency centered at 2.3 GHz. Then the IF signal is amplified, and is fed into the acoust-optical radiospectrometer (AOS) to obtain the spectrum with resolution of 1.6 MHz. The telescope system is enclosed in a space frame radome with a Gore-Tex membrane. The radome loss including the frame blockage is 10 % at 492 GHz. It is very difficult for observers to access the telescope site in the winter season, and hence, we operate the telescope in a remote way from the University of Tokyo using the commercial satellite communication system (JCSAT). The other details of the telescope are described elsewhere [10].

We first preinstalled the telescope system at Nobeyama Radio Observatory, and carried out test operations from June 1997 to May 1998. During this period, we observed the CO ($J = 3 - 2$) line at 345 GHz, since it is impossible to observe the 492 GHz line at Nobeyama. During this test operation, we were able to improve both hardware and the software effectively. After the test operation, the telescope system was once disassembled, and it was transported to the summit of Mount Fuji by a helicopter in July 1998. After that we installed various instruments, and started test observations. We first succeeded in detecting the 492 GHz line toward the Orion molecular cloud in September 1998, and started the scientific operation from November of the same year.

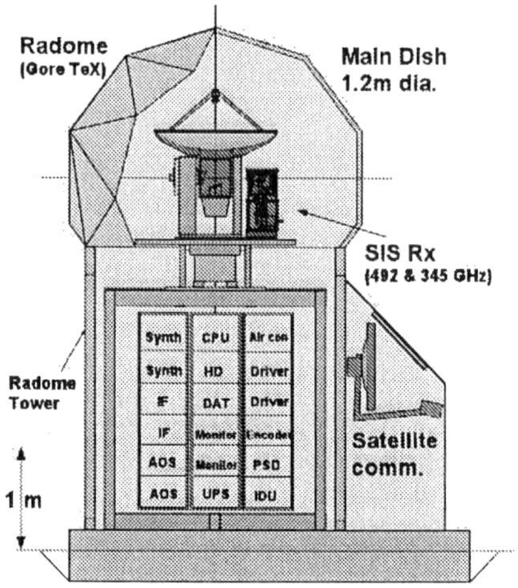

Fig. 1. A schematic illustration of the Mount Fuji submillimeter-wave telescope

Since then, we have successfully operated the telescope for 4 observing seasons. We have already surveyed more than 40 square degrees of the sky with the 492 GHz line. We have also mapped a few representative sources with the 809 GHz line. The atmospheric opacity toward zenith often becomes as low as 0.5 at 492 GHz. This good observing condition made it possible to make an efficient survey observation with the CI lines.

3. Observational Results

3.1. The CI emission in the molecular-cloud scale

We have obtained the CI (492 GHz) distributions of many giant molecular clouds including Orion A, Orion B, DR15, DR21, NGC2264, M17, W3, W44, and W51. Here I would like to introduce the result of the Orion A molecular cloud, as an example.

The Orion A molecular cloud is the nearest giant molecular cloud, and is subject to extensive studies in various wavelength regions. The Orion A molecular cloud is composed of a very active star-forming region associated with the Orion nebula at the northern part and a rather quiescent region extended to the south of the cloud. Ikeda et al. [11] [12] observed the CI distribution toward the entire region of the Orion A molecular cloud for the first time. An extent of the CI distribution resembles that of the ^{13}CO ($J = 1 - 0$) distribution observed by Bally et al.(1987), which indicates that the CI emission essentially comes from the molecular cloud itself. Ikeda et al. derived the column densities of C and CO from the CI and ^{13}CO data, and found that the C/CO column density ratio ranges from 0.1 to 0.2 in the central ridge of the cloud. On the other hand, the C/CO ratio is found to be significantly higher in the cloud edges.

The distribution of the CI emission has been discussed in terms of the photodissociation region (PDR) picture [13] [14]. When we think a plane-parallel cloud illuminated by the UV radiation from one side, the major form of the gas-phase carbon changes from C^+ to C, and then C to CO as the depth from the surface of the cloud. Therefore the neutral carbon atom can exist in a thin layer near the surface of the cloud. The higher C/CO ratio in the cloud edges observed in the Orion A molecular cloud is consistent with this picture. However, it is rather difficult to explain the almost constant C/CO ratio over the variety of physical conditions with this simple picture. The constant C/CO ratio means that the neutral carbon atom coexists with CO, and there would be a certain mechanism to adjust the C/CO ratio to be within such a small range.

3.2. The physical condition of the CI emitting region

In order to understand the physical condition of the CI emitting region, Kuboi et al. [15] mapped the OMC-1 region in the Orion A molecular cloud with the 809 GHz line for the first time (Figure 2). This is the largest image of the 809 GHz emission so far reported. A distribution of the 809 GHz emission is concentrated in the vicinity of the OB stars (θ^1 Ori and θ^2 Ori), whereas that of the 492 GHz emission is extended to the molecular cloud ridge running from north to south. By use of these data, Kuboi et al. determined the column density and the excitation temperature separately under the assumption of the LTE condition. The distribution of the column density is found to be very similar to that of the 492 GHz emission. On the other hand, the excitation temperature is enhanced in the small region surrounding the OB stars, probably because of the heating by the intense UV field. Therefore, there is no apparent correlation between the excitation temperature and the column density, indicating that the UV field strength is not an important factor determining the CI column density.

Now we can compare the column densities of C and CO in detail toward the OMC-1 region, as shown in Figure 3. The CII emission is known to have a peak around the OB stars. The CO column density has a peak around the Orion KL region. The column density of C has a peak behind the CO column density peak with respect to the OB stars. Therefore, the spatial order is C^+/CO/C from the OB stars to the column density peak of C. This is apparently inconsistent with the PDR picture, where the spatial order of C^+/C/CO is expected. Furthermore, the density has to be lower than 10^3 cm^{-3} in order to explain the observed column

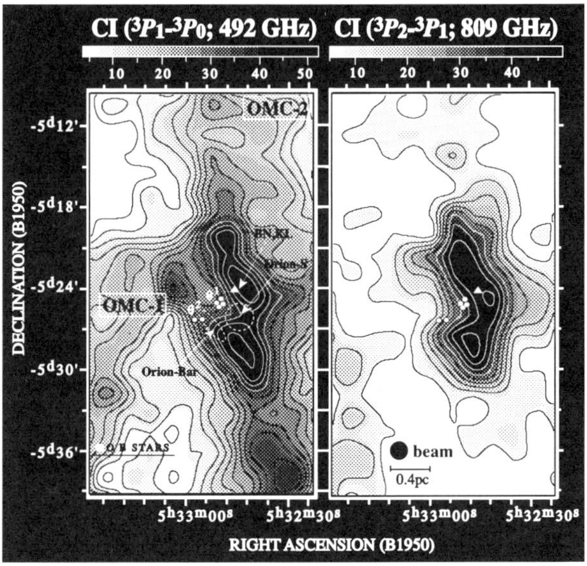

Fig. 2. Integrated intensity maps of the 492 GHz (left) and 809 GHz (right) lines of CI observed toward OMC-1

density of the C peak (10^{18} cm^{-2}) with the PDR model [14]. This is much lower than the density reported so far for this region. Therefore it seems difficult to explain the column density of C by the simple PDR model.

One possible explanation is that the cloud has a clumpy structure so as that the UV radiation can penetrate deep into the cloud. If the clump size and density are adjusted as model parameters, the observed column density would be explained. However, the C^+/CO/C arrangement cannot be explained even with this modified picture. In order to reproduce the arrangement, the clump size has to be different between the CO peak and the C peak.

3.3. The C^+/CO/C arrangement and its meanings

The C^+/CO/C arrangement seen in the OMC-1 region is generally observed in various regions associated with OB stars, such as DR15, M17, and ρOph. Particularly Oka et al. [16] clearly demonstrated this arrangement in the DR15 region. Thus this arrangement does not originate from a special situation of each source, but from the general property of the cloud illuminated by the strong UV radiation. Here I would like to introduce the ρ Oph molecular cloud

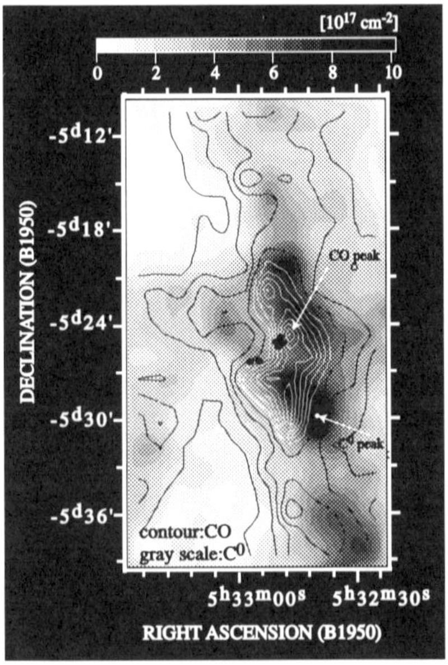

Fig. 3. Column density distributions of C (gray scale) and CO (contour) in OMC-1

observed by Kamegai et al. as an example [17].

The ρ Oph molecular cloud is a large dark cloud complex extending over more than 35 square degrees. Particularly the L1688 region is known to be an active star-forming region close to the Sun (160 pc). In addition, the B2V star (HD147889) is associated with the L1688 cloud, and hence it is known as a typical photodissociation region. Kamegai et al. conducted the 492 GHz observations toward the entire region of L1688 (Figure 6). The CI emission has a peak at the ρ Oph A position, where ^{13}CO emission has also a peak [18]. In addition, the second peak is found in the CI emission toward the southeast position of ρOph A, which is not associated with the ^{13}CO emission peak.

In Figure 4, the CI distribution is compared with the CII distribution [19] and the C^{18}O ($J = 1 - 0$) distribution [20]. The CII distribution is almost centered at HD147889, and traces an ionized cloud and a diffuse part of the photodissociation region. On the other hand, the C^{18}O distribution traces the dense core region, where active star formation is taking place. In fact, distributions of the millimeter- and submillimeter-wave continuum emissions are similar to that of the C^{18}O emission [21]. Kamegai et al. pointed out that the second CI peak is located behind the dense core region with respect to the HD147889. Hence the

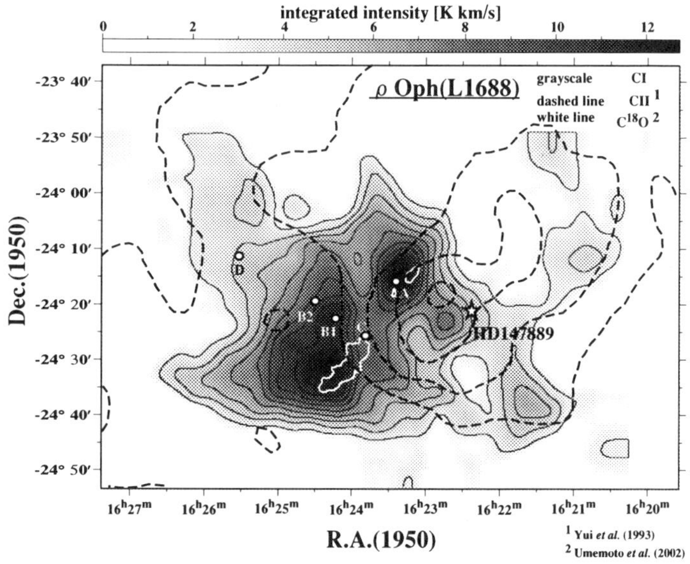

Fig. 4. Distributions of CII (dotted contour), CI (gray scale), and $C^{18}O$ (white contour) toward L1688

arrangement, C^+/CO/C, is clearly seen in this region.

It is suggested that this region is influenced by a shock wave from the Sco OB2 association, which is located at the southwest of the L1688 cloud. Loren and Wooten proposed that the flattened structure of the dense region is formed by such shock compression [22]. In this case, the second CI peak in L1688 is located just behind the compression layer, and can be regarded as the formation site of molecular cloud cores. The conversion timescale from C to CO is 10^6 yr, which is comparable to the dynamical timescale [23]. Therefore the neutral carbon atom can survive for a while in a rapidly compressed cloud. The age of the Sco OB2 association is estimated to be a few times 10^6 yr [24]. Therefore the compression time scale of the L1688 core would be much shorter, and it seems possible that the neutral carbon atom still remains at least the youngest part of the cloud.

If this mechanism works, the C^+/CO/C arrangement seen in various molecular clouds associated with OB stars can be understood as an evolutionary consequence of the cloud. The CII line traces the ionized gas in the vicinity of OB stars, the CO lines the dense core region where the star is being born, and the CI lines the formation site of the dense cores.

3.4. Molecular cloud formation without influences of external UV radiation source

Next I would like to introduce the result obtained toward a quiescent dark cloud, where no OB stars are associated. Maezawa et al. [25] observed the HCL2 region in the Taurus dark cloud complex. HCL2 involves a few protostars and T-Tauri stars, and is regarded as an active formation site of low-mass stars. Although this region has extensively been studied with various molecular lines, the CI observation was carried out toward very limited positions [26]. Maezawa et al. delineated the CI distribution in the entire region of the HCL2 region for the first time, and found an very interesting part at the south end of the HCL2 ring, where the CI emission is very intense (CI cloud).

In the CI cloud, the $C^{18}O$ emission is fairly weak [27], and the C/CO column density ratio is evaluated to be larger than 0.8. This ratio is comparable to the ratio found in the diffuse cloud (> 1), although the visual extinction toward the CI cloud is higher than 6. The average H_2 density is estimated to be 5.4×10^3 cm^{-3}, which is intermediate between a diffuse cloud and a dense core. Furthermore, the CI cloud spatially exists between the dense core (north-west) and the diffuse cloud (south-east). These results suggest that the CI cloud is in the early stage of the dense core formation. The neutral carbon atom would be abundant because of incomplete conversion of C to CO there. If this is the case, the dense core formation is taking place from northwest to southeast in this region. This direction is consistent with the result inferred for the TMC-1 ridge from the molecular abundance gradient [28]. By increasing similar samples of the CI cloud in other regions in future observations, we would be able to clarify the physical processes that govern the spontaneous formation of molecular clouds.

4. Acknowledgements

The author is grateful to Professor Katsuhiko Sato and Professor Kazuo Makishima for their continuous encouragement and support throughout this work. The author also thanks to Tomoharu Oka for his valuable discussions, and to Kazuhisa Kamegai and Nobuyuki Kuboi for their assistance in preparation of the manuscript. This paper is based on the result of the Mount Fuji submillimeter-wave project in collaboration with Tomohoaru Oka, Masafumi Ikeda, Hiroyuki Maezawa, Tetsuya Ito, Kazuhisa Kamegai, Takeshi Sakai, Nobuyuki Kuboi, Kunihiko Tanaka, Koudai Matsuo, Ken Shimbo, Masaaki Hayashida, Ken'ichi Tatematsu, Yutaro Sekimoto, Takashi Noguchi, Keisuke Miyazawa, Hiroyuki Ozeki, Junji Inatani, Msastoshi Ohishi, Hideo Fujiwara, and Shuji Saito.

References

[1] C. Heiles and H.J. Habing, Astrophys. J. Suppl. 14 (1974) 1.
[2] T.M. Dame et al., Astrophys. J. 322 (1987) 706.
[3] T.G. Phillips, P.J. Huggins, T.B.H. Kuiper, and R.E. Miller, Astrophys. J. 236 (1980) L103.
[4] E.L. Wright et al., Astrophys. J. 381 (1991) 200.
[5] R. Plume, D.T. Jaffe, and J. Keene, Astrophys. J. 425 (1994) L49.
[6] D.T. Jaffe, A.I. Harris, M. Silber, T. Genzel, and A.L. Betz, Astrophys. J. 290 (1985) L59.
[7] S. Yamamoto et al., Astrophys. J. 547 (2001) L165.
[8] Y. Sekimoto et al. Int. J. IR & MM Waves 17 (1996) 1263.
[9] T. Noguchi, H. Maezawa, S.-C. Shi, and S. Yamamoto, IEEE Trans. Appl. Supercond. 71 (2001) 167.
[10] Y. Sekimoto et al., Rev. Sci. Instr. 71 (2000) 2895.
[11] M. Ikeda et al., Astrophys. J. 527 (1999) L59.
[12] M. Ikeda, T. Oka, K. Tatematsu, Y. Sekimoto, and S. Yamamoto, Astrophys. J. Suppl. in press.
[13] A.G.G.M. Tielens and D. Hollenbach, Astrophys. J. 291 (1985) 722.
[14] M.J. Kaufman, M.G. Wolfire, D.J. Hollenbach, and M.L. Luhman, Astrophys. J. 527 (1999) 795.
[15] N. Kuboi et al., in preparation.
[16] T. Oka et al., Astrophys. J. 558 (2001) 176.
[17] K. Kamegai et al., Astrophys. J. submitted.
[18] R.B. Loren, Astrophys. J. 338 (1989) 902.
[19] Y.Y. Yui et al., Astrophys. J. 419 (1980) L37.
[20] T. Umemoto et al., in preparation.
[21] F. Motte, P. André, and R. Neri, Astron. Astrophys. 336 (1998) 150.
[22] R.B. Loren and A. Wooten, Astrophys. J. 306 (1986) 142.
[23] H. Suzuki et al., Astrophys. J. 392 (1992) 551.
[24] E.J. de Geus, Astron. Astrophys. 262 (1992) 258.
[25] H. Maezawa et al., Astrophys. J. 524 (1999) L129.
[26] P. Schilke et al., Astron. Astrophys. 294 (1995) L17.
[27] K. Sunada and Y. Kitamura, private communication.
[28] Y. Yirahara et al., Astrophys. J. 394 (1992) 539.

Low Energy Antiprotons in Cosmic Rays during a Half Solar Cycle Period

T. Maeno, T. Matsukawa, T. Mitsui[1], M. Nozaki, Y. Shikaze, K. Tanizaki, K. Yamato
Kobe University, Kobe, Hyogo 657-8501, Japan
K. Abe, K. Anraku, Y. Asaoka, M. Fujikawa, H. Fuke, S. Haino, M. Imori, M. Ishino, K. Izumi, S. Matsuda, N. Matsui, H. Matsumoto, H. Matsunaga[2], M. Motoki[1], J. Nishimura, S. Orito[3], T. Sanuki, T. Saeki, T. Sonoda, I. Ueda, Y. Yamamoto
University of Tokyo, Tokyo 113-0033, Japan
Y. Makida, J. Suzuki, K. Tanaka, A. Yamamoto, T. Yoshida, K. Yoshimura
High Energy Accelerator Research Organization (KEK), Tsukuba, Ibaraki 305-0801, Japan
J. Mitchell, A. Moiseev, J. Ormes, M. Sasaki, R. Streitmatter
National Aeronautics and Space Administration, Goddard Space Flight Center, Greenbelt, MD 20771, USA
E. S. Seo, J. Z. Wang
University of Maryland, College Park, MD 20742, USA
N. Yajima, T. Yamagami
The Institute of Space and Astronautical Science (ISAS), Sagamihara, Kanagawa 229-8510, Japan

1. Present address: Tohoku University, Japan
2. Present address: University of Tsukuba, Japan
3. deceased

Abstract

The BESS spectrometer has measured the energy spectrum of cosmic-ray antiprotons ($\bar{p}$'s) successively since 1993 to 2001, during a half solar cycle period. In total, 2090 antiprotons have been clearly identified in energy range 0.18 to 4.20 GeV based on magnetic-rigidity, time-of-flight, energy loss, and Cherenkov radiation. From these successive measurements of $\bar{p}$ spectrum at various solar activities, we discuss about the origin of cosmic-ray $\bar{p}$'s and the effect of the solar modulation.

1. Introduction

Cosmic-ray $\bar{p}$'s should certainly be produced by the interaction of galactic high energy cosmic-rays with the interstellar medium. It is considered that this "secondary" $\bar{p}$ is the dominant component of cosmic-ray $\bar{p}$'s. Since the interstellar medium consists mainly of thermalized hydrogen gas, the most of the secondary $\bar{p}$'s should be produced with non-zero momentum. Therefore, the energy spectrum of the "secondary" $\bar{p}$ is expected to show a characteristic peak around 2 GeV and sharp decreases of the flux below and above the peak.

One can also conceive more novel or exotic elementary-particle processes such as the annihilation of neutralino dark matter [1, 2] or the evaporation of primordial black hole [3]. The $\bar{p}$'s from these "primary" sources are expected to show very soft energy spectra [4], peaking toward the lower energies, and would exhibit large solar modulations [5]. Thus they are distinguishable in principle from the secondary $\bar{p}$ component.

The detection of cosmic-ray $\bar{p}$'s has been difficult to achieve, because of huge backgrounds and the extremely small flux especially at low energies. And because of the loss by interaction with atmosphere, the measurement should be carried out at high altitude, such as balloon or satellite experiments. The first [6] and subsequent [7] evidence for cosmic-ray $\bar{p}$'s were reported at relatively high energies, where it was not possible to positively identify the $\bar{p}$'s with a mass measurement. The first "mass-identified" and thus unambiguous detection of cosmic-ray $\bar{p}$'s was performed by BESS 1993 and BESS 1994 [8] in the low-energy region (8 events at 0.3 to 0.5 GeV), which was followed by IMAX [9] and CAPRICE [10] detections. The BESS 1995 measured the energy spectrum [11] for the first time, based on 43 $\bar{p}$'s over the range 0.18 to 1.4 GeV. In BESS 1997, we installed a threshold-type aerogel Cherenkov counter [12] to extract $\bar{p}$'s among the e^-/μ^- background in the high energy region. As a result, we measured the spectrum at solar minimum in the energy range from 0.18 to 3.56 GeV [13], and detected a characteristic peak at 2 GeV of secondary $\bar{p}$ in the spectrum. Since 1998 to 2000, we had observed the variations of the spectrum with solar activities [14, 15]. In BESS 2001, we carried out the experiment in New Mexico, USA, and observed the background $\bar{p}$'s which are produced in collisions between high energy cosmic-rays and atmosphere. The subtraction of this "atmospheric" $\bar{p}$ is dominant correction in the analysis.

Figure 1. shows the flight summary of BESS experiment since 1993 to 2000. The dead-time fraction has been reduced from 39% to 13% by a high speed data acquisition system which has been installed since BESS 1997. The time resolution has been improved from 110 psec to 75 psec by new time of flight

	1993	1994	1995	1997	1998	1999	2000
Flight Time (hrs)	17.5	17.0	19.5	20.5	22.0	37.5	40.3
Obs. Time (hrs)	14.0	15.0	17.5	18.3	20.0	31.3	32.5
Dead Time (%)	39.0	50.0	39.0	13.0	15.0	11.5	12.0
Recorded Events (M)	4.0	4.2	4.5	16.2	19.0	16.8	15.8
Total Data Size (GB)	4.5	6.5	8.0	31.0	38.0	41.0	37.7
Δp/p (%)	0.5						
ΔTOF (psec)	300		110	75			
Energy Range (GeV)	0.2-0.6		0.2-1.4	0.2-3.6	0.2-4.2		
# of Antiprotons	6	2	43	415	398	668	558

Fig. 1. Flight summary of BESS experiment.

counters [16]. And detectable energy range has been extended up to 4.2 GeV by the aerogel Cherenkov counter [12]. Because of these successive improvements, the number of observed $\bar{p}$'s has been increased ten times as many.

2. BESS spectrometer

The BESS spectrometer was designed [17, 18] and developed [12, 16, 19, 20] as a high-resolution spectrometer. A uniform field of 1 Tesla is produced by a thin superconducting solenoid [21], and the field region is filled with tracking detectors. This geometry results in an acceptance of 0.3 $\mathrm{m^2sr}$. Tracking is performed by fitting up to 28 hit-points in drift chambers, resulting in a magnetic-rigidity ($R \equiv Pc/Ze$) resolution of 0.5% at 1 GV. Energy deposit in the drift chamber gas is also obtained as a truncated mean of the integrated charges of hit pulses. The upper and lower scintillator-hodoscopes (TOF) [16] provide time-of-flight and two $\mathrm{d}E/\mathrm{d}x$ measurements. Time resolution of each counter is 55 ps, resulting in a $1/\beta$ resolution of 0.014, where β is defined as particle velocity normalized by the speed of light. The instrument also incorporates a threshold-type Cherenkov counter [12] with a silica-aerogel radiator (n=1.02) that can reject e^- and μ^- backgrounds by a factor of 4000 and identify $\bar{p}$'s from such backgrounds up to 4.2 GeV. In addition to biased trigger modes [14, 20] selecting negatively-charged particles preferentially, one of every 60 (30) first-level triggered events were recorded to provide unbiased samples. Note that the first-level trigger was provided by a coincidence between the top and bottom scintillators, with the threshold set at 1/3 of the pulse height from minimum ionizing particles.

Since 1993 to 2000, the experiments were carried out in northern Canada, Lynn Lake, where the geomagnetic cutoff rigidity is 0.3 GV. In 2001, the experiment was carried out in southern USA, Ft. Summner, where the geomagnetic cutoff rigidity is 4.0 GV. Data for flux measurements were taken at altitudes above 34 km. Data during ascent were also collected at various altitudes, in order to estimate a background of "atmospheric" p's.

3. Result

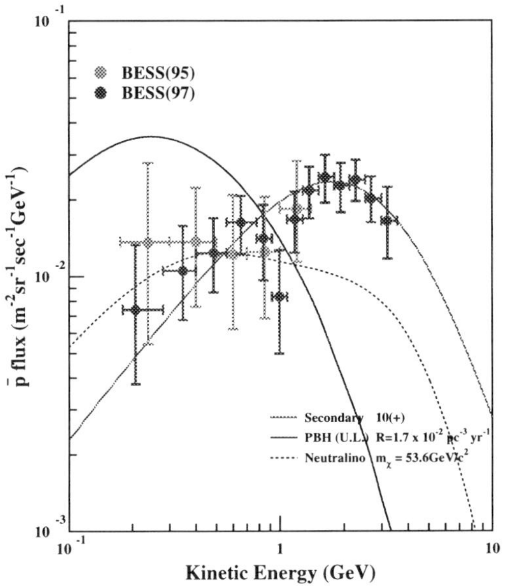

Fig. 2. The $\bar{p}$ spectrum measured by BESS in recent solar-minimum period [13, 14]. A clear peak at 2 GeV of "secondary" $\bar{p}$'s was detected for the first time. The purple solid curve shows the "secondary" $\bar{p}$ spectrum [22], which well reproduces the measured spectrum in the peak region. The blue solid curve shows the upper limit on the expected $\bar{p}$ spectrum produced from primordial black hole [4]. The black dashed curve shows one example of the expected spectrum produced from annihilation of neutralino dark matter [2].

Figure 2. shows the $\bar{p}$ spectrum measured in recent solar-minimum period [13, 14]. We have detected for the first time a characteristic peak at 2 GeV of "secondary" $\bar{p}$'s. The peak spectrum is reproduced by a theoretical calculation [22], implying that the dominant component of the cosmic-ray $\bar{p}$'s is the "secondary" $\bar{p}$'s. From this, we confirmed that the propagation models are basically correct and that different cosmic-ray species undergo a universal propagation.

Based on the BESS 1995 data [11], the upper limit on the explosion rate ($\mathcal{R}$) of primordial black hole was derived [4]:

$$\mathcal{R} \leq 1.7 \times 10^{-2} pc^{-3} yr^{-1} (90\% \, C.L.),$$

which is almost 8 orders of magnitude more stringent than the present upper limit on the rate of 50-TeV γ-ray bursts [23], and the practical sensitivity for 100-MeV γ-ray bursts [24]. Note that, because the cosmic-ray $\bar{p}$'s escape out from the Galaxy, by using BESS measurement we can constrain "local" distribution of primordial black holes within a few kpc away from the solar system [4].

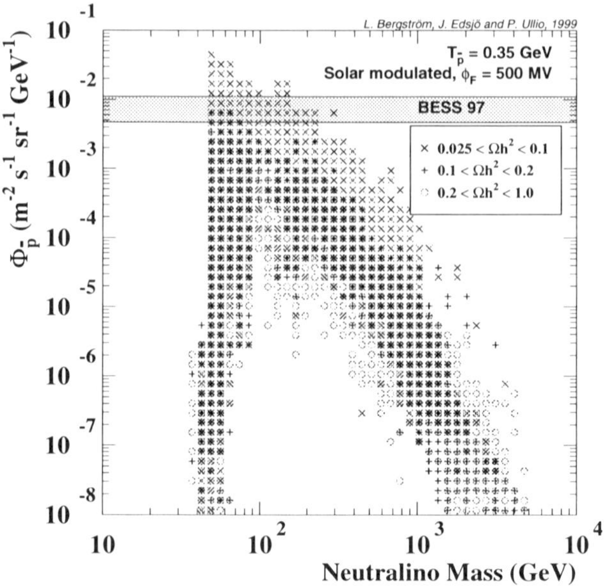

Fig. 3. The expected $\bar{p}$ fluxes [2] at kinetic energy of 0.35 GeV compared with BESS 1997 data [13]. The models have been coded according to relic density of neutralino dark matter, $\Omega_\chi h^2$.

Moreover, our measurements were compared with expected $\bar{p}$ spectra from SUSY dark matter [1, 2]. Shown In Figure 2. is one example of expected $\bar{p}$ spectrum. As this example, the expected spectra of $\bar{p}$'s produced from neutralino annihilation have rather broader distribution. Therefore, it is difficult to find signals of neutralino annihilation in the low energy region. Figure 3. shows expected $\bar{p}$ fluxes [2] at kinetic energy of 0.35 GeV, compared with BESS 1997 data. The highest values of the fluxes in Figure 3. are four times larger than the our data. However, uncertainties in the calculations are too large in the present; the halo profile, the propagation model, and the solar modulation. For this reason it is impossible to exclude any SUSY model by using $\bar{p}$ measurements presently [2].

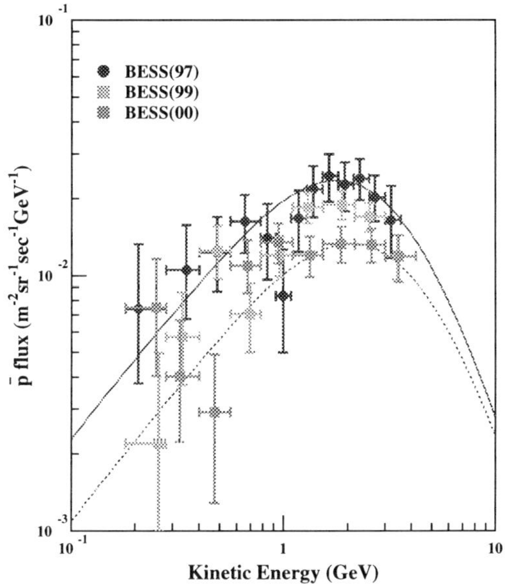

Fig. 4. The variations of $\bar{p}$ spectrum [13, 15] with solar activities. The solid and dashed curves show the "secondary" $\bar{p}$ spectra [22] in the solar-minimum and solar-maximum, respectively. According to increase of solar activities from 1997 to 2000, clear decrease was observed in the peak region. On the other hand, in the low energy region poor statistics prevent us to draw firm conclusion of primary $\bar{p}$ components.

Figure 4. shows the variations of $\bar{p}$ spectrum with solar activities. According to increase of solar activities, the spectrum shows distinct decrease in the peak region, which is well reproduced by theoretical calculation [22]. From this, we can confirm the solar modulation model is basically correct. On the other hand, in the low energy region we cannot draw firm discussion because of poor statistics.

As mentioned before, if very soft "primary" $\bar{p}$ components exist, the spectrum is expected to show large variation in the low energy region [5]. Therefore, more precise measurements are desired to observe spectrum variations.

For this reason, a long duration balloon flight in Antarctica, BESS-Polar is being prepared to extend the BESS scientific objectives in an ideal ballooning environment in Antarctica [25]. This experiment aims at extremely sensitive measurement of low energy $\bar{p}$ to search for any novel sources. Flight time will be about twenty days, which is twenty times as long as present BESS experiment.

The first flight in Antarctica is expected to be realized in 2003 $\sim$ 2004 and second flight in the coming solar minimum period. More than 10^4 $\bar{p}$ may be observed in the energy range of 0.1 $\sim$ 4.2 GeV.

4. Conclusion

The BESS experiment has performed eight balloon flights since 1993. In total, 2090 antiprotons were clearly identified in energy range 0.18 to 4.20 GeV. Based on these successive measurements, various discussions of the origin of cosmic-ray $\bar{p}$ were derived. However, in order to draw firm conclusion of the primary $\bar{p}$ components, we need more precise measurement, which will be performed by BESS-Poler. The first BESS-Polar flight in Antarctica is expected in 2003/2004.

5. Acknowledgements

We thank NASA and NSBF for the balloon expedition, and KEK and ISAS for continuous support. This work was supported by Grant-in-Aid for Science Research, MEXT in Japan; and by NASA in the USA. Analysis was performed using the computing facilities at ICEPP, the University of Tokyo.

References

[1] A. Bottino, et al., Phys. Rev. D **58**, 123503 (1998); A. Bottino, et al., Phys. Rev. D **62**, 56006 (2000).
[2] L. Bergström, et al., Astrophys. J. **215**, 526 (1999).
[3] S. W. Hawking, Commun. Math. Phys. **43**, 199 (1975).
[4] K. Maki, T. Mitsui and S. Orito, Phys. Rev. Lett. **76**, 3474 (1996).
[5] T. Mitsui, K. Maki and S. Orito, Phys. Lett. B **389**, 169 (1996).
[6] R. L. Golden *et al.*, Phys. Rev. Lett. **43**, 1196 (1979).
[7] E. A. Bogomolov *et al.*, Proc. 16th Int. Cosmic Ray Conf. (Kyoto) **1**, 330 (1979).

[8] K. Yoshimura *et al.*, Phys. Rev. Lett. **75**, 3792 (1995); A. Moiseev *et al.*, Astrophys. J. **474**, 479 (1997).

[9] J. W. Mitchell *et al.*, Phys. Rev. Lett. **76**, 3057 (1996).

[10] M. Boezio, et al., Astrophys. J. **487**, 415 (1997); D. Bergström, et al., Astrophys. J. **177**, 534 (2000).

[11] H. Matsunaga, et al., Phys. Rev. Lett. **81**, 4052 (1998).

[12] Y. Asaoka *et al.*, Nucl. Instr. Meth. A **416**, 236 (1998).

[13] S. Orito, et al., Phys. Rev. Lett. **84**, 1078 (2000).

[14] T. Maeno, et al., Astroparticle Phys. **16**, 121 (2001).

[15] Y. Asaoka, et al., Phys. Rev. Lett. (to be printed).

[16] Y. Shikaze *et al.*, Nucl. Instr. Meth. A **455**, 596 (2000).

[17] S. Orito, Proc. of the ASTROMAG Workshop, KEK Report 87-19, 111 (1987).

[18] A. Yamamoto *et al.*, IEEE Trans. Magn. **24**, 1421 (1988).

[19] A. Yamamoto *et al.*, Adv. Space Res. **14(2)**, 75 (1994).

[20] Y. Ajima *et al.*, Nucl. Instr. Meth. A **443**, 71 (2000).

[21] Y. Makida, *et al.*, IEEE Trans. Applied Supercon. **5(2)**, 638 (1995).

[22] J. W. Bieber, *et al.*, Phys. Rev. Lett. **83**, 674 (1999).

[23] D. E. Alexandreas, *et al.*, Phys. Rev. Lett. **71**, 2524 (1993).

[24] D. V. Semikov, Astrophys. J. **436**, 254 (1994).

[25] A. Yamamoto, *et al.*, Adv. in Space Res. (2000).

Results from the Super-Kamiokande Experiment

Kyoshi NISHIJIMA for the Super-Kamiokande collaboration
Department of Physics, School of Science, Tokai University, 1117 Kita-Kaname, Hiratsuka, Kanagawa 259-1292, Japan

Abstract

The latest results from the Super-Kamiokande experiment are presented. We have obtained precise value of ^8B solar neutrino flux by the elastic scattering and confirm the strong deficit of it. We found no significant distortion of recoil electron energy spectrum or no significant difference between day and nighttime flux. For atmospheric neutrinos, we updated results on the flavor ratio and the zenith angle distribution. $\nu_\mu \rightarrow \nu_\tau$ oscillation suffices to explain all the atmospheric neutrino results in hands. We also present the results of search for supernova neutrino bursts and search for high energy neutrino sources.

1. Introduction

The Super-Kamiokande detector is a 50 kton large imaging water Cherenkov detector located at a depth of 2,700 m.w.e. in the Kamioka mine in Japan. The central 32,000 tons is optically separated from surrounding veto layer of water, and 11,146 50 cm diameter photomultiplier tubes (PMTs) are mounted on the inner wall. This detector is designed for a multi-purpose experiment, especially for study of neutrino mass and mixing from precise meassurements of the flux of solar neutrinos and atmospheric neutrinos.

Recently, SNO group reported the neutrino flux measured by charged current reaction[1]. Comparing their flux with the precise value of flux obtained by Super-Kamiokande via elastic scattering[2], the deficit of solar neutrino flux is strongly demonstrated to be due to neutrino oscillation. On the other hand, atmospheric neutrino anomaly was first reported by Kamiokande experiment in 1988[3], and the evidence of neutrino oscillation has been already confirmed by the atmospheric neutrino observation with the Super-Kamiokande[4].

In this paper, I present the updated results on solar neutrinos and atmospheric neutrinos observed from Super-Kamiokande since 1996. I also present the results of supernova search and the search for astronomical neutrino sources.

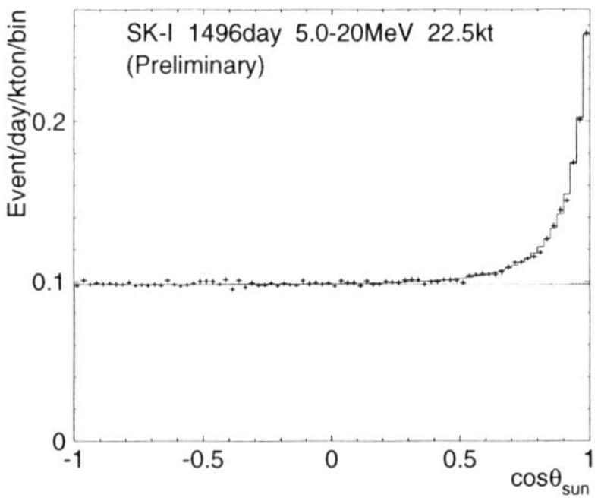

Fig. 1. $\cos\theta_{Sun}$ distribution, where θ_{Sun} is the angle between the direction of reconstructed recoil electron and the Sun's direction.

2. Solar neutrinos

In Super-Kamiokande solar neutrinos are observed via elastic neutrino-electron scattering $\nu_e + e^- \rightarrow \nu_e + e^-$ in water. The direction of the recoil electron well follows the direction of the neutrinos.

Fig.1. shows the $\cos\theta_{Sun}$ distribution, where θ_{Sun} is the angle between the direction of recoil electron and the Sun's direction. By subtracting the background we obtain $22,385 \pm 226(stat.)^{+783}_{-716}(sys.)$ signal events above 5.0 MeV for 1,496 live days and 22,500 tons target water. Assuming the ^8B spectrum shape[5] the ^8B neutrino flux can be calculated to be $2.35 \pm 0.02(stat.)^{+0.013}_{-0.012}(sys.) \times 10^6 cm^{-2}s^{-1}$. Comparing to the SSM calculated by Bahcall et al.[6], the ratio is

$$\frac{Data}{SSM_{BP2000}} = 0.465 \pm 0.005(stat.)^{+0.013}_{-0.012}(sys.)$$

The result clearly indicates the deficit of solar neutrinos, and can be well explained by $\nu_e \rightarrow \nu_{\mu,\tau}$ oscillation.

The Super-Kamiokande measures the energy spectrum of recoil electrons, which shape reflects the shape of the spectrum of incoming neutrinos. Fig.2. shows the observed energy spectrum of the recoil electrons normalized to the SSM_{BP2000}. Observed data is consistent with flat spectrum. The χ^2/dof for flat normalozed spectrum is 17.4/18, corresponding to 50% confidence level.

Since the Super-Kamiokande experiment is a real time one, day-night and seasonal variation of the solar neutrino flux can be studied. Day-night flux difference is caused by the regeneration of ν_e due to the matter effect on neutrinos when it passes through the earth[7][8]. Then the MSW effect predicts an enhancement of the solar neutrino flux during nights. Fig.3. shows the neutrino flux rate

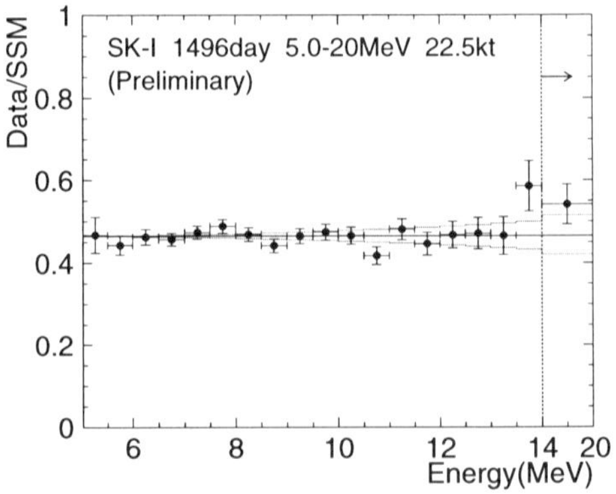

Fig. 2. Observed electron energy spectrum normalized to the SSM prediction.

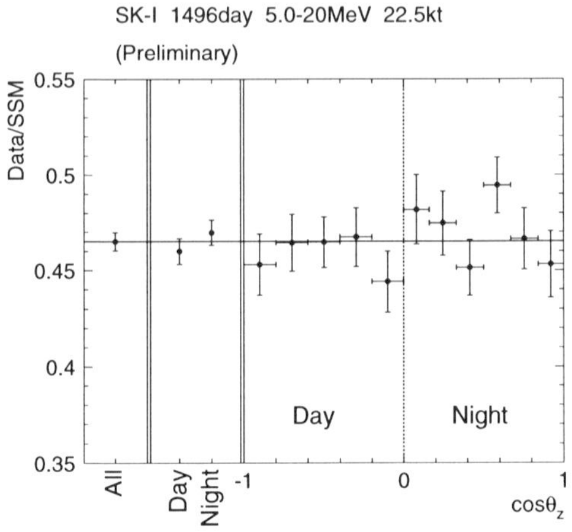

Fig. 3. Normalized neutrino rate as a function of the zenith angle of the Sun's position.

normalized to SSM_{BP2000} for five day bins and six night bins. Summed up day flux and night flux are also plotted. The daytime and nighttime flux difference is solar model independent. The result of the measured day-night asymmetry is

$$\frac{N-D}{(N+D)/2} = 0.021 \pm 0.022(stat.)^{+0.013}_{-0.012}(sys.),$$

where D and N are the day flux and the night flux, respectively. The difference between the day flux and the night flux is only less than 1σ level.

Using the recoil electron energy spectrum for each zenith angle bin, the zenith angle spectrum is analized in a two-neutrino oscillation scenario, $\nu_e \leftrightarrow \nu_{\mu,\tau}$, and $\nu_e \leftrightarrow \nu_{sterile}$. Fig.4. and Fig. 5. show the excluded regions obtained from flux independent analysis and the allowed regions obtained by the flux constraint analysis at 95% C.L. for each scenario, respsctively. Four allowed regions by global fit are also overlaid in these figures.

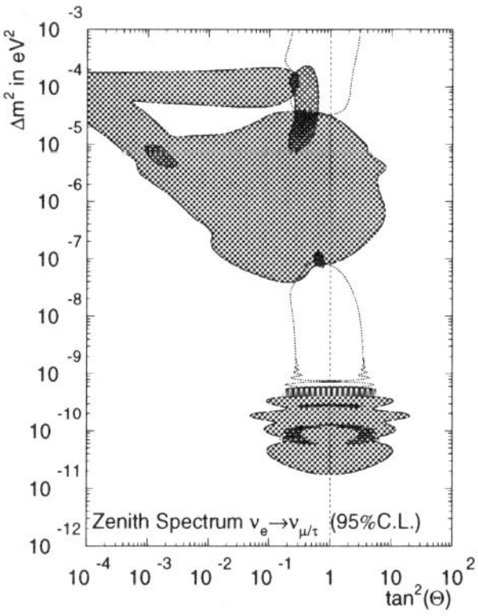

Fig. 4. Exclusion area for two-flavor oscillation $\nu_e \leftrightarrow \nu_{\mu,\tau}$ from flux independent zenith angle spectrum analysis at 95% C.L.(grey area). The allowd areas at 95% C.L. obtained by the flux-constraint analysis using the zenith angle spectrum and the SSM flux prediction (inside dotted lines) and by analysis of the combined absolute flux data of Homestake, GALLEX, SAGE, and Super-Kamiokande(crosshatched area) are overlaid.

3. Atmospheric neutrinos

We classified observed atmospheric neutrino data as four types. Fully-contained(FC) events have all their tracks in the inner detector and partially-contained(PC) events have escaping tracks from the inner detector. Upward-going

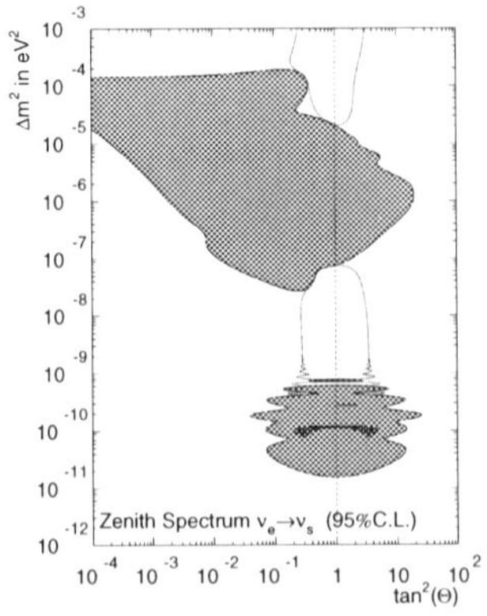

Fig. 5. Exclusion and allowed areas for two-flavor oscillation $\nu_e \leftrightarrow \nu_{sterile}$. The definition is the same as in 4.

muons(UGM) are produced in the rock beneath the detector and go through the inner detector (upward through-going muons,UTM) or stop inside of it(upward stopping muons,USM). Range of neutrino energy covers from 0.5 GeV to above 100 GeV. Analysis of these three samples are described below.

3.1. Analysis of FC+PC events

We have observed 10,313 FC events and 754 PC events for 1,289 live days. The FC events are divided into single and multi ring events. These single ring events are further divided into $e - like$ and $\mu - like$ events based on likelihood of shape around Cherenkov ring, where mis-ID probability is less than 2%. The data sample is summarized in Table 1. for "sub-GeV"(E_{vis} < 1.33 GeV) and "multi-GeV"(E_{vis} > 1.33 GeV), respectively, where E_{vis} is the visible energy. Also included are the prediction from the model of Honda et al[9]. The number of $\mu - like$ events is significantly less than the expected one.

In order to cancel experimental and theoretical uncertainties, the measured flavor ratio is conventionally compared to expectation as the "double ratio", $R \equiv (\mu/e)_{data}/(\mu/e)_{MC}$, where μ and e are numbers of $\mu - like$ and $e - like$ events for data and Monte Carlo simulation. The double ratio for sub-GeV single ring samples is then,

$$R \equiv \frac{(\mu/e)_{DATA}}{(\mu/e)_{MC}} = 0.638 \pm 0.017(stat.) \pm 0.050(sys.).$$

Table 1. Summary of the Sub-GeV, Multi-GeV and PC event samples observed in Super-Kamiokande. Data are compared with the Monte Carlo prediction

	Sub-GeV(FC)		*Multi-GeV(FC)*		*Multi-GeV(PC)*	
	Data	MC(Honda)	Data	MC(Honda)	Data	MC(Honda)
single-ring	5652	6740.4	1184	1451.1		
$e - like$	2864	2667.6	626	612.8		
$\mu - like$	2788	4072.8	558	838.3		
multi-ring(μ)	2159	2585.1	1318	1648.1		
total	7811	9325.5	2502	3099.1	754	1065.0

For multi-GeV single ring and PC samples the double ratio is,

$$R \equiv \frac{(\mu/e)_{DATA}}{(\mu/e)_{MC}} = 0.675^{+0.034}_{-0.032}(stat.) \pm 0.080(sys.),$$

where PC events are included in the multi-GeV muon sample.

The zenith angle is a direct measure of its flight length, and hence neutrino oscillation manifest itself as a possible distortion of the zenith angle distribution which does not depend on uncertainties in the absolute flux or cross sections. Fig.6.(a)-(f) shows the zenith angle distribution of $e - like$ and $\mu - like$ events for sub-GeV and multi-GeV samples. In what follows $\cos\theta = 1$ corresponds to vertically downward direction and $\cos\theta = -1$ corresponds to vertically upward direction. Comparing with MC prediction from no oscillation, a strong deficit of $\mu - like$ events in the upward direction is clearly seen, while prediction from the best fit oscillation parameters quite well reproduces the observed zenith angle distribution. For the single ring multi-GeV $\mu - like$ events, up-down asymmetry is $(U - D)/(U + D) = -0.303 \pm 0.030(stat.) \pm 0.004(sys.)$, where U is the number of upward-going events($-1 < \cos\theta < -0.2$), D is the number of downward-going events ($0.2 < \cos\theta < 1$) and $(U - D)/(U + D)$ is expected to be $0.005 \pm 0.006(stat.) \pm 0.009(sys.)$ from MC. This up-down asymmetry for $\mu - like$ events is one of the evidence for neutrino oscillation.

3.2. Upward-going muons

The typical energy of neutrinos which produce UTMs and USMs are around 100 GeV and around 10 GeV, respectively. Now we have 1,416 UTMs for 1,268 live days and 345 USMs for 1,247 live days. The observed flux of UGMs compared with expectation is summarized in Table 2.. The flux ratio, $\mathcal{R} \equiv (UTMs/USMs)$ is more sensitive to neutrino oscillations since most of sys-

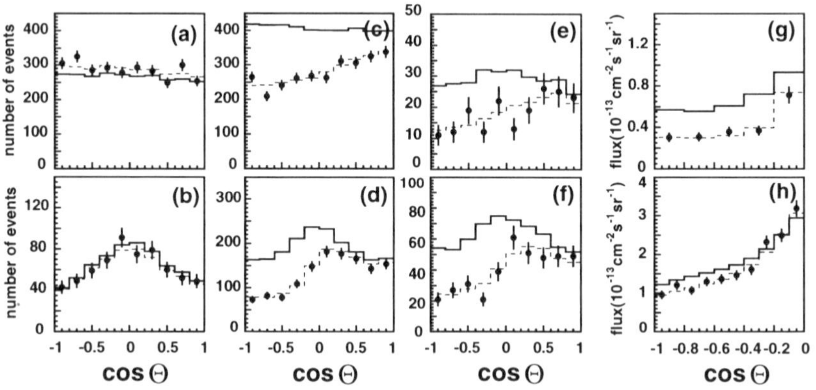

Fig. 6. Zenith angle distribution of FC+PC single-ring (a)sub-GeV $e - like$, (b)multi-GeV $e - like$, (c)sub-GeV $\mu - like$, (d)multi-GeV $\mu - like$, multi-ring (e)sub-GeV $\mu - like$, (f)multi-GeV $\mu - like$, (g)upward stopping, and (h)upward through-going muon events. Solid histograms show the prediction from no oscillation, while dotted histograms show the prediction from the $\nu_\mu \leftrightarrow \nu_\tau$ oscillation for the best fit parameters.

Table 2. Summary of upward-going muon samples.

Sample	Number of events	Flux($\times 10^{-13} cm^{-2} s^{-1} sr^{-1}$) Measured	MC(Honda)	MC(Bartol)
Through-going	1416	$1.70 \pm 0.05 \pm 0.02$	1.84 ± 0.41	1.97 ± 0.44
Stopping	345	$0.41 \pm 0.02 \pm 0.02$	0.68 ± 0.15	0.73 ± 0.16

tematic errors due to the uncertainties of absolute cosmic ray flux cancel. The obtained ratio is $\mathcal{R} = 0.242 \pm 0.017(stat.)^{+0.013}_{-0.011}(sys.)$, which is significantly smaller than the expected $\mathcal{R} = 0.37 \pm 0.05(theor.)$ for both Bartol[10] and Honda[9] flux calculations.

Fig.6.(g),(h) show the zenith angle distribution of the USM flux and UGM flux, respsctively, The agreement between the data and the expectation from no oscillation hypothesis is very poor for the shape of the distribution.

3.3. Oscillation analysis

Neutrino oscillation analysis assuming $\nu_\mu \rightarrow \nu_\tau$ oscillation has been performed using all the atmospheric data. The best fit for physical solution occurs at

($\sin^2 2\theta = 1.00, \Delta m^2 = 2.5 \times 10^{-3} eV^2$) with $\chi^2_{min} = 157.5/170\ dof$. In contrast, assuming no oscillation, $\chi^2_{min} = 393.5/170\ dof$ and no oscillation hypothesis is ruled out. Fig.7. shows the allowed regions in $\sin^2 2\theta$ and Δm^2 plane at 68%, 95%, and 99% confidence level.

The oscillation analysis described above assumes ν_μ oscillates into ν_τ. However, $\nu_\mu \rightarrow \nu_{sterile}$ oscillation also can well repoduce the above atmospheric neutrino data. In order to discriminate between $\nu_\mu \rightarrow \nu_\tau$ and $\nu_\mu \rightarrow \nu_{sterile}$ oscillation, we study the difference in zenith angle distribution due to neutral currents and matter effects [11]. Obtained the excluded regions in $\sin^2 2\theta - \Delta m^2$ plane at the 90(99)% C.L. are shown in Fig.8.. There is no evidence favoring sterile neutrinos.

4. Astronomical neutrinos

In 1987, Kamiokande detector succeeded to detect neutrinos from SN1987A [12]. We search for another supernova neutrino burst using Super-Kamiokande data accumulated during 1,704 days. There is no clear evidence of neutrino burst from supernova. Assuming supernova with twelve solar mass, full detection efficiency of our detector keeps up to about 80 kpc. Then we obtain 90% C.L. upper limits for supernova explosions in our galaxy as 0.26 SN/yr(including the Kamiokande data), and also obtain for extra galaxy up to 80 kpc as 0.49 SN/yr.

We also search for high energy neutrino sources of astronomical origin, using the entire sample of UGM data. One of the methods is to scan entire visible region of the sky from the detector and examine the significance. Fig.9. shows cumulative Pisson probability sky map which is the probability of having the observed number of UGMs and above from the background noise. The result shows it is consistent with the case of no neutrino source.

5. Conclusion

We have presented the latest results on solar neutrinos, atmospheric neutrinos, and astrnomical neutrinos from Super-Kamiokande.

For solar neutrinos, the energy spectrum of recoil electrons is consistent with flat, and day/night difference is as small as 2.1%. Oscillation analysis shows that the small mixing angle and the just-so solutions in the global flux analysis are disfavored with 95% C.L. by Super-Kamiokande zenith-spectrum analysis for $\nu_e \rightarrow \nu_{\mu,\tau}$ oscillations. The large mixing regions are favored with flux constraint analysis. $\nu_e \rightarrow \nu_{sterile}$ solutions are disfavored with 95% C.L. by zenith-spectrum.

For atmospheric neutrinos, observation clearly shows the deficit and up-down asymmetry of muons in the sub-GeV and multi-GeV energy range. The

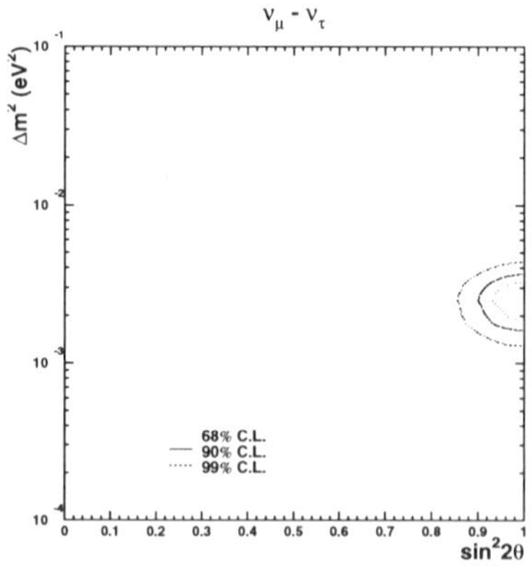

Fig. 7. Allowed region in the $(\sin^2 2\theta, \Delta m^2)$ space from the combined data of FC,PC and UGMs assuming $\nu_\mu \to \nu_\tau$ oscillation.

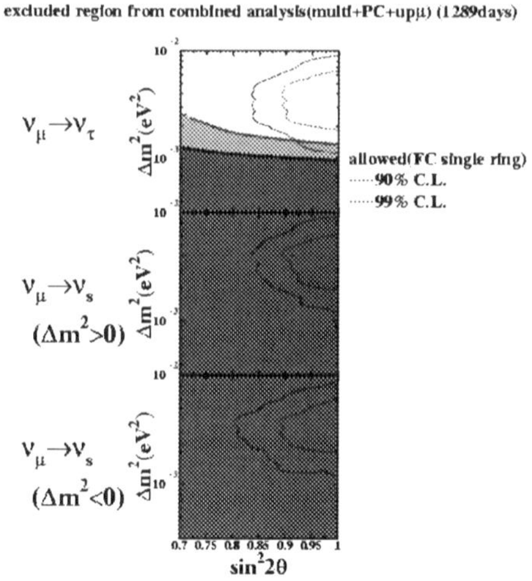

Fig. 8. Excluded regions for three oscillation nodes.The light(dark) gray regions are excluded at 90(99)% C.L. The thin dotted(solid) lines indicate the 90(99)% C.L. allowed regions from the analysis of FC single-ring events.

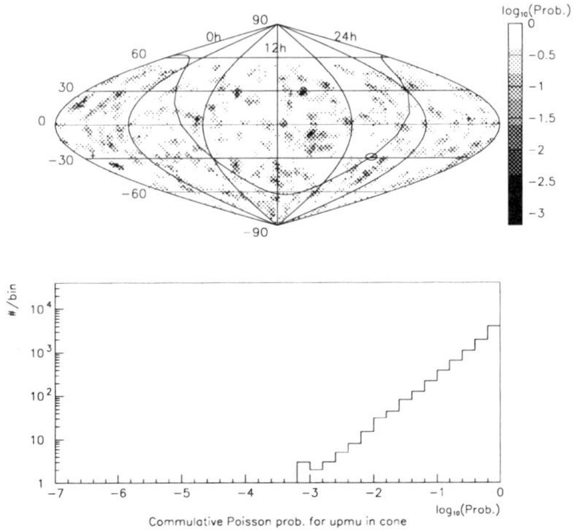

Fig. 9. Excluded regions for three oscillation nodes.The light(dark) gray regions are excluded at 90(99)% C.L. The thin dotted(solid) lines indicate the 90(99)% C.L. allowed regions from the analysis of FC single-ring events.

deficit of UTMs compared with horizontal direction and the deficit of upward stopping muons are also clearly seen. Detailed analyses show oscillation between ν_μ and ν_τ suffices to reproduce all the atmospheric neutrino data in hand. On the other hand, sterile neutrino hypothesis is rejected at the 99% C.L.

References

[1] Q. R. Ahmad et al., Phys. Rev. Lett. 87 (2001) 071301.
[2] S. Fukuda et al., Phys. Rev. Lett. 86 (2001) 5651.
[3] K. S. Hirata et al., Phys. Lett. B205 (1988) 416.
[4] Y. Fukuda et al., Phys. Rev. Lett. 81 (1998) 1562.
[5] C. E. Ortiz et al., Phys. Rev. Lett. 85 (2000) 2909.
[6] J. N. Bahcall, M. H. Pinsonneault, and S. Basu, Astrophys. J. 555 (2001) 990.
[7] S. P. Mikheyev and A. Y. Smirnov, Sov. J. Nucl. Phys. 42 (1985) 913.
[8] L. Wolfenstein, Phys. Rev. D17 (1978) 2369.
[9] M. Honda et al., Phys. Rev. D52 (1995) 4985.
[10] V. Agrawal et.al., Phys. Rev. D53 (1996) 1314.
[11] S. Fukuda et al., Phys. Rev. Lett. 85 (2000) 3999.
[12] K. Hirata et al., Phys. Rev. Lett. 58 (1987) 1490.

Diversity of Type Ia Supernovae: Rapidly Rotating Progenitors

Tatsuhiro UENISHI, Ken'ichi NOMOTO
Department of Astronomy, School of Science, The University of Tokyo, 7-3-1 Hongo, Bunkyo-ku, Tokyo 113-0033, Japan
Izumi HACHISU
Department of Earth Science and Astronomy, College of Art and Science, The University of Tokyo, 3-8-1 Komaba, Meguro-ku, Tokyo 153-0041, Japan

Abstract

Type Ia supernovae are used as standard candle because they are believed to have uniform light curves and spectral evolution. However, their observed brightness have a little diversity. Explaining this diversity is very important to determine cosmological parameters. We simulate the evolution of accreting white dwarfs and computed their structures. Their rapid rotation affects the Chandrasekhar's limiting mass and the final structure of the white dwarf just before the supernova expolosion. Difference in their initial masses at the onset of accretion results in the diversity in the final masses, which can be large enough to explain the observed diversity of maximum brightness of Type Ia supernovae.

1. Introduction

Type Ia supernovae are widely used as a standard candle because of their uniform light curve and spectrum(e.g. Perlmutter 2002 for a review). However, they are not the perfect standard candle. Their light curve have slight diversity and caribrated by empirical law obtained from nearby SNe Ia. The origin of this diversity is important when we investigate high-z SNe Ia because the same empirical law is applied for them.

The purpose of the present study is to explain this diversity with the rotation effect of accreting white dwarfs which eventually explode as SNe Ia. The white dwarf gains not only mass but angular momentum from its companion star. The centrifugal force reduces net gravity and increases the limiting mass for a supernova explosion. Brightness of SNe Ia may not be uniform if the masses of their progenitors have some diversity.

2. Methods

To calculare the structures of rapidly rotating white dwarfs, we used Hachisu's axisymmetric code (Hachisu 1986). We adopt the equation of state with zero temperature for white dwarf matter. As rotation law, j-constant law is adopted. No Coulomb interaction is included.

We assume that a white dwarf is at rest when it starts to gain mass and angular momentum, i.e. $J = 0$ at $M = M_0$. At each step of the calculation, the white dwarf is kept in hydrostatic equilibrium by the balance between the gravitational force, centrifugal force(equation (1)), and the pressure gradient and gains mass and specific angular momentum.

$$\int \rho^{-1} dP + \Phi - \int \Omega^2 w dw = C(\text{constant}) \tag{1}$$

$$j = \alpha j_{\text{Kepler}} = \alpha R \sqrt{\Phi_{\text{surface}}}, \tag{2}$$

where accreting gas is supposed to obey Keplerian rotation law and gives its angular momentum to the white dwarf with a certain efficiency α(equation (2)). The obtained new potentials are used for the calculation of the next step of accretion.

We expect two possible fates of the accreting white dwarfs: (1) the mass of the white dwarf gets close to the Chandrasekhar's limiting mass and a supernova explosion occurs when its maximum density reaches 2.0×10^9 g $\cdot$ cm^{-3}; and (2) the rotation of the white dwarf becomes too fast for itself to be stable against to secular instability($T/W > 0.14$ reference, Hachisu 1986).

3. Results

Results are shown in Figure 1. In this case, the accreting white dwarf reaches the upper edge with $T/W = 0.14$ regardless of its initial mass(upper panel). It means that whether SNe Ia occur under this condition is not clear.

For decreasing α, the final point of each track moves along the edge of the data area. For $M_0 = 0.6 M_\odot$, the white dwarf can become a supernova if $\alpha < 0.20$. In the lower panel, evolutionary tracks for $\alpha = 0.20$ are shown.

It is seen that a smaller initial mass results in a larger final mass. This means that an initially smaller white dwarf accretes more matter and rotates faster. Faster rotation allows more massive white dwarf to be stable.

To estimate brightness, we applied the $\rho - X(^{56}\text{Ni})$ relation, where $X(^{56}\text{Ni})$ is the mass fraction of the synthesized ^{56}Ni, obtained from the sphericalcarbon deflagration model W7 (Nomoto et al. 1984). The more massive progenitor produces more ^{56}Ni, thus becoming a brighter SN Ia since the light curve of SN

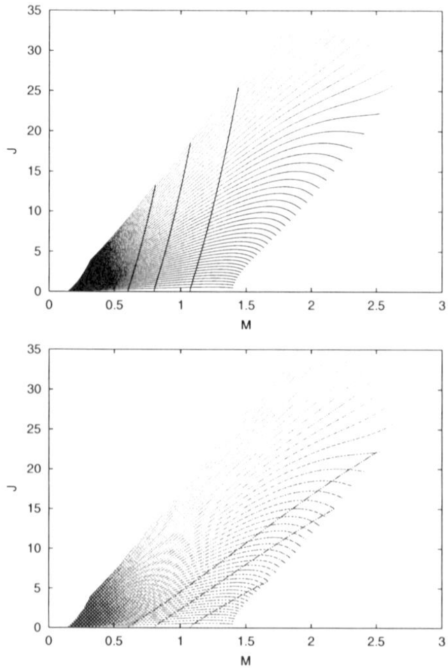

Fig. 1. Evolutionary tracks of white dwarfs whose initial mass are $0.6M_\odot$, $0.8M_\odot$, $1.07M_\odot$. Upper panel is for $\alpha = 1.0$ and lower panel is for $\alpha = 0.2$.

Ia is powered by the radioactive decay of ^{56}Ni. Difference in the synthesized ^{56}Ni mass between 0.6 case and 1.07 case corresponds to 0.4 magnitude. We can conclude from this result that the effect of rotation of progenitors can explain the diversity of SN Ia to some extent.

4. Discussion

To test whether this is a reasonable conclusion, we applied the result to the explanation of the dependence of SN Ia brightness on their parent galaxies. It is reported that SNe Ia have the dependence on their parent galaxies(Hamuy et al. 2000). That is, the brightest SNe Ia are observed only in spiral galaxies, where the dimmer SNe Ia are found in both spiral and elliptical galaxies.

In elliptical galaxies, star formation has stopped long time ago, and massive stars have already evolved. Therefore, a white dwarf with small M_0 cannot accrete enough mass to make a supernova explosion because no companion stars can supply enough matter to the white dwarf. On the other hand, in spiral galaxies, star formation is still active, so that the companion stars of the white dwarf can be massive. Therefore, white dwarfs of small M_0 can get enough mass to become SNe Ia. Accreting white dwarfs with smaller M_0 grow to large $M_{\rm fin}$ at the explosion, thus being expected to become brighter SN Ia. This implies that the upper limit of SNe Ia brightness is higher in spirals than in ellipticals. This explains the observation that the most luminous SNe Ia seem to be observed only in spirals, while dimmer SNe Ia are observed in both spirals and ellipticals.

5. Conclusion

Our main conclusion is that the rotation of progenitors can explain the diversity of the brightness of SNe Ia to some extent. However, this study needs to be extended by two means. First is an axsymmetric explosion simulation. We need to know how a nuclear flame propagates in axisymmetric explosion. Second is more precise treatment of angular momentum transportation.

In spite of these, we believe that our result is qualitatively true, since we can explain dependence on parents galaxies by using this result.

6. List of Symbols/Nomenclature

M=Mass, $M_\odot$ J=Angular momentum, 10^{49} erg · sec
T=Total rotational energy, erg W=Total gravitational energy, erg
j= Specific angular momentum, erg · sec/g
Φ= Gravitational potential Ω= Angular velocity, sec^{-1}
ρ= Density g/cm^3 P= Pressure, g/cm^3
R = Radius, cm w= Radius in axisymmetric coordinates, cm
M_0=Initial mass of the white dwarf, $M_\odot$
α= Efficiency of angular momentum during accretion

References

[1] S. Perlmutter, in this volume(2002)
[2] I. Hachisu, Astrophys. J. 61 (1986) 479.
[3] K. Nomoto et al., Astrophys. J. 286 (1984) 644.
[4] M. Hamuy et al., Astrophys. J. 120 (2000) 1479.

Probing the Formation and Evolution of Galaxies and Clusters with Gamma-rays and Cosmic Rays

Susumu INOUE
Division of Theoretical Astrophysics, National Astronomical Observatory, 2-21-1 Osawa, Mitaka, Tokyo 181-8588, Japan

Abstract

High energy, nonthermal phenomena are invaluable probes of shock heating processes which are crucial to the formation and evolution of large scale structure in the universe, and should offer new insight into studies of galaxies, groups, clusters, etc. Here we discuss the following two subjects: 1) heating of intracluster gas by high power radio galaxies and associated 'prompt' gamma-ray emission from nonthermal electrons [4], and 2) cosmic ray synthesis of ^{6}Li by structure formation shocks in the early Galaxy, and its use as a probe of dissipative processes during Galaxy formation [8].

1. Heating of Intracluster Gas by Powerful Radio Galaxies and Associated High Energy Gamma-ray Emission

The luminosity-temperature and entropy-temperature relations of clusters and groups of galaxies strongly suggest that the the hot gas in these systems has been mechanically heated by some energy source in addition to gravitational shock heating. While supernovae (SNe) are the most obvious candidate heating sources, a number of recent studies have indicated that they may be energetically insufficient [2]. Here we investigate the possibility of shock heating of the intracluster medium (ICM) by powerful FRII radio galaxies (RGs) in relation to these problems, making use of recent observational data.

We utilize 3 types of observational information on RGs: the observed correlation between jet kinetic power L_j and radio luminosity L_r [7]; the redshift evolution of the luminosity function (LF) of RGs [10], and the local LF of RGs inside clusters [5]. By integrating the LFs over L_r and redshift z, the total energy output of RG jets in the universe U_j can be evaluated with reasonable confidence, $U_j \sim 1 - 3 \times 10^{57}$ergMpc^{-3}, the bulk of which is attained at redshifts $z \sim 1 - 3$ Combining this with the local RG LF in clusters [5], the total energy input into the ICM can be estimated to be 1–2 keV per particle, precisely in the range recent models require to reproduce the observed deviations from self-similarity in the X-ray scaling relations.

RGs have a crucial advantage over SNe in that their large-scale jets deposit energy directly into the low density ICM outside galaxies, and are thus much less susceptible to radiative losses. The amount of heating can be even higher than the above if the cocoons of RGs additionally dissipate their energy into the ICM.

This picture of ICM shock heating by RGs can be most clearly tested and distinguished from SNe by observing the high energy, inverse Compton (IC) gamma-rays emitted by shock-accelerated electrons during the era of ICM heating, which is feasible with the *GLAST* satellite (Fig.1.). At sufficiently high energies, relativistic electrons have IC cooling times much shorter than the RG lifetime, so that the resultant gamma-ray emission is 'prompt' and directly traces the period of strong shock activity. The recently detected excess hard X-rays from groups [3] may be interpreted as relic non-thermal emission of past shock activity by RGs. This work is described in more detail in [4].

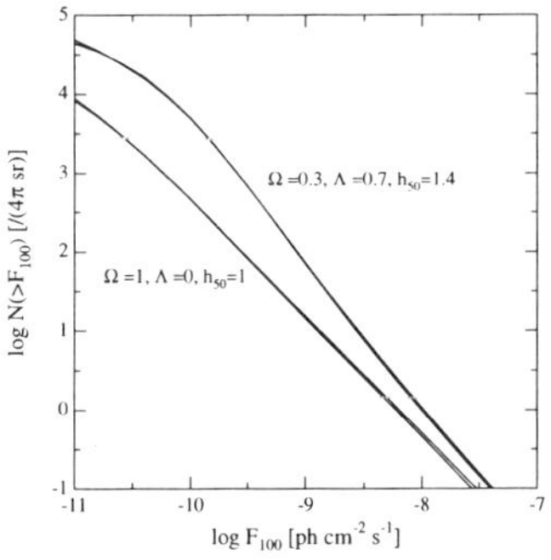

Fig. 1. The cumulative all-sky number of RGs with gamma-ray flux larger than F_{100}, the integral gamma-ray flux above 100 MeV. Curves are shown for model LFs (A, B and C) of [10] but are almost degenerate within either the Einstein-de Sitter (bottom) or the standard lambda cosmologies (top).

2. Cosmic Ray Production of ^{6}Li by Structure Formation Shocks in the Early Galaxy

The bulk of the light elements Li, Be and B are believed to arise from nonthermal nuclear reactions induced by cosmic rays (CRs). The abundances of Be and B observed in metal-poor halo stars (MPHS) are well explained as spallation products of CNO-enriched CRs accelerated by supernova (SN) shocks [9]. However, the origin of ^{6}Li observed in MPHS is more puzzling, as models based on SN CRs require an unrealistically high CR injection efficiency or an additional low energy CR component lacking observational support [9]. Here we investigate a more natural ^{6}Li production scenario: nuclear reactions induced by CRs accelerated at structure formation (SF) shocks, i.e. gravitational virialization shocks driven by the infall and merging of sub-Galactic gas clumps during the hierarchical build-up of structure in the early Galaxy. Such shocks are inevitable consequences in the currently standard picture of hierarchical structure formation.

The main SF shock during Galaxy formation should be due to the major merger of the two most massive clumps occurring at $z \sim 2$ [1]. Assuming a standard lambda cosmology and the total merged halo mass of $M_t \sim 3 \times 10^{12} \mathrm{M}_\odot$, the attained virial temperature is $k_B T_{\rm v} \simeq 0.26$keV, corresponding to a total dissipated specific energy $\epsilon_{\rm SF} \sim 3 k_B T_v/2 \simeq 0.4$ keV. This is to be compared with that estimated for SNe in the early halo, $\epsilon_{\rm SN} \sim 0.1$ keV [8], implying that SF shocks can be more energetic than SNe at early epochs.

A previous model of light element evolution by [9] is extended to incorporate CRs from SF shocks. SN CRs are modeled with standard assumptions and parameters: an injection efficiency $\xi_{\rm SN} = 0.15$ for each SN of kinetic energy $E_{SN} = 10^{51}$erg, with a power-law momentum spectrum of injection index $\gamma_{\rm SN} = 2.1$. As the evolution of Galactic structure with respect to [Fe/H] involves a number of uncertainties, SF CRs are included in a simple, parameterized way: they begin to be injected from a certain time $t_{\rm SF}$, maintains a constant flux for a duration $\tau_{\rm SF}$, and then returns to zero. The time-integrated CR energy is normalized to give $\epsilon_{\rm SF} \simeq 0.4$ keV. The injection efficiency and spectral shape are taken to be similar to SN CRs, but the injection index $\gamma_{\rm SF}$ is chosen to be steeper, as SF shocks should generally have low Mach numbers. Some specific parameter sets considered for $t_{\rm SF}$(Gyr), $\tau_{\rm SF}$(Gyr) and $\gamma_{\rm SF}$ are: I (0.12, 0.1, 3), II (0.22, 0.1, 3), III (0.32, 0.1, 3), IV (0.22, 0.1, 2), and V (0.1, 0.5, 3). The model results vs. observations are displayed in Fig.2..

It is seen that SN CRs succeed for Be but fail for ^{6}Li, whereas SF CRs can explain the present ^{6}Li data quite nicely. A crucial difference compared to SNe is that SF shocks do not eject any freshly synthesized CNO nor Fe, so that $\alpha - \alpha$ fusion is the dominant production channel at low [Fe/H], providing the observed amounts of ^{6}Li with little Be/B production and no direct correlation with Fe. As is evident in Fig.2., the SF CR models predict clearly testable trends in ^{6}Li/H vs. [Fe/H] and ^{6}Li/Be vs. [Fe/H] which are quite distinct from SN CR models.

A further unique expectation is of correlations between ^{6}Li/H and the kinematic properties of the MPHS. Recent studies of halo star kinematics show the Galactic halo to be of a 2-component nature, consisting of an inner, flattened and rotating halo and an outer, spherical and non-rotating halo. This suggests

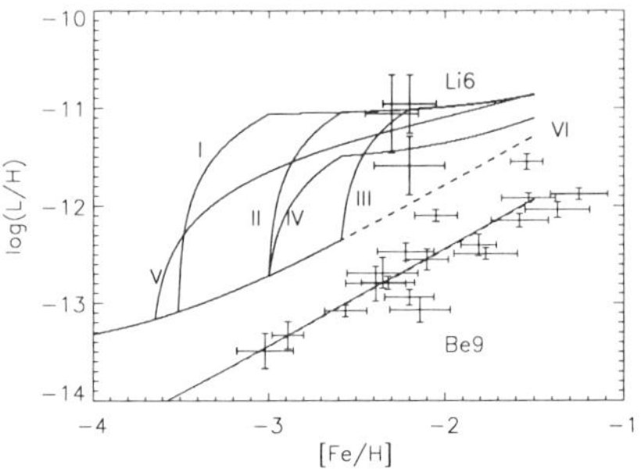

Fig. 2. Model results vs. current observational data for ^{6}Li (thick) and Be (thin) production by SN CRs only (dashed curves) and SN plus SF CRs (solid curves), each roman numeral corresponding the to the parameter set discussed in the text.

that during formation of these structures, gaseous dissipative processes have been crucial to the former whereas dissipationless stellar dynamics determine the latter. Since ^{6}Li production in our picture is a direct consequence of the main dissipation mechanism of gravitational shock heating, its abundance may be interpreted as a fossil record of gas dynamical processes during Galaxy formation. One prediction is that ^{6}Li/H should be systematically larger in stars of the inner halo compared to those of the outer halo. If such trends are indeed observed, it would not only confirm the SF CR origin of ^{6}Li, but may potentially point to a new field of '^{6}Li Galactic archaeology', whereby extensive observations of ^{6}Li in MPHS, e.g. by the Subaru HDS, can be exploited as unique probes of the past dynamical history of our Galaxy. See [8] for more details of this work.

References

[1] K. Bekki & M. Chiba, ApJ 558 (2001) 666
[2] R. G. Bower et al., MNRAS 325 (2001) 497
[3] Y. Fukazawa et al., ApJ 546 (2001) L87
[4] S. Inoue & S. Sasaki, ApJ 562 (2001) 618
[5] M. J. Ledlow & F. N. Owen, AJ 112 (1996) 9
[6] R. Ramaty et al., ApJ 534 (2000) 747
[7] S. Rawlings, in *Extragalactic Radio Sources* J. Roland et al. (1992) Cambridge
[8] T. K. Suzuki & S. Inoue, submitted to ApJ Lett. (2001)
[9] T.K. Suzuki & Y. Yoshii, ApJ 549 (2001) 303
[10] C. J. Willott et al., MNRAS 322 (2001) 536

Iron line emission in the Afterglow of GRB000214 and the environment around the GRB

Shigehiro NAGATAKI
Department of Physics, School of Science, The University of Tokyo, 7-3-1 Hongo, Bunkyo-ku, Tokyo 113-0033, Japan, nagataki@utap.phys.s.u-tokyo.ac.jp

Abstract

We present a model for explaining the Fe Kα line and the continuum in the afterglow of GRB000214. For reproducing the observation, we need a ring-like remnant around the progenitor, like that of SN 1987A produced by the mass-loss of the progenitor. The continuum spectrum can be fitted by the afterglow emission of the fireball pointing toward the observer, which does not collide with the ring. On the other hand, the line can be fitted by the recombinations of the ionized electrons in the ring illuminated by the X-ray afterglow. The significance of this study is that our model may strongly constrain the GRB models, that is, a massive star will be the progenitor of the GRB.

1. Introduction

There are four Gamma-Ray Bursts (hereafter GRBs) displaying the Fe Kα emission lines in their afterglows [1]. Of course, it is difficult to construct a model that reproduces the GRB and Fe line at the same time. This is because clean environment with high entropy is required for making a GRB, in which iron nuclei will be hardly synthesized.

In this study, we present a model to reproduce the observation of GRB 000214, in which we consider two systems for the line emission and the continuum. That is, the Fe Kα line is produced by the recombinations of ionized electrons in the surrounding ring illuminated by the X-ray afterglow. On the other hand, the continuum power-law flux is obtained by the afterglow emitted from the fireball toward the observer (see. Fig.1a). In this study, we show that many observational facts such as intensity of the Fe line emission and spectrum shaping are explained by our model. Detailed explanation is presented in our paper [2].

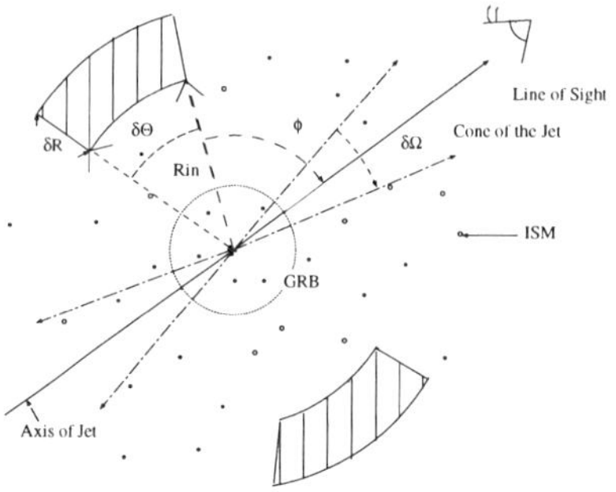

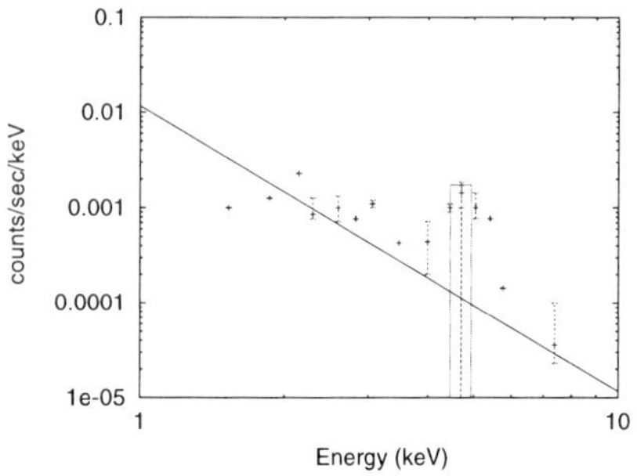

Fig. 1. (a) Schematic diagram of the GRB's environment in our model. (b) Integrated spectrum, solid line : afterglow emission toward the observer, dashed line : fluorescent line emission of Fe Kα originated from the ring illuminated by the X-ray afterglow in the energy bin of 0.48 keV, crosses is the observation of the afterglow and the Fe line in Antonelli et al. (2000)

2. Model and Results

2.1. continuum

We analyze the afterglow of GRB000214 using a fireball model. Seven free parameters required to determine the spectrum in their model are determined below. $E_{\text{spherical}}$ is given as,

$$E_{\text{spherical}} = \frac{4\pi}{\delta\Omega}\, E_{\text{jet}} \tag{1}$$

(see Fig.1a), which is the kinetic energy of the fireball, estimated as if the explosion were spherical. ϵ_B is equipartition parameter for magnetic field, ϵ_e is equipartition parameter for electric field, n_{ism} is number density for the interstellar matter, t_d is days from the Burst, γ_0 is initial fireball's Lorentz factor, D is the distance from the GRB center to the observer. From observations, $t_d = 0.5$, $D = 1.2 \times 10^{28}$ cm, while n_{ism} is set to 1/cc. We fix $E_{\text{spherical}} = 2.8 \times 10^{54}$ ergs, $\epsilon_B = \epsilon_e = 0.5$, $\gamma_0 = 200$. The radiation of GRB afterglows is divided into two phases, adiabatic and radiative phases. To determine under which processes the system is, we compare two quantity, γ_c, and γ_m. Both of them are given like below.

$$\gamma_c = \frac{3 m_e}{16\, \epsilon_B\, \sigma_T\, m_p\, c\, t\, \gamma_0^3\, n_{\text{ism}}}, \tag{2}$$

$$\gamma_m \propto 610\, \epsilon_e\, \gamma_0. \tag{3}$$

At γ_c synchrotron cooling is most effective. On the other hand, above γ_m, the number density of electrons: $N(\gamma)$ accelerated behind the shock obeys power law as

$N(\gamma) \propto \gamma^{-p} d\gamma$, $(\gamma \geq \gamma_m)$. Since γ_c is larger than γ_m with the above parameters, the situation is concluded to be the adiabatic phase. For adiabatic blast wave,

$$\begin{aligned} \nu_c &= 2.7 \times 10^{11} {\epsilon_B}^{-3/2} \left(\frac{E_{\rm spherical}}{10^{54}{\rm ergs}}\right)^{-1/2} \left(\frac{n_{\rm ism}}{1/{\rm cc}}\right)^{-1} \\ & \times {t_d}^{-1/2} \ {\rm Hz}. \end{aligned} \quad (4)$$

$$\nu_m = 5.7 \times 10^{15} {\epsilon_B}^{1/2} {\epsilon_e}^{2} \left(\frac{E_{\rm spherical}}{10^{54}{\rm ergs}}\right)^{1/2} t_d^{-3/2} \ {\rm Hz}. \quad (5)$$

$$\begin{aligned} F_{\nu,{\rm max}} &= 1.1 \times 10^{7} \ {\epsilon_B}^{1/2} \left(\frac{E_{\rm spherical}}{10^{54}{\rm ergs}}\right) \left(\frac{n_{\rm ism}}{1/{\rm cc}}\right)^{1/2} \times \\ & \left(\frac{D}{10^{28}{\rm cm}}\right)^{-2} \ \mu{\rm Jy}. \end{aligned} \quad (6)$$

where ν_c and ν_m correspond to the relevant Lorenz factor γ_c and γ_m. The flux of the afterglow, $F_{\rm X}$ can be estimated as

$$F_{\rm X} = \left(\frac{\nu_c}{\nu_m}\right)^{-3/2} \left(\frac{\nu}{\nu_c}\right)^{-2} \nu \, F_{\nu,{\rm max}} \ {\rm ergs \ cm^{-2} \, s^{-1}}. \quad (7)$$

In Fig.1b, with the above parameters, we can fit the observed continuum spectrum well.

2.2. *emission line*

The number of the line photons from GRB000214 is estimated to be 2×10^{56}. Thus, if one iron nuclei is assumed to emit one line photon, the mass of iron can be estimated to be $12.6 M_\odot$, which is too much to be accepted. In order to solve the difficulty, we should consider that one iron nuclei emit k line photons in average, which can be estimated as

$$k = \frac{t_{\rm ill}}{t_{\rm ion} + t_{\rm rec}} q \sim \frac{t_{\rm ill}}{t_{\rm rec}} q, \quad (8)$$

where $t_{\rm ill}$, $t_{\rm ion}$, $t_{\rm rec}$ and q are illuminated timescale, ionization timescale, recombination timescale and the fraction of the absorbed ionizing fluence and reprocessed into the line, respectively. $t_{\rm ill}$ is about 10^5s from the observations and q is estimated to be 0.1 theoretically. $t_{\rm rec}$ can be estimated as

$$t_{\rm rec} = \frac{1}{\alpha({\rm T}) n_{\rm e}}, \quad (9)$$

where $n_{\rm e}$ is electron number density and

$$\alpha({\rm T}) = 1.4 \times 10^{-12} \lambda^{1/2} \left(0.429 + 0.5 \ln \lambda + \frac{0.496}{\lambda^{1/3}}\right). \quad (10)$$

Here λ is defined as $1.1\times10^8/\mathrm{T}$. From the observations, it is reported that $\mathrm{T}\leq 10^7$[K] [4]. Thus k can be estimated as

$$k \geq 7.1 \times 10^3 \left(\frac{q}{0.1}\right)\left(\frac{n_\mathrm{e}}{10^{11}}\right). \tag{11}$$

As a result, the required mass of iron is

$$M_\mathrm{Fe} \sim \frac{12.6}{k} M_\odot \leq 1.8 \times 10^{-3} \left(\frac{0.1}{q}\right)\left(\frac{10^{11}}{n_\mathrm{e}}\right). \tag{12}$$

Finally, the total mass of the ring around the GRB can be estimated to be

$$M_\mathrm{tot} \leq 1.5 \left(\frac{1}{A_\odot}\right)\left(\frac{0.1}{q}\right)\left(\frac{10^{11}}{n_\mathrm{e}}\right) M_\odot, \tag{13}$$

where $A_\odot = X_\mathrm{Fe}/X_{\mathrm{Fe},\odot}$. Thus it is concluded that the required mass is in the range (1.5 - 15)$M_\odot$ if n_e is in the range $(10^{10} - 10^{11})$ cm^{-3}. The concluded mass will be able to be supplied by a mass loss process as in our model. Also, although the required electron number density seems to be very large, it is reported that $n_\mathrm{e} \sim 2.4 \times 10^{10}$ for SN 1997cy.

3. Conclusion

We have shown that mass loss model is natural and may be the solution for the iron emission line problem. This indicates that the progenitor of a GRB may be a massive star. The reasons are as follows. (1) a ring-like shape can be formed as a result of mass loss process. (2) SN1998bw is Type Ic SN, which is the evidence that the progenitor lost a lot of mass before the explosion. (3) The intensity of iron line from GRB000214 can be reproduced when the density of CSM is in the range $(10^{10} - 10^{11})$, which is about same order with that observed in SN1997cy. Further observations by HETE II, Swift and Astro E2 will present us valuable informations which determine which model is correct in the near future.

References

[1] L. Piro, et al. Science 290 (2000) 955.
[2] K. Kotake, S. Nagataki PASJ 53 (2001) 579.
[3] L.A. Antonelli, et al. ApJ 545 (2000) L39.
[4] D. Yonetoku, et al. ApJ 545 (2001) L23.

The Implications of $\bar{\nu}_e e^- \to W^-\gamma$ for the Detection of High-Energy $\bar{\nu}_e$

H. Athar[1,2] and Guey-Lin Lin[2]
[1] *Physics Division, National Center for Theoretical Sciences, Hsinchu 300, Taiwan*
[2] *Institute of Physics, National Chiao-Tung University, Hsinchu 300, Taiwan*

Abstract

We discuss some motivations for detecting high-energy neutrinos through the pure electroweak processes such as $\bar{\nu}_e e^- \to W^-$ and $\bar{\nu}_e e^- \to W^-\gamma$. We argue that the latter process can be viewed as an enhancement to the former one. The event-rate enhancement is estimated.

1. Introduction

An observation of high-energy neutrinos beyond the atmospheric background will mark the beginning of high-energy neutrino astronomy. Several high-energy neutrino detectors, commonly known as high-energy neutrino telescopes, are currently in an advanced stage of their deployments. These detectors operate according to the high-energy neutrino interactions, particularly those occurring inside the high-energy neutrino telescopes. The studies facilitated by the high-energy neutrino telescopes will complement the high-energy gamma ray astronomy for understanding the origin of cosmic high-energy radiation [1].

In this talk, we shall discuss some motivations for detecting the high-energy neutrinos through the pure electroweak interactions. In particular, we focus on the resonant absorption process $\bar{\nu}_e e^- \to W^-(\to \bar{\nu}_\mu \mu^-)$, now referred to as *Glashow resonance* [2, 3] and the scattering process $\bar{\nu}_e e^- \to W^-\gamma$ [4]. The role of the latter process will be discussed.

2. The Detection Modes of High-Energy Neutrinos

The typical detection of high-energy neutrinos is through the charged-current neutrino-nucleon scattering $\nu_l(\bar{\nu}_l) + N \to l^-(l^+) + X$. For $10^7\text{GeV} < E_\nu < 10^{12}\text{GeV}$, the cross section of this process can be parameterized as $\sigma^{CC}(\nu N) = 5.5 \cdot 10^{-36}\text{cm}^2(E_\nu/1\text{ GeV})^{0.363}$ [5]. In general, the cross section of neutrino-lepton scattering is suppressed except for the $\bar{\nu}_e e^-$ scattering at the resonance energy $E_{\bar{\nu}_e} = 6.3 \cdot 10^6$ GeV. At this energy, the resonant absorption $\bar{\nu}_e e^- \to W^- \to$

hadrons occurs with a peak cross section $\sigma^{res}(s = m_W^2) = 0.3\ \mu$b. Although the resonant peak spreads only over the W-boson width, the overall event rate in the high-energy neutrino telescope is non-negligible after convoluting the absorption cross section with the incoming neutrino flux[5].

It is interesting to note that the signature of $\bar{\nu}_e e^- \to W^-\gamma$ is similar to that of the resonant absorption provided the outgoing photon is soft. If the photon is hard, one might be able to separate this process from the resonant absorption by detecting the hard photon. There are a few reasons for identifying the resonant absorption and $\bar{\nu}_e e^- \to W^-\gamma$ processes in the high-energy neutrino telescopes. First of all, unlike the neutrino-nucleon scattering, the resonant absorption and $\bar{\nu}_e e^- \to W^-\gamma$ do not contain any hadronic uncertainties. Therefore, they can be used to probe the absolute flux of $\bar{\nu}_e$. Secondly, the $\bar{\nu}_e$ flux measured in the high-energy neutrino telescope is in fact identical to the intrinsic $\bar{\nu}_e$ flux from a cosmologically distant source, in spite of the flavor oscillations taking place between the source and the high-energy neutrino telescope[6]. Finally, we should point out that only the $\bar{\nu}_e$ flux near the energy $E_{\bar{\nu}_e} = 6.3 \cdot 10^6$ GeV can be probed by the resonant absorption and $\bar{\nu}_e e^- \to W^-\gamma$. For other neutrino energies, the cross section of neutrino-nucleon scattering dominates.

3. Results and Discussion

The cross sections for resonant absorption and $\bar{\nu}_e e^- \to W^-\gamma$ are both well known. We have further calculated the photon spectrum of the latter process[7]:

$$\frac{\mathrm{d}\sigma}{\mathrm{d}y} = A(\lambda)\Big[(\lambda-1)(\lambda^2+1)y^{-1} + 4\lambda^2(\lambda-1)y - 2\lambda^3 y^2 - \lambda(3\lambda^2 - 4\lambda + 3)\Big], \quad (1)$$

where $A(\lambda) = \sqrt{2}\alpha G_F/(\lambda^2(\lambda-1)^2)$, with $\lambda = s/m_W^2$ and $y = E_\gamma/E_{\bar{\nu}_e}$. The appearance of $1/y$ term in the spectrum $\mathrm{d}\sigma/\mathrm{d}y$ indicates that the outgoing photon tends to be soft. Indeed, we have computed $\langle y \rangle$, the average value of y, for different incoming neutrino energies. We have found $\langle y \rangle \simeq 1.3 \cdot 10^{-3}$ for $E_{\bar{\nu}_e} = 6.6 \cdot 10^6$ GeV, $\langle y \rangle \simeq 6.0 \cdot 10^{-3}$ for $E_{\bar{\nu}_e} = 8.6 \cdot 10^6$ GeV, whereas $\langle y \rangle \simeq 9.7 \cdot 10^{-2}$ for $E_{\bar{\nu}_e} = 1.1 \cdot 10^7$ GeV. The corresponding averaged photon energies in these cases are $8.6 \cdot 10^3$ GeV, $4.1 \cdot 10^4$ GeV, and 10^6 GeV respectively. With the above energies, the photon radiation length is about 30 cm in the water or ice [8]. It turns out that the detection of such photons is impractical given the large separation of photomultiplier tubes in the current design of high-energy neutrino telescopes.

Without detecting the outgoing photon in $\bar{\nu}_e e^- \to W^-\gamma$, one should treat this process as an enhancement to the resonant absorption. It is important to compare $\bar{\nu}_e e^- \to W^-\gamma$ cross section with those of conventional channels. The behaviors of these cross sections as functions of $E_{\bar{\nu}_e}$ are shown in Fig. 1. In

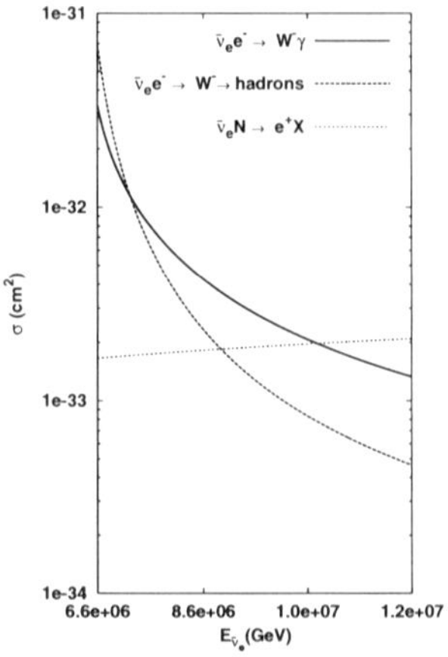

Fig. 1. High-energy $\bar{\nu}_e$ absorption cross section σ (cm^2), over two different target particles as a function of electron anti neutrino energy $E_{\bar{\nu}_e}$ (GeV). The minimum value of $E_{\bar{\nu}_e}$ corresponds to $(M_W + \Gamma_W)^2/2m_e$.

this figure, we have included, besides the cross section of the current process, the resonant cross section, $\sigma_{\bar{\nu}_e e^- \to W^- \to \text{hadrons}}$, taken from Ref. [9], as well as the charged current deep-inelastic $\bar{\nu}_e$ scattering cross section over nuclei, $\sigma^{CC}_{\bar{\nu}_e N \to e^+ X}$, taken from Ref. [5] with CTEQ4-DIS parton distribution functions. From Fig. 1, we note that $\sigma_{\bar{\nu}_e e^- \to W^- \gamma}$ dominates over the other two for less than half an order of magnitude in $E_{\bar{\nu}_e}$ $(7 \cdot 10^6 \leq E_{\bar{\nu}_e}/\text{GeV} \leq 1 \cdot 10^7)$. The dominance is however within a factor of 2.

In Ref. [5], the hadron shower event rate due to the Glashow resonance formation in the $\bar{\nu}_e e^-$ scattering was estimated. We have computed an enhancement to this event rate by incorporating the contribution of $\bar{\nu}_e e^- \to W^-\gamma$ (see Fig. 1). The enhancement is typically about 10%. This enhancement, if become measurable, can be considered as a signature of hard photon emission in the absorption process $\bar{\nu}_e e^- \to W^-\gamma$ in a high-energy neutrino telescope. In this estimate, we have used the $\mathrm{d}N/\mathrm{d}E_{\bar{\nu}_e}$ given by

$$\frac{\mathrm{d}N}{\mathrm{d}E_{\bar{\nu}_e}} \simeq 10^{-8} \left(\frac{E_{\bar{\nu}_e}}{1\text{GeV}} \right)^{-2} \text{cm}^{-2}\text{s}^{-1}\text{sr}^{-1}\text{GeV}^{-1}, \tag{2}$$

for the relevant $E_{\bar{\nu}_e}$ range. In Eq. (2), the $\mathrm{d}N/\mathrm{d}E_{\bar{\nu}_e}$ is the downward going differential $\bar{\nu}_e$ flux arriving at the high-energy neutrino telescope. Briefly, it is for the gamma ray burst fireball model proposed in Ref. [10], where $p\gamma$ interactions are suggested to produce the high-energy $\bar{\nu}_e$ flux at the gamma ray burst fireball site.

In conclusion, we have provided some motivations for detecting the high-energy $\bar{\nu}_e$ through the resonant absorption process $\bar{\nu}_e e^- \to W^-$ and the scattering $\bar{\nu}_e e^- \to W^-\gamma$, which are purely electroweak in nature. The latter process is shown to enhance the event rate of the former process by about 10%.

H.A. is supported by Physics Division of National Center for Theoretical Sciences. G.L.L. is supported by the National Science Council under the grant number NSC90-2112-M009-023 and Research Excellence Project on Cosmology and Particle Astrophysics (CosPA) funded by Taiwan's Ministry of Education.

References

[1] For recent review articles, see, for instance, J. G. Learned and K. Mannheim, Ann. Rev. Nucl. Part. Sci. 50 (2000) 679; F. Halzen, Phys. Rept. 333 (2000) 349 and references therein.

[2] S. L. Glashow, Phys. Rev. 118 (1960) 316.

[3] J. N. Bahcall and S. C. Frautschi, Phys. Rev. 135B (1960) 788; V. S. Berezinskiǐ and A. Z. Gazizov, JETP Lett. 25 (1977) 254 [Pisma Zh. Eksp. Teor. Fiz. 25 (1977) 276]; V. S. Berezinsky, D. Cline and D. N. Schramm, Phys. Lett. B78 (1978) 635; I. M. Zheleznykh and É. A. Taǐnov, Sov. J. Nucl. Phys. 32 (1980) 242 [Yad. Fiz. 32 (1980) 468]; V. S. Berezinsky and V. L. Ginzburg, Mon. Not. R. Astr. Soc. 194 (1981) 3; R. W. Brown and F. W. Stecker, Phys. Rev. D26 (1982) 373; G. Domokos and S. Kovesi-Domokos, Phys. Lett. B346 (1995) 317.

[4] R. W. Brown, D. Sahdev and K. O. Mikaelian, Phys. Rev. D20 (1979) 1164; K. O. Mikaelian and I. M. Zheleznykh, Phys. Rev. D22 (1980) 2122; V. S. Berezinskiǐ and A. Z. Gazizov, Sov. J. Nucl. Phys. 33 (1981) 120 [Yad. Fiz. 33 (1981) 230]; F. Wilczek, Phys. Rev. Lett. 55 (1985) 1252; D. Seckel, Phys. Rev. Lett. 80 (1998) 900.

[5] R. Gandhi, C. Quigg, M. H. Reno and I. Sarcevic, Phys. Rev. D58 (1998) 093009.

[6] H. Athar, M. Jeżabek and O. Yasuda, Phys. Rev. D62 (2000) 103007 and references therein.

[7] H. Athar and G.-L. Lin, hep-ph/0108204.

[8] T. Stanev, C. Vankov, R. E. Streitmatter, R. W. Ellsworth and T. Bowen, Phys. Rev. D25 (1982) 1291.

[9] T. K. Gaisser, F. Halzen and T. Stanev, Phys. Rept. 258 (1995) 173 [Erratum-ibid. 271 (1996) 355].

[10] E. Waxman and J. Bahcall, Phys. Rev. Lett. 78 (1997) 2292.

The 2dF Galaxy Redshift Survey: Cosmological Parameters and Galaxy Biasing

Ofer LAHAV
Institute of Astronomy, University of Cambridge, Madingley Road, Cambridge CB3 0HA, UK

Abstract

The 2dF Galaxy Redshift Survey (2dFGRS) has already measured over 220,000 redshifts of nearby ($z \sim 0.1$) galaxies. It allows us to estimate fundamental cosmological parameters and to subdivide the survey into subsets according to the galaxy intrinsic properties. The large-scale structure analysis of the survey has already yielded several significant results: (i) the shape of the power spectrum of fluctuations is consistent with the Λ-Cold Dark Matter concordance model; (ii) from joint analysis of the 2dFGRS data with the Cosmic Microwave Background anisotropies the linear-theory rms mass fluctuations is $\sigma_{8\mathrm{m}} \approx 0.73$, lower than the COBE-alone normalization and previous estimates from cluster abundance and cosmic shear; (iii) The biasing parameter of bright galaxies on scales $\gtrsim 10\,h^{-1}$ Mpc is unity; and (iv) on scales smaller than $\lesssim 10\,h^{-1}$ Mpc red galaxies are more strongly clustered than blue galaxies.

1. Introduction

Multifibre technology now allows us to measure redshifts of millions of galaxies. The Anglo-Australian 2 degree Field Galaxy Redshift Survey* (2dFGRS) has already measured redshifts for 220,000 galaxies selected from the APM catalogue (as of March 2002). The median redshift of the 2dFGRS is $\bar{z} \sim 0.1$. It aims to acquire a complete sample of ~250,000 galaxy spectra, down to an extinction corrected magnitude limit of $b_J < 19.45$ (Colless et al. 2001). A sample of this size allows large-scale structure statistics to be measured with very small random errors. In this review we summarize some recent results from 2dFGRS

*The 2dFGRS Team comprises: I.J. Baldry, C.M. Baugh, J. Bland-Hawthorn, T.J. Bridges, R.D. Cannon, S. Cole, C.A. Collins, M. Colless, W.J. Couch, N.G.J. Cross, G.B. Dalton, R. DePropris, S.P. Driver, G. Efstathiou, R.S. Ellis, C.S. Frenk, K. Glazebrook, E. Hawkins, C.A. Jackson, O. Lahav, I.J. Lewis, S.L. Lumsden, S. Maddox, D.S. Madgwick, S. Moody, P. Norberg, J.A. Peacock, B.A. Peterson, W. Sutherland, K. Taylor. For more details on the survey see http://www.mso.anu.edu.au/2dFGRS/

on clustering and galaxy biasing.

2. The Power spectrum of 2dF Galaxies

An initial estimate of the convolved, redshift-space power spectrum of the 2dFGRS has already been determined (Percival et al. 2001; hereafter P01) for a sample of 160,000 redshifts. On scales $0.02 < k < 0.15\,h\,\mathrm{Mpc}^{-1}$, the data are robust and the shape of the power spectrum is not affected by redshift-space or non-linear effects, though the amplitude is increased by redshift-space distortions. P01 and Efstathiou et al. (2001; hereafter E01) have mainly compared the *shape* of the 2dFGRS and CMB power spectra, and concluded that they are consistent with each other (see also Tegmark, Hamilton & Xu 2001), within the Λ-CDM framework.

Lahav et al. (2001) have estimated the *amplitudes* of the linear-theory rms fluctuations in mass $\sigma_{8\mathrm{m}}$ and in galaxies $\sigma_{8\mathrm{g}}$. More precisely, consider the ratio of galaxy to matter power spectra, and use the ratio of these to define the bias parameter:

$$b^2 \equiv \frac{\sigma_{8\mathrm{g}}^2}{\sigma_{8\mathrm{m}}^2} \equiv \frac{P_{\mathrm{gg}}(k)}{P_{\mathrm{mm}}(k)} \tag{1}$$

On scales of $0.02 < k < 0.15\,h\,\mathrm{Mpc}^{-1}$ the fluctuations are close to the linear regime, and there are good reasons (e.g. Benson et al. 2000) to expect that b should tend to a constant. Here we do not test the assumption that the biasing is scale-independent, but we do allow it to be function of luminosity and redshift. Another necessary complication is that we need to distinguish between the apparent values of $\sigma_{8\mathrm{g}}$ as measured in redshift space ($\sigma_{8\mathrm{g}}^S$) and the real-space value that would be measured in the absence of redshift-space distortions ($\sigma_{8\mathrm{g}}^R$). It is the latter value that is required in order to estimate the bias. We emphasize that here $\sigma_{8\mathrm{g}}$ is the linear-theory normalization, not the observed non-linear $\sigma_{8\mathrm{gNL}}$. For example, the 2dFGRS correlation function of Norberg et al. (2001a) can be translated to a non-linear $\sigma_{8\mathrm{gNL}}^R(L_*) = 0.87 \pm 0.07$, at an effective redshift of approximately 0.07. In practice, nonlinear corrections to σ_8 are expected to be relatively small for CDM-like spectra.

The 2dFGRS power spectrum (Fig. 2) is fitted in P01 over the above range in k, assuming scale-invariant primordial fluctuations and a Λ-CDM cosmology, for four free parameters: $\Omega_{\mathrm{m}}h$, $\Omega_{\mathrm{b}}/\Omega_{\mathrm{m}}$, h and the redshift space $\sigma_{8\mathrm{g}}^S$. Assuming a Gaussian prior on the Hubble constant $h = 0.7 \pm 0.07$ (based on Freedman et al. 2001) the shape of the recovered spectrum within the above k-range was used to yield 68 per cent confidence limits on the shape parameter $\Omega_{\mathrm{m}}h = 0.20 \pm 0.03$, and the baryon fraction $\Omega_{\mathrm{b}}/\Omega_{\mathrm{m}} = 0.15 \pm 0.07$, in accordance with the popular

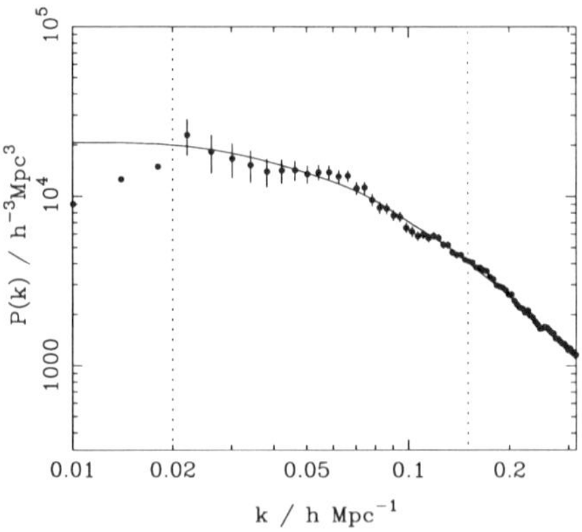

Fig. 1. The observed (i.e. convolved with the window function) 2dFGRS power spectrum (as derived in Percival et al. 2001). The solid line shows a linear theory Λ-CDM fit (convolved with the window function) with $\Omega_\mathrm{m}h = 0.2, \Omega_\mathrm{b}/\Omega_\mathrm{m} = 0.15, h = 0.7, n = 1$ and best-fitting $\sigma^S_{8\mathrm{g}}(z_s, L_s) = 0.94$. Only the range $0.02 < k < 0.15\,h\,\mathrm{Mpc}^{-1}$ is used at the present analysis (roughly corresponding to CMB harmonics $200 < \ell < 1500$ in a flat $\Omega_\mathrm{m} = 0.3$ Universe). The good fit of the linear theory power spectrum at $k > 0.15\,h\,\mathrm{Mpc}^{-1}$ is due to a conspiracy between the non-linear growth and finger-of-god smearing.

'concordance' model. For fixed 'concordance model' parameters $n = 1, \Omega_\mathrm{m} = 1 - \Omega_\Lambda = 0.3$, $\omega_\mathrm{b} = 0.02$ and a Hubble constant $h = 0.70$, we find that the amplitude of 2dFGRS galaxies in redshift space is $\sigma^S_{8\mathrm{g}}(L_s, z_s) \approx 0.94$. As shown in P01, the likelihood analysis gives a second (non-standard) solution, with $\Omega_\mathrm{m}h \sim 0.6$, and the baryon fraction $\Omega_\mathrm{b}/\Omega_\mathrm{m} = 0.4$, which generates baryonic 'wiggles'. We ignore this case at the present analysis. We also note that even if there are features in the primordial power spectrum, they would get washed out by the 2dFGRS window function (Elgaroy, Gramann & Lahav 2001).

In reality, the effective redshift for the P01 analysis is not zero, but $z_s \sim 0.17$. This is higher than the median redshift of 2dFGRS ($z_m \sim 0.11$) due to the weighting scheme used in estimating the power spectrum. Similarly, $L_s \simeq 1.9L_*$, rather than the $L_s \simeq L_*$ that would apply for a flux-limited sample. The effects of redshift-space distortion, evolution of biasing and luminosity bias on the determination of b are quite significant, at the level of ~ 15 per cent each.

3. Combining 2dFGRS & CMB

A simultaneous analysis of the constraints placed on cosmological parameters by different kinds of data is essential because each probe (e.g. CMB, SNe Ia, redshift surveys, cluster abundance, and peculiar velocities) typically constrains a different combination of parameters (e.g. Bahcall et al. 1999; Bridle et al. 2001a; E01). A particular case of joint analysis is that of galaxy redshift surveys and the CMB. While the CMB probes the fluctuations in matter, the galaxy redshift surveys measure the perturbations in the light distribution of particular tracer (e.g. galaxies of certain type). Therefore, for a fixed set of cosmological parameters, a combination of the two can tell us about the way galaxies are 'biased' relative to the mass fluctuations (e.g. Webster et al. 1998).

The CMB fluctuations are commonly represented by the spherical harmonics C_ℓ. The connection between the harmonic ℓ and k is roughly

$$\ell \simeq k \, \frac{2c}{H_0 \, \Omega_{\rm m}^{0.4}} \tag{2}$$

for a flat Universe. For $\Omega_{\rm m} = 0.3$ the 2dFGRS range $0.02 < k < 0.15\, h\,{\rm Mpc}^{-1}$ corresponds approximately to $200 < \ell < 1500$, which is well covered by the recent CMB experiments.

A well-known problem in estimating cosmological parameters is the degeneracy of parameters, and the choice of free parameters. Here we assume a flat Universe (i.e. zero curvature), and no tensor component in the CMB (for discussion of the degeneracy with respect to these parameters see E01). We consider five free parameters: the matter density parameter $\Omega_{\rm m}$, the linear-theory amplitude of the mass fluctuations $\sigma_{8{\rm m}}$, the present-epoch linear biasing parameter $b(L_s, z=0)$ (for the survey effective luminosity $L_s \simeq 1.9 L_*$), the Hubble constant $h \equiv H_0/(100\,{\rm km\,s^{-1}})$, and the baryon density parameter $\omega_{\rm b} \equiv \Omega_{\rm b} h^2$. As we are mainly interested in combinations of $\sigma_{8{\rm m}}$, b and $\Omega_{\rm m}$, we shall marginalize over the remaining parameters. We also check the robustness of the results to some 'extra parameters', the optical depth τ due to reionization and the primordial spectral index n.

The latest CMB measurements from Boomerang (Netterfield et al. 2001, de Bernardis et al. 2001), Maxima (Lee et al. 2001; Stomper et al. 2001) and DASI (Halverson et al. 2001; Pryke et al. 2001) suggest three acoustic peaks. Parameter fitting to a Λ-CDM model indicates consistency between the different experiments, and a best-fit Universe with zero curvature, and an initial spectrum with spectral index $n \simeq 1$ (e.g. Wang et al. 2001, E01 and references therein). Unlike the earlier Boomerang and Maxima results, the new data also show that the baryon contribution is consistent with the Big Bang Nucleosynthesis value $\omega_{\rm b} \simeq$

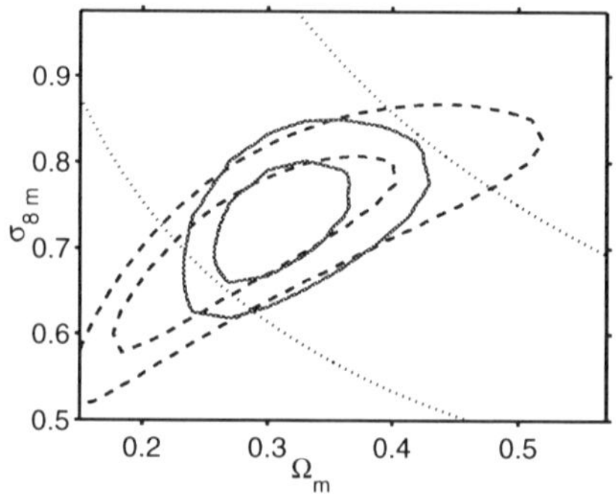

Fig. 2. The likelihood function of CMB alone (dashed lines) in terms of the mass fluctuation amplitude $\sigma_{8\mathrm{m}}$ and the present epoch Ω_{m}. The marginalization over the Hubble constant is done with a Gaussian centred at $h = 0.7$ and standard deviation of 0.07. Other parameters are held fixed ($n = 1, \omega_{\mathrm{b}} = 0.02, \tau = 0.0$). The contours are for (two-parameter) 68 per cent and 95 per cent confidence intervals. The solid lines show the contours (68 per cent and 95 per cent) for the joint 2dFGRS+CMB analysis, after marginalization over h, $b(L_s, 0)$ and ω_{b}. Other parameters are held fixed ($n = 1, \tau = 0.0$). Note that the contours of 2dFGRS+CMB are much tighter than when using CMB alone. Two recent extreme cluster abundance determinations are overlayed as the upper dotted line (Pierpaoli et al. 2001) and the lower dotted line (Viana et al. 2001). From Lahav et al. (2001).

0.02 (O'Meara et al. 2001). We have used a compilation of COBE, Boomerang, Maxima and DASI data, after marginalization over calibration errors (Bridle et al. 2001b).

When combining 2dFGRS and CMB data the parameterization for the log-likelihoods is then:

$$\begin{aligned} \ln \mathcal{L}_{\mathrm{tot}} &= \ln \mathcal{L}_{\mathrm{2dFGRS}}[\Omega_{\mathrm{m}}, h, \omega_{\mathrm{b}}, \sigma_{8\mathrm{m}}, b(L_s, 0)] \\ &+ \ln \mathcal{L}_{\mathrm{CMB}}[\Omega_{\mathrm{m}}, h, \omega_{\mathrm{b}}, \sigma_{8\mathrm{m}}], \end{aligned} \quad (3)$$

where $\mathcal{L}_{\mathrm{2dFGRS}}$ and $\mathcal{L}_{\mathrm{CMB}}$ are the likelihood functions for 2dFGRS and the CMB.

The 2dFGRS likelihood function takes into account the redshift-space distortions, an epoch-independent galaxy clustering biasing scheme, and the redshift evolution of $\Omega_{\mathrm{m}}(z)$. Other parameters are held fixed ($n = 1, \tau = 0$).

Fig. 2. (solid lines) shows the 2dFGRS+CMB likelihood as a function of ($\Omega_{\mathrm{m}}, \sigma_{8\mathrm{m}}$), after marginalization over $h, b(L_s, 0)$ and ω_{b}. The peak of the distribution is consistent with the result for the CMB alone (shown by the dashed lines in Fig. 2.), but we see that the contours are tighter due to the addition of the 2dFGRS data. Further marginalization over Ω_{m} gives $\sigma_{8\mathrm{m}} = 0.73 \pm 0.05$.

To study the biasing parameter we marginalize the 2dFGRS likelihood over $h, \omega_{\rm b}$, $\sigma_{8\rm m}$ and $\Omega_{\rm m}$ (with fixed parameters are held fixed $n = 1, \tau = 0$) and we get $b(L_s, 0) = 1.00 \pm 0.09$ (1-sigma).

To translate the biasing parameter from L_s to e.g. L_* galaxies one can either assume (somewhat ad-hoc) no luminosity segregation on large scales, or divide by the factor 1.14 (Norberg et al. 2001a) that applies on small scales. We also tested sensitivity to the spectral index n and the optical depth τ. Overall, our results can be described by the following formula:

$$b(L_*, z = 0) = (0.88 \pm 0.08) \exp[-\tau + 0.5(n - 1)]. \quad (4)$$

4. Comparison with other measurements

4.1. Other estimates of 2dFGRS amplitude of fluctuations

An independent measurement from 2dFGRS comes from redshift-space distortions on scales $\lesssim 10\,h^{-1}\,$Mpc (Peacock et al. 2001). This gives $\beta(L_s, z_s) = 0.43 \pm 0.07$. Using the full likelihood function in the $(b, \Omega_{\rm m})$ plane we derive a somewhat larger (but consistent) value, $\beta(L_s, z_s) \simeq 0.52 \pm 0.18$.

A study of the bi-spectrum of the 2dFGRS (Verde et al. 2001) on smaller scales ($0.1 < k < 0.5\,h\,{\rm Mpc}^{-1}$) sets constraints on deviations from linear biasing, and it gives a best-fit solution consistent with linear biasing of unity. The agreement with the 2dF+CMB result is impressive, given that the methods used are entirely different. In fact, by matching the two results one can get constraints on e.g. the optical depth $\tau \lesssim 0.2$.

4.2. Comparison with other independent measurements

Our derived values for $\sigma^R_{8\rm g}(L_*, z = 0)$ is in accord with with the values derived from the early Sloan Digital Sky Survey (SDSS) by Szalay et al. (2001).

Cluster abundance is a popular method for constraining $\sigma_{8\rm m}$ and $\Omega_{\rm m}$ on scales of $\sim 10\,h^{-1}\,$Mpc. Four recent analyses span a wide range of values, but interestingly they are all orthogonal to our CMB and 2dF constraints (Fig. 3). For example, for $\Omega_{\rm m} = 0.3$ Pierpaoli, Scott & White (2001), Seljak (2001), Reiprich & Boehringer (2001), and Viana, Nichol & Liddle (2001) found: $\sigma_{8\rm m} \simeq 1.02; 0.75; 0.68; 0.61$ respectively (with typical errors of 10 per cent). The discrepancy between the different estimates is in part due to differences in the assumed mass-temperature relation. The cluster physics still needs to be better understood before we can conclude which of the above results is more plausible. We see in Fig. 2 that the lower cluster abundance results are actually in good agreement with our value from the 2dFGRS+CMB, $\sigma_{8\rm m} \simeq 0.73 \pm 0.05$.

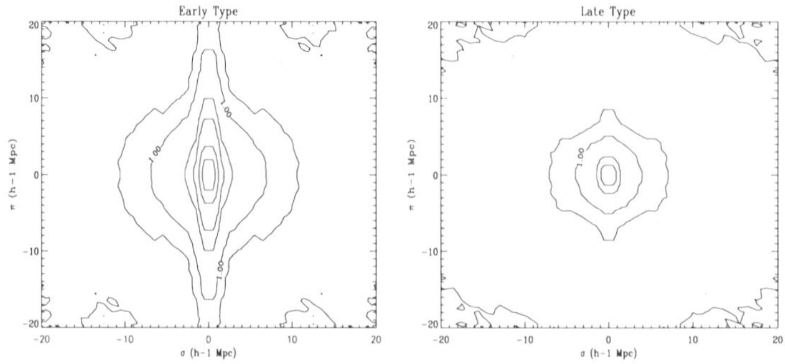

Fig. 3. The two point correlation function $\xi(\sigma, \pi)$ plotted for passively (left) and actively (right) star-forming galaxies. The contour levels are 15,10,5,3,1 (labelled),0.5,0. From Madgwick & Lahav (2001) and Hawkins et al. (in preparation).

The measurements of weak gravitational lensing (cosmic shear) are sensitive to the amplitude of the matter power spectrum on mildly non-linear scales. For example, for $\Omega_{\rm m} = 0.3$ Bacon et al. (2002) and Van Waerbeke et al. (2002) find $\sigma_{8\rm m} \simeq 0.97$ (with errors of about 20 per cent). These estimates are higher than the $\sigma_{8\rm m}$ value that we obtain from 2dFGRS+CMB, but note the large error bars in this recently developed method.

5. Clustering per spectral type

Although biasing was commonly neglected until the early 1980s, it has become evident that on scales $\lesssim 10\,h^{-1}$ Mpc different galaxy populations exhibit different clustering amplitudes, the so-called morphology-density relation (e.g. Dressler 1980; Hermit et al. 1996). Biasing on small scales is also predicted in the simulations of hierarchical clustering from CDM initial conditions (e.g. Benson et al. 2000). It is important therefore to pay attention to the scale on which biasing operates.

Madgwick et al. (2001) have utilized the method of Principal Component Analysis (PCA) to compress each galaxy spectrum into one quantity, $\eta \approx 0.5\,pc_1 + pc_2$. Qualitatively, η is an indicator of the current star formation in a galaxy. This allows us to divide the 2dFGRS into η-types, and to study e.g. luminosity functions and clustering per type.

Norberg et al. (2001b) found that for L_* galaxies, the real space correlation function amplitude of η early-type galaxies is $\sim 50\%$ higher than that of late-type galaxies. Fig 3 (from Madgwick & Lahav 2001) shows the redshift space correlation function in terms of the line-of-sight and perpendicular to the line-of-sight separation $\xi(\sigma, \pi)$. In this plot we show the correlation function calculated from the most passively ('red') and actively ('blue') star-forming galaxies. It can be seen that the clustering properties of the two samples are quite distinct on scales $\lesssim 10\,h^{-1}\,\mathrm{Mpc}$. The 'red' galaxies display a prominent 'finger-of-god' effect and also have a higher overall normalization than the 'blue' galaxies. This is a manifestation of the well-known morphology-density relation. This diagram allows us to determine the combination of the mass density and biasing parameter $\beta = \Omega_m^{0.6}/b$ (e.g. Peacock et al. 2001; Hawkins et al., in preparation).

6. Discussion

We have combined the latest 2dFGRS and CMB data. The first main result of this joint analysis is the normalization of the mass fluctuations, $\sigma_{8\mathrm{m}} = 0.73 \pm 0.05$. This normalization is lower than the COBE normalization and previous estimates from cluster abundance, but it is actually in agreement with recently revised cluster abundance normalization. The results from cosmic shear are still somewhat higher, but with larger error bars.

The second result is for the biasing parameter for optically-selected L_s galaxies, $b(L_s, 0) = 1.00 \pm 0.09$, which is consistent with no biasing ('light traces mass') on scales of tens of Mpc. When translated to L_* via a correction valid for small scales we get a slight anti-bias, $b(L_*, 0) \simeq 0.9$. Our result of linear biasing of unity on scales ($\gtrsim 10\,h^{-1}\,\mathrm{Mpc}$) is actually in agreement with predictions of simulations (e.g. Blanton et al. 2000 Benson et al. 2000; Somerville et al. 2001). It was also demonstrated by Fry (1996) that even if biasing was larger than unity at high redshift, it would converge towards unity at late epochs. Yet on scales $\lesssim 10\,h^{-1}\,\mathrm{Mpc}$ different galaxy populations have different clustering amplitudes.

It may well be that in the future the cosmological parameters will be fixed by CMB, SNe etc. Then, for fixed reasonable cosmological parameters, one can use redshift surveys to study biasing, evolution, etc. The above analysis is a modest illustration of this approach. Future work along these lines will include exploring non-linear biasing models (e.g. Dekel & Lahav 1999; Sigad, Branchini & Dekel 2001; Verde et al. 2001) per spectral type or colour (Madgwick et al 2001; Norberg et al. 2001b; Hawkins et al., in preparation; Zehavi et al. 2001) and the detailed variation of other galaxy properties with local mass density.

Overall, the results from 2dFGRS fit well into the 'concordance' model which has emerged from various cosmological data sets. The Λ-CDM model with comparable amounts of dark matter and dark energy is rather esoteric, but it is remarkable that different measurements converge to the 'concordance model' with parameters:

- $\Omega_k \approx 0$,
- $\Omega_m = 1 - \Omega_\Lambda \approx 0.3$,
- $\Omega_b h^2 \approx 0.02$,
- $h \approx 0.7$
- the age of the Universe $t_0 \approx 14$ Gyr,
- the spectral index $n \approx 1$,
- $\sigma_{8\mathrm{m}} \approx 0.7$.

Perhaps the least accurate estimates on that list are for Ω_{m} and $\sigma_{8\mathrm{m}}$.

While phenomenologically the Λ-CDM model has been successful in fitting a wide range of cosmological data, there are some open questions:

- Both components of the model, Λ and CDM, have not been directly measured. Are they 'real' entities or just 'epicycles' ?
- Why is $\Omega_m \sim \Omega_\Lambda$ at the present-epoch ? Do we need to introduce a new physics or invoke the Anthropic Principle to explain it ?
- There are still open problems in Λ-CDM on the small scales e.g. galaxy profiles and satellites.
- The age of the Universe is uncomfortably close to some estimates for the age of the Globular Clusters, when their epoch of formation is also taken into account (Gnedin, Lahav & Rees 2001).
- Could other (yet unknown) models fit the data equally well ?
- Where does the field go from here ? Would the activity focus on refinement of the cosmological parameters within Λ-CDM, or on introducing entirely new paradigms ?

These issues will no doubt be revisited soon with larger and more accurate data sets. We will soon be able to map the fluctuations with scale and epoch, and to analyze jointly redshift surveys (e.g. 2dF, SDSS) and CMB (e.g. MAP, Planck) data. These high quality data sets will allow us to study a wider range of models and parameters.

ACKNOWLEDGMENTS

I thank Sarah Bridle, Darren Madgwick and the 2dFGRS team members for their contribution to the work described here. The 2dF Galaxy Redshift Survey was made possible through the dedicated efforts of the staff of the Anglo-Australian Observatory, both in creating the 2dF instrument and in supporting it on the telescope. I also thank the conference organizers for the hospitality in Tokyo.

References

[1] Bacon D.J., Massey R.J., Refregier A.R., Ellis, R.S., 2002, astro-ph/0203134

[2] Bahcall N.A., Ostriker J.P., Perlmutter S., Steinhardt P.J., 1999, Science, 284, 148

[3] Benson A.J., Cole S., Frenk C.S., Baugh C.M., Lacey C.G., 2000, MNRAS, 311, 793

[4] Blanton M., Cen R., Ostriker J.P., Strauss M.A., Tegmark M., 2000. ApJ, 531, 1

[5] Bridle S.L., Zehavi I., Dekel A., Lahav O., Hobson M.P., Lasenby A.N., 2001a, MNRAS, 321, 333

[6] Bridle S.L., Crittenden R., Melchiorri A., Hobson M.P., Kneissl R., Lasenby A., 2001b, astro-ph/0112114

[7] Colless M. & the 2dFGRS team, 2001, MNRAS, 328, 1039

[8] de Bernardis P., et al. 2001, astro-ph/0105296

[9] Dekel A., Lahav O., 1999, ApJ, 520, 24

[10] Dressler A., 1980, ApJ, 236, 351

[11] Efstathiou G. & the 2dFGRS team, 2001, astro-ph/0109152

[12] Elgaroy O., Gramann M., Lahav O., 2001, MNRAS, in press, astro-ph/0111208

[13] Freedman W.L., et al., 2001, ApJ, 553, 47

[14] Fry J.N., 1996, ApJ, 461, L65

[15] Gnedin, O., Lahav, O., Rees, M.J., 2001, M.J., astro-ph/0108034

[16] Halverson N.W., et al. 2001, astro-ph/0104489

[17] Hermit S., Santiago B.X., Lahav O., Strauss M.A., Davis M., Dressler A., Huchra J.P., 1996, MNRAS, 283, 709

[18] Lahav O., Bridle S.L., & the 2dF Team, 2001, MNRAS, in press, astro-ph/0112162

[19] Lee A.T. et al., 2001, astro-ph/0104459

[20] Madgwick D.S. & the 2dFGRS team, 2001, MNRAS, in press, astro-

ph/0107197

[21] Madgwick D.S., Lahav O., 2001, astro-ph/0110220

[22] Netterfield C.B. et al., 2001, astro-ph/0104460

[23] Norberg P. & the 2dFGRS team, 2001a, MNRAS, 328, 64

[24] Norberg P. & the 2dFGRS team, 2001b, MNRAS, astro-ph/0112043

[25] O'Meara J.M. et al., 2001, astro-ph/0011179

[26] Peacock J.A. & the 2dFGRS team, 2001, Nature, 410, 169

[27] Percival W.J. & the 2dFGRS team, 2001, MNRAS, 327, 1297

[28] Pierpaoli E., Scott D., White M., 2001, MNRAS, 325, 77

[29] Pryke C. et al., 2001, astro-ph/0104490

[30] Reiprich T.H., Boehringer H., 2001, astro-ph/0111285

[31] Seljak U., 2001, astro-ph/0111362

[32] Sigad Y., Branchini E., Dekel A., 1999, ApJ, 520, 24

[33] Somerville R., Lemson G., Sigad Y., Dekel A., Colberg J., Kauffmann G., White S.D.M., 2001, 320, 289

[34] Stompor R. et al., 2001, astro-ph/0105062

[35] Szalay A.S. & the SDSS team, 2001, astro-ph/0107419

[36] Tegmark M., Hamilton A.J.S., Xu Y., 2001, astro-ph/0111575

[37] Van Waerbeke L., et al. 2002, astro-ph/0202503

[38] Verde L. et al. & the 2dFGRS team, submitted to MNRAS, astro-ph/0112161

[39] Viana P.T.P., Nichol R.C., Liddle A.R., 2001, astro-ph/0111394

[40] Wang X., Tegmark M., Zaldarriaga M., 2001, astro-ph/0105091

[41] Webster M., Bridle S.L., Hobson M.P., Lasenby A.N., Lahav O., Rocha, G., 1998, ApJ Lett, 509, L65

[42] Zehavi I., et al., 2001, astro-ph/0106476

Density Profiles and Clustering of Dark Matter Halos

Yasushi Suto
Department of Physics and Research Center for the Early Universe (RESCEU), School of Science, University of Tokyo, Tokyo 113-0033, Japan

Abstract

We report our work concerning the density profiles and clustering of dark matter halos. First we describe a phenomenological model for the clustering of dark matter halos on the light-cone with particular emphasis on the scale-, mass-, and time-dependence of halo biasing. A comparison of the model predictions against the light-cone output from the Hubble Volume N-body simulation indicates that our model is fairly accurate for scales above $\sim 5h^{-1}$Mpc. Next we derive a constraint on the density profiles of dark matter halos on the basis of the gravitational arc statistics. Our preliminary comparison with the available sample of the lensing arcs suggests that dark matter halos should have steep inner profiles as reported in the recent high-resolution N-body simulations.

1. Introduction

The standard picture of cosmological structure formation suggests that any visible object forms in a gravitational potential of *dark matter halos*. Therefore, a detailed description of dark matter halos is the most fundamental step toward understanding the formation and evolution of visible objects in the universe.

First, we describe a theoretical model of clustering of *dark matter halos* expected along the past light-cone of an observer (Hamana, Yoshida, Suto & Evrard 2001). A special attention is payed to the scale-, time-, and mass-dependence of halo biasing. We also incorporate the light-cone effect (Matareese et al. 1997; Yamamoto & Suto 1998; Moscardini et al. 1998), which is crucial when one analyzes data distributing over a broad redshift range, in addition to several existing theoretical prescriptions (Peacock & Dodds 1996; Kaiser 1987; Sheth & Tormen 1999; Yoshida, Sheth & Diaferio 2001). Then we show that our model predictions are in good agreement with the result from a light-cone output of the Hubble volume simulation (Evrard et al. 2001). Next we discuss a constraint on the slope of the inner density profile of dark matter halos via the gravitational arc statistics (Oguri, Taruya & Suto 2001).

2. Theoretical model for two-point correlation functions of dark matter halo on the light-cone

As emphasized by Suto et al. (1999), for instance, observations of high-redshift objects are carried out only through the past light-cone defined at $z = 0$, and the corresponding theoretical modeling should properly take account of a variety of physical effects which are briefly summarized below.

2.1. Nonlinear gravitational evolution of dark matter density fluctuations

Assuming the cold dark matter (CDM) paradigm, the linear power spectrum of the mass density fluctuations is computed by solving the Boltzmann equation for systems of CDM, baryons, photons and (usually massless for simplicity) neutrinos. The resulting spectrum $P^{\rm R}_{\rm linear}(k, z)$ in real space is specified by a set of cosmological parameters including the density parameter Ω_0, the baryon density parameter Ω_b, and the Hubble constant h in units of 100/km/s/Mpc, and the cosmological constant λ_0. Then one can obtain its nonlinear counterpart in real space, $P^{\rm R}_{\rm nl}(k, z)$ by adopting a fitting formula of Peacock & Dodds (1996).

2.2. Empirical model of halo biasing

The most important ingredient in describing the clustering of halos is their biasing properties. We construct an improved halo bias model of the two-point statistics which reproduces the scale-dependence of the Taruya & Suto (2000) bias correcting the mass-dependent but scale-independent bias of Sheth & Tormen (1999) on linear scales as follows:

$$b_{\rm halo}(M, R, z) = b_{\rm ST}(M, z)\left[1.0 + b_{\rm ST}(M, z)\sigma_M(R, z)\right]^{0.15}, \qquad (1)$$

$$b_{\rm ST}(M, z) = 1 + \frac{\nu - 1}{\delta_c(z)} + \frac{0.6}{\delta_c(z)(1 + 0.9\nu^{0.3})}, \qquad (2)$$

for $R > 2R_{\rm vir}(M, z)$, where $R_{\rm vir}(M, z)$ is the virial radius of the halo of mass M at z. In order to incorporate the halo exclusion effect approximately, we set $b_{\rm halo}(M, R, z) = 0$ for $R < 2R_{\rm vir}(M, z)$. In the above expressions, $\sigma_M(R, z)$ is the mass variance smoothed over the top-hat radius $R \equiv (3M/4\pi\rho_0)^{1/3}$, ρ_0 is the mean density, $\delta_c(z) = 3(12\pi)^{2/3}D(z)/20$, $D(z)$ is the linear growth rate of mass fluctuations (normalized as unity at $z = 0$), and $\nu = [\delta_c(z)/\sigma_M(R, z = 0)]^2$.

2.3. Redshift-space distortion

In linear theory of gravitational evolution of fluctuations, any density fluctuations induce the corresponding peculiar velocity field, which results in the systematic distortion of the pattern of distribution of objects in redshift space

(Kaiser 1987). In addition, virialized nonlinear objects have an isotropic and large velocity dispersion. This *finger-of-God* effect significantly suppresses the observed amplitude of correlation on small scales. With those effects, the nonlinear power spectrum *in redshift space* is given as

$$P^{\rm S}(k, \mu, z) = P^{\rm R}_{\rm nl}(k, z)[1 + \beta_{\rm halo}\mu^2]^2 D_{\rm vel}[k\mu\sigma_{\rm halo}], \tag{3}$$

where $D_{\rm vel}$ is the Fourier transform of the pairwise peculiar velocity distribution function (e.g., Magira et al. 2000), μ is the direction cosine in k-space, $\sigma_{\rm halo}$ is the one-dimensional *pair-wise* velocity dispersion of halos, and

$$\beta_{\rm halo} \equiv -\frac{1}{b_{\rm halo}} \frac{d \ln D(z)}{d \ln z}. \tag{4}$$

While both $\sigma_{\rm halo}$ and $b_{\rm halo}$ depend on the halo mass M, separation R, and z in reality, we neglect their scale-dependence in computing the redshift distortion, and adopt the halo number-weighted averages:

$$\sigma^2_{\rm halo}(> M, z) \equiv \frac{\displaystyle\int_M^\infty 2D^2(z)\sigma_{\rm v}^2(M, z=0)\, n_{\rm J}(M, z) dM}{\displaystyle\int_M^\infty n_{\rm J}(M, z) dM}, \tag{5}$$

$$b_{\rm halo}(> M, z) \equiv \frac{\displaystyle\int_M^\infty b_{\rm ST}(M, z)\, n_{\rm J}(M, z) dM}{\displaystyle\int_M^\infty n_{\rm J}(M, z) dM}, \tag{6}$$

where we adopt the halo mass function $n_{\rm J}(M, z)$ fitted by Jenkins et al. (2001). The value of $\sigma_{\rm v}(M, z = 0)$, the halo center-of-mass velocity dispersion at $z = 0$, is modeled following Yoshida, Sheth & Diaferio (2001). Then our empirical halo bias model can be applied to the two-point correlation function of halos at z in redshift space as

$$\xi_{\rm halo}(M, R, z) = b^2_{\rm halo}(M, R, z) \int_0^\infty P^{\rm S}(k, z) \frac{\sin kR}{kR} \frac{k^2 dk}{2\pi^2}. \tag{7}$$

2.4. Cosmological light-cone effect

All cosmological observations are carried out on a light-cone, the null hypersurface of an observer at $z = 0$, and not on any constant-time hypersurface. Thus clustering amplitude and shape of objects should naturally evolve even *within* the survey volume of a given observational catalogue. Unless restricting the objects at a narrow bin of z at the expense of the statistical significance, the proper understanding of the data requires a theoretical model to take account of the average over the light cone (Matsubara, Suto, & Szapudi 1997; Mataresse et al. 1997; Moscardini et al. 1998; Nakamura, Matsubara, & Suto 1998; Yamamoto & Suto 1999; Suto et al. 1999). According to the present prescription, the two-point correlation function of halos on the light-cone is computed by averaging over the appropriate halo number density and the comoving volume element between the survey range $z_{\rm min} < z < z_{\rm max}$:

$$\xi_{\rm halo}^{\rm LC}(>M,R) = \frac{\displaystyle\int_M^\infty dM \int_{z_{\rm min}}^{z_{\rm max}} dz \frac{dV_{\rm c}}{dz}\, n_{\rm J}^2(M,z)\xi_{\rm halo}(M,R,z)}{\displaystyle\int_M^\infty dM \int_{z_{\rm min}}^{z_{\rm max}} dz \frac{dV_{\rm c}}{dz}\, n_{\rm J}^2(M,z)}, \tag{8}$$

where $dV_{\rm c}/dz$ is the comoving volume element per unit solid angle. While the above expression assumes a mass-limited sample for simplicity, any observational selection function can be included in the present model fairly straightforwardly (Hamana, Colombi & Suto 2001) once the the relation between the luminosity of the visible objects and the mass of the hosting dark matter halos is specified.

3. Comparison with the light-cone output from the Hubble volume simulation

Figure 1 plots the projected distribution of identified halos from "light-cone output" of the Hubble Volume ΛCDM simulation (Evrard et al. 2001) with $\Omega_{\rm b} = 0.04$, $\Omega_{\rm CDM} = 0.26$, $\sigma_8 = 0.9$, $\Omega_\Lambda = 0.7$ and $h = 0.7$. The total number of halos at $0 < z < 3$ is 21,090, 5,554 and 1,543, on the light-cone for $M > 2.2 \times 10^{13} h^{-1} M_\odot$, $5.0 \times 10^{13} h^{-1} M_\odot$, and $1.0 \times 10^{14} h^{-1} M_\odot$, respectively. The stronger clustering for the more massive halos is visible in this plot.

Figure 2 compares our model predictions with the clustering of those simulated halos. For the dark matter correlation functions, our model reproduces the simulation data almost perfectly at $R > 3h^{-1}$Mpc (see also Hamana, Colombi & Suto 2001). This scale corresponds to the mean particle separation of this particular simulation, and thus the current simulation might systematically underestimate the real clustering below this scale especially at $z > 0.5$ (see Hamana, Yoshida & Suto 2002). Our model and simulation data also show quite good agreement for dark halos at scales larger than $5h^{-1}$Mpc. Below that scale, they start to deviate slightly in a complicated fashion depending on the mass of halo and the redshift range. This discrepancy may be ascribed to both the numerical limitations of the current simulations and our rather simplified model for the halo biasing. Nevertheless the clustering of *clusters* on scales below $5h^{-1}$Mpc is difficult to determine observationally anyway, and our model predictions differ from the simulation data only by ~ 20 percent at most. Therefore we conclude that in practice our empirical model provides a successful description of halo clustering

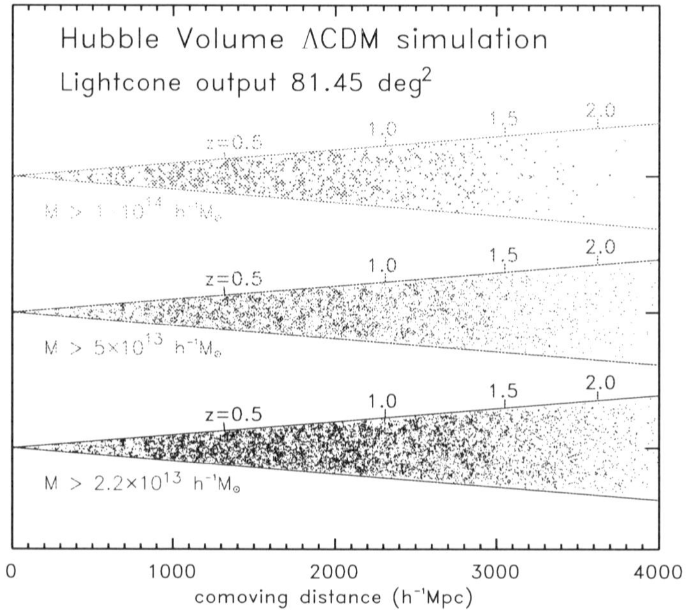

Fig. 1. Projected distribution of dark matter halos with $M_{\rm halo} > M_{\rm min}$ in the light-cone output of the Hubble volume simulation; $M_{\rm min} = 1 \times 10^{14} h^{-1} M_\odot$, $5 \times 10^{13} h^{-1} M_\odot$ and $2.2 \times 10^{13} h^{-1} M_\odot$ from top to bottom.

on the light-cone.

4. Constraint on the central density profiles of dark matter halos using gravitational lensing arcs

Turn next to the density profiles of dark matter halos. Several recent high-resolution simulations indicate that dark halos have the central cusp approximated by $\rho(r) \propto r^{-1\sim-1.5}$ (Navarro, Frenk & White 1996; Fukushige & Makino 1997; Moore et al. 1999; Jing & Suto 2000). On the other hand, both th rotation curves of the low surface brightness galaxies (Salucci et al. 2000; de Blok et al. 2001) and the inner region of the cluster CL0024-1654 reconstructed from gravitational lensing images (Tyson et al. 1998) indicate the presence of the flat core instead of the central cusp.

Since the current situation concerning the core structure of dark halo is somewhat puzzling, it is important to develop another independent methodology to probe the profiles. For this purpose, we consider the arc statistics of the

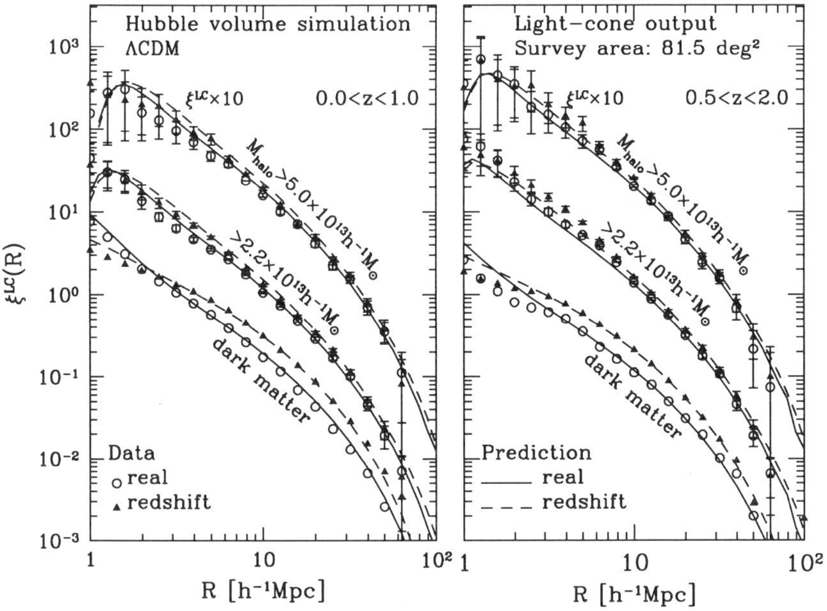

Fig. 2. Two-point correlation functions of halos on the light-cone; simulation results (symbols; open circles and filled triangles for real and redshift spaces, respectively) and our predictions (solid and dotted lines for real and redshift spaces, respectively). The error bars denote the standard deviation computed from 200 random re-samplings of the bootstrap method. The amplitudes of ξ^{LC} for $M_{\rm halo} \geq 5.0 \times 10^{13} h^{-1} M_{\odot}$ are increased by an order of magnitude for clarity.

gravitational lensing (Oguri, Taruya & Suto 2001). We adopt a parameterized profile for dark matter halos:

$$\rho(r) = \frac{\rho_{\rm crit}\delta_{\rm c}}{(r/r_{\rm s})^{\alpha}(1+r/r_{\rm s})^{3-\alpha}}, \tag{9}$$

where $r_{\rm s}$ is a scale radius and $\delta_{\rm c}$ is a characteristic density (Navarro, Frenk & White 1996; Jing & Suto 2000). The shape of halos is also characterized by the concentration parameter, which is defined as the ratio of the virial radius, $r_{\rm vir}$, of the halo to the scaled radius $r_{\rm s}$. In practice, we properly take account of the probability distribution of the concentration parameter reported from numerical simulations as well as their mass and redshift dependence (Bullock et al. 2001). Thus the halo shape is simply characterized by the two parameters; the inner slope α and the normalization amplitude of the median concentration parameter $c_{\rm norm}$ (Keeton & Madau 2001).

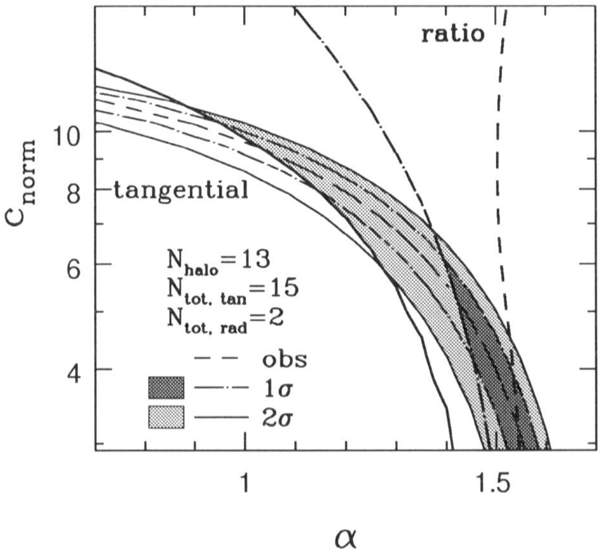

Fig. 3. Tentative constraints on α and $c_{\rm norm}$ from a sample of 13 clusters with $0.1 < z < 0.4$ and $S_{\rm lim} = 10^{-12}$erg/s/cm^{-2}. The low-density spatially flat cold dark matter model is assumed. Dashed lines represent the relation of α and $c_{\rm norm}$ which reproduce the observed number for tangential arcs and the number ratio of radial to tangential arcs. Dark shaded and light shaded regions indicate the allowed regions combined from the two constraints taking account of the 1σ and 2σ statistical errors, respectively.

Then we compute the numbers of tangential and radial arcs as a function of the inner slope α and the concentration parameter $c_{\rm norm}$. We incorporate several realistic effects including the magnification bias, the finite size of the source galaxies, and the luminosity distribution and evolution of source galaxies. We find that the numbers of arcs sensitively depend on the values of α and $c_{\rm norm}$. In addition, the numbers of arcs, if divided by the corresponding number of halos, are almost insensitive to the underlying cosmological parameters. Therefore they prove to be a powerful discriminator of a family of halo density profiles suggested by recent numerical simulations (see also Molikawa & Hattori 2001).

Figure 3 shows our tentative comparison with the currently available data from the imaging survey for gravitational lensing of X-ray selected clusters of galaxies (Luppino et al. 1999). From these clusters we choose a subsample of 13 clusters which satisfy the condition $0.1 < z < 0.4$ and $S(0.5-2.0{\rm keV}) > S_{\rm lim} = 10^{-12}$erg/s/cm^2. Within these 13 clusters, 15 tangential and 2 radial arcs with $\epsilon_{\rm th} = 4$ are reported. We assume the ΛCDM cosmology and neglect possible systematic errors (e.g., intrinsic ellipticities of source galaxies, non-sphericities in lensing halos, etc.; these are taken into account in Oguri 2002). This result implies that dark matter halos should have steep inner profiles as those predicted by simulations, $\alpha \sim 1.5$, but rather smaller concentration, $c_{\rm norm} \sim 4$. While it is premature to draw definite conclusions from the present comparison, this analysis clearly exhibits the extent to which our current methodology puts useful constraints on α and $c_{\rm norm}$ separately.

Acknowledgements

The present work is based on my collaboration with S.Colombi, A.E.Evrard, T.Hamana, M.Oguri, A.Taruya N.Yoshida. This research was supported in part by the Grant-in-Aid by the Ministry of Education, Science, Sports and Culture of Japan (07CE2002, 12640231).

References

[1] Bullock, J. S., Kolatt, T. S., Sigad, Y., Somerville, R. S., Kravtsov, A. V., Klypin, A. A., Primack, J. R., & Dekel, A. 2001, MNRAS, 321, 559
[2] de Blok, W. J. G., McGaugh, S. S., Bosma, A., & Rubin, V. C. 2001, ApJ, 552, L23
[3] Evrard, A.E. et al. 2002, ApJ, in press (astro-ph/0110246).
[4] Fukushige, T., & Makino, J. 1997, ApJ, 477, L9
[5] Hamana, T., Colombi, S., & Suto, Y. 2001, A& A, 367, 18
[6] Hamana, T., Yoshida, N., & Suto, Y. 2002, ApJ, in press (astro-ph/0111158).
[7] Hamana, T., Yoshida, N., Suto, Y., & Evrard, A.E. 2001, ApJ, 561, L143
[8] Jenkins, A. et al. 2001, MNRAS, 321, 372
[9] Jing, Y.P. 1998, ApJ, 503, L9
[10] Jing, Y. P., & Suto, Y. 2000, ApJ, 529, L69
[11] Kaiser, N. 1987, MNRAS, 227, 1
[12] Keeton, C. R., & Madau, P. 2001, ApJ, 549, L25
[13] Luppino, G. A., Gioia, I. M., Hammer, F., Le Févre, O., & Annis, J. A. 1999, A&AS, 136, 117
[14] Magira, H., Jing, Y.P., & Suto, Y. 2000, ApJ, 528, 30
[15] Matarrese, S., Coles, P., Lucchin, F., & Moscardini, L. 1997, MNRAS, 286, 115
[16] Matsubara, T., Suto, Y., & Szapudi, I. 1997, ApJ, 491, L1
[17] Mo, H.J., & White, S.D.M 1996,MNRAS, 282, 347
[18] Molikawa, K., & Hattori, M. 2001, ApJ, 559, 544

[19] Moore, B., Quinn, T., Governato, F., Stadel, J., & Lake, G. 1999, MNRAS, 310, 1147

[20] Moscardini, L., Coles, P., Lucchin, & F., Matarrese, S. 1998, MNRAS, 299, 95

[21] Nakamura, T. T., Matsubara, T., & Suto, Y. 1998, ApJ, 494, 13

[22] Navarro, J.F., Frenk, C.S., & White, S. D. M. 1996, ApJ, 462, 563

[23] Oguri, M., Taruya, A., & Suto, Y. 2001, ApJ, 559, 572

[24] Oguri, M., 2002, Master's thesis, submitted to University of Tokyo, and ApJ, in press.

[25] Peacock, J.A., & Dodds, S.J. 1996, MNRAS, 280, L19

[26] Salucci, P., & Burkert, A. 2000, ApJ, 537, L9

[27] Sheth, R.K., & Tormen, G. 1999, MNRAS, 308, 119

[28] Suto, Y., Magira, H., Jing, Y. P., Matsubara, T., & Yamamoto, K. 1999, Prog.Theor.Phys.Suppl., 133, 183

[29] Taruya, A. & Suto,Y. 2000, ApJ, 542, 559

[30] Tyson, J. A., Kochanski, G. P., & Dell'Antonio I. P. 1998, ApJ, 498, L107

[31] Yamamoto, K., & Suto, Y. 1999, ApJ, 517, 1

[32] Yoshida, N., Sheth, R., & Diaferio, A. 2001, MNRAS, 325, 803

High-Resolution Numerical Simulations for Galaxy Formation

Y. P. Jing
Shanghai Astronomical Observatory, the Partner Group of MPI für Astrophysik, Nandan Road 80, Shanghai 200030, China
Yasushi Suto
Department of Physics and Research Center for the Early Universe (RESCEU) School of Science, University of Tokyo, Tokyo 113-0033, Japan

Abstract

Understanding of the structure evolution in the Universe has been greatly advanced with a rapid progress in high-resolution cosmological simulations. In this contribution, we report a new set of cosmological simulations with 512^3 ($\sim 1.4 \times 10^8$) particles in a simulation box of $100\,h^{-1}$Mpc. With this unprecedented resolution, we have successfully overcomed the over-merging problem while retaining a cosmological volume size that is crucial for the statistical analysis. The simulations are being applied to studying important astrophysical problems that are sensitive to the simulation resolution.

1. Introduction

Galaxy formation is one of the most important and most challenging problems in astrophysics. In the standard theories of galaxy formation, it is generally believed that the Universe is dominated by dark matter (DM), the early density perturbations generated in the inflationary era are amplified by gravitational instability, the DM halos form via the collapse of the high density regions, and galaxies are formed from the cooled gas within those DM halos.

The formation of DM halos, more generally the DM distribution on small scales, however, is a strongly non-linear problem. Direct numerical simulations provide the only reliable tool to approach this problem. Although there have been many exciting developments in analytical, semi-analytical or empirical modeling of the DM distribution, these results are generally calibrated with N-body simulations and their validity still depends heavily on the resolution of the simulations used.

In this contribution, we report a new set of cosmological N-body simulations that were obtained recently under a collaboration of Shanghai Astronomical

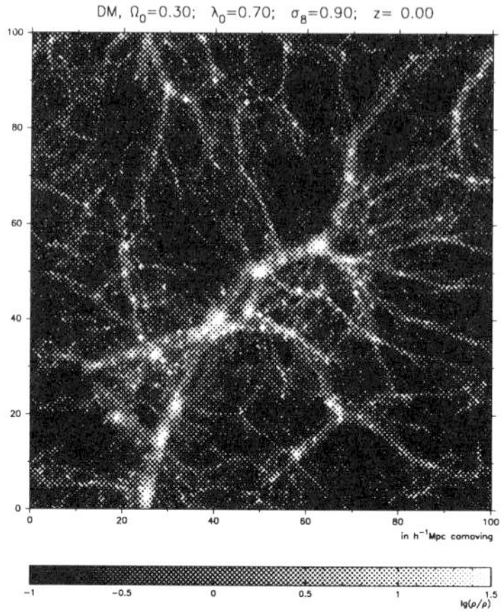

Fig. 1. Dark matter distribution around the most massive halo in the LCDM100a simulation. The slice is $100\,h^{-1}$Mpc wide and $10\,h^{-1}$Mpc thick.

Observatory and the University of Tokyo. We use 512^3 particles to simulate a Cold Dark Matter (CDM) model in a cosmological volume of $100(h^{-1}\mathrm{Mpc})^3$. The simulations have achieved a mass resolution which is an order of magnitude better than the advanced simulations of the Virgo Consortium (Jenkins, A. et al. 1998) as well as our previous simulations (Jing & Suto 1998). They are now being applied to studying a number of astrophysical problems that are sensitive to the simulation resolutions. As their first application, we (Jing & Suto 2002) have studied the internal matter distributions within virialized halos and have successfully found an ellipsoid description for the halo density profile that is more accurate than the conventional spherical description.

2. N-body simulations

The simulations are generated with our Particle-Particle-Particle-Mesh (P^3M) code on the supercomputer VPP5000 at the National Astronomical Observatory of Japan. The code adopts the standard P^3M algorithm (Hockney & Eastwood 1981; Efstathiou et al. 1985), and is fully vectorized (Jing & Suto 1998), and has been recently parallelized. Each simulation uses 512^3 simulation particles. Five simulations are generated for two representative cold dark matter (CDM) models of galaxy formation. The linear DM distribution in CDM models is completely specified by the density parameter Ω_0, the cosmological constant λ_0, the shape Γ and the normalization σ_8 of the linear power spectrum. Table 1

Fig. 2. Dark matter distribution within the most massive halo in the LCDM100a simulation. The physical size of the figure is 2 times of the halo virial radius

summarizes the physical and simulation parameters used for these simulations. The physical parameters are chosen so that the models are consistent with most observations of the real Universe. A box size of $100\,h^{-1}$Mpc implies that the particle mass in this system is only $m_p \approx 10^9 h^{-1} M_\odot$. The force resolution is $\eta = 3 \sim 6\,h^{-1}$kpc (the Plummer form), so typical galactic halos can be well resolved. Each model (with one box size) will have a few different realizations in order to properly account for the cosmic variance (still in progress). One LCDM model uses a small force softening $\eta = 3\,h^{-1}$kpc and is evolved with 5000 time steps, so how the force softening affects non-linear structures of DM distribution can be checked. Therefore, these simulations will become an important source for studying galaxy formation, internal structures of DM halos, and nonlinear clustering of dark matter.

Table 1. List of simulations

Model	N	Ω_0	λ_0	σ_8	Γ	m_p ($h^{-1}M_\odot$)	L (h^{-1}Mpc)	steps	samples
LCDM100a	512^3	0.3	0.7	0.9	0.2	6.2×10^8	100	5000	1
LCDM100	512^3	0.3	0.7	0.9	0.2	6.2×10^8	100	1200	2
SCDM100	512^3	1.0	0.0	0.55	0.5	2.1×10^9	100	1200	2

3. Preliminary results and future work

Figure 1 shows a dark matter distribution of the LCDM100a simulation centered at the most massive halo. The plot corresponds to a slice of $100\,h^{-1}$Mpc wide and $10\,h^{-1}$Mpc thick. Filamentary structures are very visible, as many previous studies of such simulations have discovered. More interesting features of the simulations are seen in Figure 2 which shows the DM distribution of the most massive halo. Substructures within the virialized region are preserved well, similar to high-resolution simulations for individual halos (Moore et al. 1999), therefore the over-merging problem encountered in previous studies of cosmological N-body simulations has now been successfully resolved in our simulations. This will be very important to studying the dynamical properties and spatial distribution of galaxies when using the simulations (e.g. based on the semi-analytical modeling).

With this large set of high-resolution simulations, many interesting cosmological problems will be investigated. As an "incomplete" list, we are 1) studying the DM distribution in the strongly clustering regimes; 2) analyzing internal structures of DM halos; 3) investigating galaxy formation by hand-inputing gas physics; 4) applying the simulations to real observations.

Acknowledgements

We thank Kohji Yoshikawa for invaluable advice on parallelizing the code. The work is supported in part by the One-Hundred Talent Program, by NKBRSF (G19990754) and by NSFC (No.10043004). The Numerical simulations were carried out at ADAC (the Astronomical Data Analysis Center) of the National Astronomical Observatory, Japan.

References

[1] Efstathiou, G., Davis, M., Frenk, C.S. & White, S.D.M. 1985, ApJS, 57, 241
[2] Hockney, R.W. & Eastwood, J.W. 1981, Computer simulations using particles. Mc Graw-Hill
[3] Jenkins, A. et al. 1998, ApJ, 499, 20
[4] Jing, Y.P., Suto, Y., 1998, ApJ, 494, L5
[5] Jing, Y.P., Suto, Y., 2002, ApJ, to be submitted
[6] Moore, B., Ghigna, S., Governato, F., Lake, G., Quinn, T., Stadel, J., & Tozzi, P., 1999, ApJ, 524, L19

Structure of Dark Matter Halos From Hierarchical Clustering

Toshiyuki FUKUSHIGE
Department of General Systems Studies, College of Arts and Sciences, University of Tokyo, 3-8-1 Komaba, Meguro-ku, Tokyo 153-8902, Japan

Abstract

We investigated the structure of the dark matter halo formed from hierarchical clustering by high-resolution N-body simulations. In all simulation runs, the density cusps proportional to $r^{-1.5}$ developed at the center. This result was independent of the halo mass and cosmological models. We could not reproduce the cusp shallower than $r^{-1.5}$, which was obtained in some of previous studies.

1. Introduction

The structure of dark matter halos formed through disspationless hierarchical clustering from cosmological initial setting has been explored by many researchers since the "finding" of the universal profile by Navarro, Frenk and White [9][10](hereafter NFW). NFW performed N-body simulations of the halo formation and found that the density profiles of dark matter halos were expressed well by a simple formula, $\rho = \rho_0/[(r/r_0)(1 + r/r_0)^2]$, where ρ_0 is a characteristic density and r_0 is a scale radius. They argued that the profile of dark matter halos have the same shape, independent of the halo mass, the initial density fluctuation spectrum or the value of the cosmological parameters.

Although the NFW results have been confirmed regarding the logarithmic slope shallower than isothermal near the center and -3 in the outskirt, disagreements concerning the slope of the central region were reported by subsequent studies in which higher-resolution simulations were performed [1][7][8][4][5](see also a review by Yasushi Suto in this volume).

We performed two series of N-body simulations of formation of dark matter halos using Barnes-Hut treecode and a special-purpose computer GRAPE-5 [6]. In the first series [2](hereafter Paper I), we simulated 12 halos with the mass of $6.6 \times 10^{11} M_\odot$ to $8.0 \times 10^{14} M_\odot$ in SCDM model . In the second series [3] (hereafter Paper II), we simulated cluster-sized 13 halos with the mass in three CDM model (SCDM, OCDM, and LCDM). For more details on the simulation models, see [2][3].

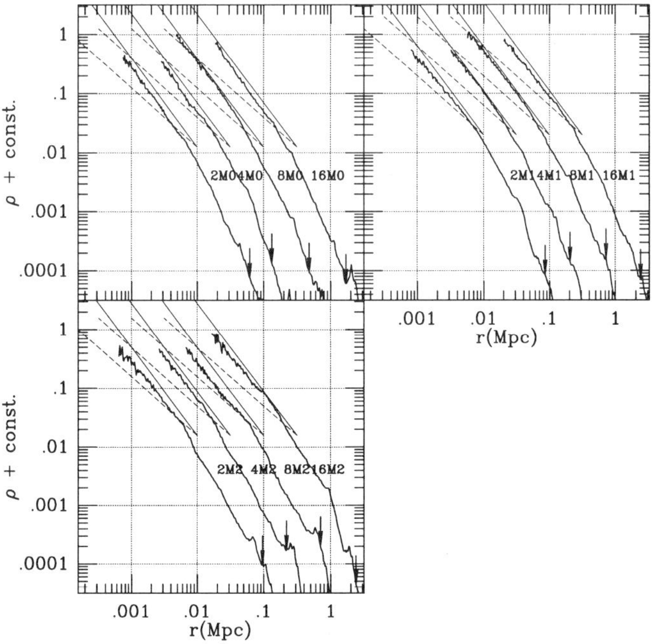

Fig. 1. Density profile of the halo for all runs in the series of simulations among which the mass scale differs (Paper I). The unit of density is $M_\odot/\mathrm{pc}^3$. The label beside the profile indicates the run name. The numbers before M in the run names mean mass scale, and the numbers after M mean the random number used for the initial distribution. The profiles are vertically shifted for illustrative purpose. The arrows indicate r_{200}. The thin solid and dashed lines indicate the densities proportional to $r^{-1.5}$ and r^{-1}.

2. Simulation Results

In this section, we present density profiles of the dark matter halos obtained in the simulations. In this study, we have to take care about numerical artifacts, such as two-body relaxation (gravothermal expansion) effects, which affects the profile in the central region. We here plot the density only for the radii unaffected by numerical artifacts. We used the two criteria introduced in Paper I; (1) $t_{\rm rel}(r)/t > 3$ and (2) $t_{\rm dy}(r)/\Delta t > 40$, where $t_{\rm rel}(r)$ is the local two-body relaxation time and $t_{\rm dy}(r)$ is the local dynamical time. We plot the densities only if both criteria are satisfied.

Figure 1. shows the evolution of density profiles at the end of simulation

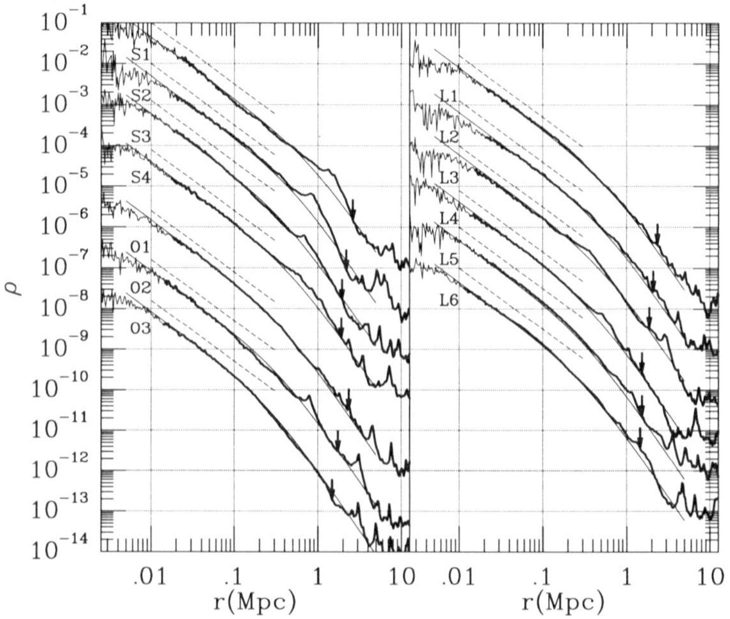

Fig. 2. Density profiles of the halos for all Runs at $z \simeq 0$ in the series of simulations among which the cosmological model differs (Paper II). Only the densities plotted in the thick lines satisfy the accuracy criteria. Those plotted in the thin lines are influenced by numerical artifacts. The unit of density is $M_{\odot}/\mathrm{pc}^3$. The labels on the left of the profiles indicate the run name. The characters, S, O, and L in the run name mean runs in SCDM, OCDM, and LCDM models, respectively. The arrows indicate r_{V}. The thin dashed lines indicate the densities proportional to $r^{-1.5}$. The thin solid curves indicate the density profile given by Moore et al. [8].

for all runs in the series of simulations among which the mass scale differs (Paper I). The position of the center of the halo was determined using the potential minimum and the density is averaged over each spherical shell.

Figure 2. shows the density profiles at $z = 0$ for all runs in the series of simulations among which the cosmological model differs (Paper II). In this figure, we plot the densities in the thick lines if the two accuracy criteria are satisfied. This means that the densities in the central region plotted in the thin lines are influenced by the numerical artifacts (mainly two-body relaxation), and they are not reliable.

In all runs we can see the central density cusps approximately proportional to $r^{-1.5}$. In other words, the power of the cusp is -1.5 and is independent of the halo mass and cosmological models we simulated.

3. Discussion

We found that, in all runs, the final halos have density cusps proportional to $r^{-1.5}$, regardless of the halo mass and cosmological models.

There are some claims that the innermost slope should converge not to $r^{-1.5}$ but to a shallower one (e.g. [11]). Indeed, if we do not pay attention to numerical artifacts, we may see in our simulation results (Figure 2.) that the innermost slopes of all runs become shallower than -1.5. However, Figure 2. shows that the inner region where the slope are shallower than $r^{-1.5}$ is not reliable because in this region the accuracy criteria are not satisfied. The numerical artifact which makes the cusp shallower is mainly the two-body relaxation effect in this region. Therefore, we emphasize that discussions based on simulation results without careful analysis of the influence by the two-body relaxation effects are misleading.

I am grateful to Junichiro Makino for the collaboration in the studies on which my talk is based, and Atsushi Kawai for their helps in preparing the hardware and software environment of the GRAPE-5 system.

References

[1] T. Fukushige, and J. Makino, ApJ 477 (1997) L9.
[2] T. Fukushige, and J. Makino, ApJ 557 (2001) 533. (Paper I)
[3] T. Fukushige, and J. Makino, ApJ submitted. (astro-ph/0108014) (Paper II)
[4] S. Ghigna, B. Moore, F. Governato, G. Lake, T. Quinn, and J. Statal, ApJ 544 (2000) 616.
[5] Y. P. Jing, and Y. Suto, ApJ 529 (2000) L69.
[6] A. Kawai, T. Fukushige, J. Makino, and M. Taiji, PASJ 52 (2000) 659.
[7] B. Moore, F. Governato, T. Quinn, J. Statal and G. Lake, ApJ 499 (1998) L5.
[8] B. Moore, T. Quinn, F. Governato, J. Statal and G. Lake, MNRAS 310 (1999) 1147.
[9] J. F. Navarro, C. S. Frenk and S. D. M. White, ApJ 462 (1996) 563.
[10] J. F. Navarro, C. S. Frenk and S. D. M. White, ApJ 490 (1997) 493.
[11] J. E. Taylor, and J. F. Navarro, ApJ submitted. (astro-ph/0104002)

Galaxies and Intergalactic Matter at Redshift $Z \sim 3$

Kurt ADELBERGER
Harvard-Smithsonian Center for Astrophysics, 60 Garden St MS 51, Cambridge, MA 02138, USA

Abstract

1. Introduction

To those concerned mainly with fundamental physics, no part of history is more interesting than the first three minutes. A theory able to explain the composition of the universe after three minutes—the relative abundances of photons, hydrogen, helium, and other particles, the spatial variations in density, the tiny value of the cosmological constant, and so on—would be a tremendous triumph for physics, and attempts to understand the early evolution of the universe have rightly been a major concern of this conference so far. But the first three minutes are only a small part of history, and only a small part of observational cosmology. Though the stage, actors, and script may have been chosen, the drama of creation had hardly begun. Stars did not yet exist, nor galaxies, nor planets, nor even many of the elements that compose them. The slow action of (possibly) well-known physical laws over billions of years was responsible for transforming the universe from a smooth sea of particles into its present complex state, but our understanding of how this happened is still poor. Observing the universe during the 13 billion years that follow the first three minutes may not teach us anything new about fundamental physics, but it can help us understand why and how the laws of physics caused matter to arrange itself into the familiar collection of stars and galaxies that we observe around us today. This, in broad terms, is the subject of my talk.

The most dramatic change in the universe over the last 13 billion years is the formation of structures on large spatial scales. Two hundred thousand years after the big bang, the universe's deviations from homogeneity were tiny, roughly one part in 10^5. Today on spatial scales $\sim 10^{25}$cm the relative variations in matter density are of order unity: $\sigma_M/\bar{M} \sim 1$. Even the coarsest observations reveal that the universe today bears little resemblance to the smooth plasma produced by the big bang. One of the major achievements in observational cosmology during the past ~ 15 years has been the demonstration that this amplification of density contrasts is a natural consequence of gravitational attraction. Parts of the universe that were slightly denser than average in the beginning would have pulled more matter towards themselves, and numerical simulations and analytic approximations have convincingly shown that this unstable process, acting over the age of the universe, could have rearranged the initially smooth distribution of matter into the clumpy distribution we observe around us now.

But this is only part of the story. Gravity on its own produces an uninteresting universe. Neither stars nor galaxies form. The structures of one epoch are erased and subsumed into the structures of the next; the rich variety of objects familiar to astronomers—the remnants, mainly, of earlier stages of cosmic evolution—these are nowhere to be found in purely gravitational simulations. To complete our understanding we need to take account of how baryons behave as the universe evolves, of what would happen to a small amount of normal matter placed on top of the underlying distribution of dark matter that is (presumably) affected solely by gravity.

Here progress has been much slower. The many complicated interactions that affect baryons and electrons are not easy to model numerically or analytically. Our understanding of the large scale distribution of baryons is reasonably good only at early times, less than $\sim 200,000$ years after the big bang, when the universe was still nearly homogeneous and baryons were tightly coupled to photons through radiation pressure. In this case the behavior of normal matter can be accurately modeled with the simple physics of linear fluid dynamics, and straightforward calculations show that the competition between photon pressure and gravitational attraction ought to fill the universe with small acoustic waves. The recent observation of these waves [2] provides a spectacular confirmation of the theoretical predictions. But this simple situation did not last long. The universe expanded and cooled; electrons vanished into atomic orbits; the photons of the microwave background streamed free; and the baryons left behind began to evolve according to more intricate rules. They fell towards overdensities in the matter distribution, shocked each other in collisions, cooled, ignited in nuclear reactions, tumbled into black holes, emitted radiation that reionized the universe, and so on. Though we cannot hope to understand every detail of this chaos, we may be able to answer general questions. Thirteen billion years ago the baryons of the universe formed small acoustic waves; today a small fraction has been crushed into stars and most of the remainder have been blasted to temperatures $T \sim 10^5$–10^7K and left to drift for a sterile eternity in intergalactic space. What physical processes were most responsible for transforming the large scale distribution of baryons in this way?

I will have time to focus on only one aspect of this question: how large a role did supernova explosions in young galaxies play? A number of arguments suggest that the energy released in supernova explosions must have had a profound impact on the general evolution of baryons in the universe. Elliptical galaxies contain little interstellar material but tend to be enveloped by gas that is rich in metals, as if the supernova explosions that accompanied their formation expelled much of their matter far into their surroundings. Rapidly star-forming galaxies in the local universe are observed to send large amounts of gas careering towards intergalactic space [5]. Metals have been found far from known galaxies at high redshift [3][4]. Numerical hydrodynamic simulations that do not include strong "feedback" from supernovae are unable to produce disk galaxies as large as those we observe around us, but the problem might be solved if heat input from supernovae reduced the clumpiness of the baryon distribution and (consequently) the transfer of angular momentum from baryons to dark matter through dynamical friction [6]. Extreme releases of heat from supernovae may be required to explain the small fraction of baryons that cooled at early times despite the universe's high density [11]. The faintness of the soft X-ray background [7] and the broken self-similarity of the cluster X-ray luminosity/temperature relationship [8] appear to indicate that something imparted ~ 1 keV of energy to each of the young universe's nucleons.

But these arguments are indirect, uncertain, and appear to be contradicted by some comparisons of observations with numerical simulations (see, e.g., [9] and references therein). Ideally one would like to directly observe supernova-driven winds emerging from young galaxies and affecting their surroundings. This is what we set out to do. Because observations suggest that most stars formed in the distant past [10], under conditions significantly different than those that currently obtain, attempts to constrain the influence of supernova explosions require one to observe the universe at high redshift. Redshift $z \sim 3$ seemed like a particularly good choice. At this redshift the comoving density of star-formation in the universe was near its peak, the number density of galaxies with extremely high star-formation rates was orders of magnitudes higher than in the local universe, and galaxies had only comparatively shallow gravitational potentials to restrain their putative supernova-driven winds.

2. Observations

Our goal was to observe the universe at redshift $z \sim 3$ and look for evidence that supernova in young galaxies had affected the intergalactic material in their vicinity. This required us to find galaxies at $z \sim 3$ and measure the characteristics of the intergalactic gas near them. Galaxies at this redshift can be found through their distinctive $U_n G\mathcal{R}$ colors, which arise due to absorption

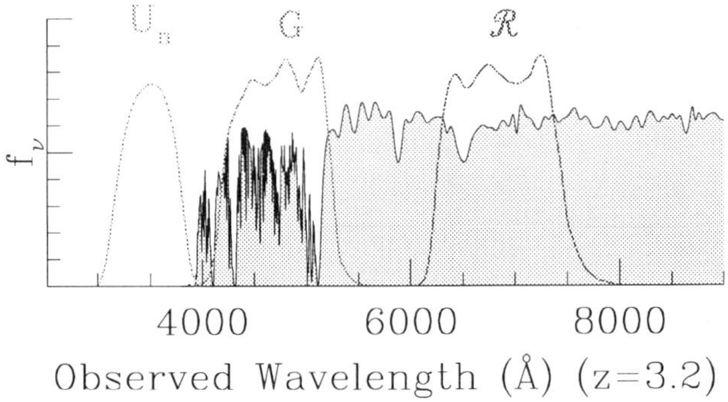

Fig. 1. Finding galaxies at $z \sim 3$. The spectra of galaxies show a strong break at (rest-frame) 912Å due to absorption of ionizing photons by hydrogen in the atmospheres of stars, in the gas that lies between stars in the galaxy, and in the intergalactic gas that lies between the galaxy and earth. For galaxies at redshifts $z > 2.8$ this break can be observed with ground-based telescopes. Its presence gives high-redshift galaxies distinctive colors that allow them to be easily identified with broadband imaging through (for example) the U_n, G, $\mathcal{R}$ filters.

from the abundant neutral hydrogen that lies between the earth and the nuclear engines buried within a distant galaxy's stars. See figure 1. The neutral hydrogen and metal content of gas in the galaxies' vicinity can be constrained by analyzing absorption lines in the spectra of background QSOs. See figure 2. These data let us measure correlations between the gas content of the intergalactic medium and the presence (or absence) of nearby galaxies. They may indicate that galaxies affect the intergalactic medium to surprisingly large radii, $r \sim 0.5h^{-1}$ comoving Mpc (4×10^{23} proper cm), as we now describe.

3. Interpretation

Our data set is still small, but we may already see some indications that supernova explosions within the observed galaxies are driving winds deep into the galaxies' surroundings. These come from four sources.

First, the spectra of individual galaxies show that the material they contain is moving at a range of velocities $\Delta v \sim 600\mathrm{kms}^{-1}$ significantly greater than expected for gas in equilibrium within the shallow potential wells of the young universe. The cool gas at larger distances from the stars produces absorption lines that tend to be blueshifted by large amounts compared to the emission lines from hot gas closer to the stars (figure 3.). One interpretation is that a signifi-

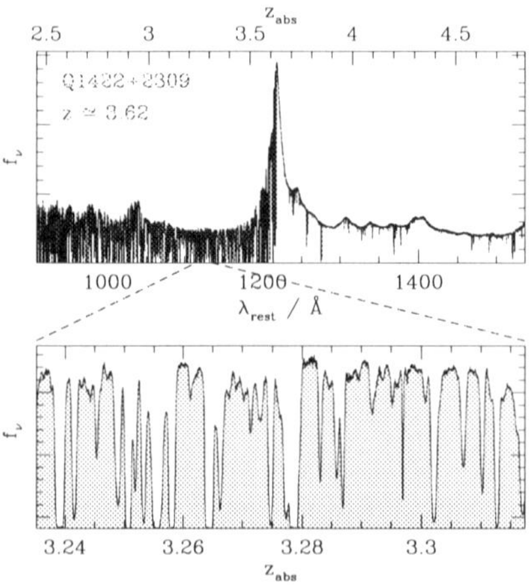

Fig. 2. Example spectrum of a high-redshift QSO. Spectra of these bright objects provide a view of the intergalactic material that lies between them and the earth. The strong emission line in the middle of the top panel is Lyman-α 1216Å at redshift $z_{\rm em} \sim 3.62$. The wavelengths of the QSO's photons gradually increased as they traveled towards earth through the expanding universe, and at a redshift $1 + z = \lambda_{\rm em}(1 + z_{\rm em})/1216$Å photons with initial wavelength $\lambda_{\rm em} < 1216$Å were stretched to $\lambda = 1216$Å and were (in many cases) absorbed by intervening neutral hydrogen at that redshift. This accounts for the complex absorption pattern at wavelengths shorter than Lyman-α (see magnified view in lower panel), and provides a detailed picture of the spatial distribution of neutral hydrogen along a single sightline through the young universe. Weaker lines due to absorption by intervening "metals" (e.g., C IV) are also present in the spectrum, though they are more difficult to see in this reproduction.

cant fraction of each galaxy's baryonic mass is being driven outwards at several hundred km $\rm s^{-1}$.

Second, comparatively little neutral hydrogen lies within $\sim 0.5h^{-1}$ comoving Mpc (4×10^{23} proper cm) of the galaxies. This is implied by the low observed level of Lyman-α absorption along the small segments of the QSOs' sightlines that pass so close to star-forming galaxies (figure 4.). The galaxies' hydrogen-ionizing radiation would not be sufficient to destroy the neutral hydrogen out to such large radii, but their winds, traveling at several hundred km $\rm s^{-1}$ throughout the galaxies' estimated 10^8 yr lifetimes, might [1].

Third, metals that are found in the high-redshift intergalactic medium tend to lie close to observed star-forming galaxies. This is illustrated in a qualitative

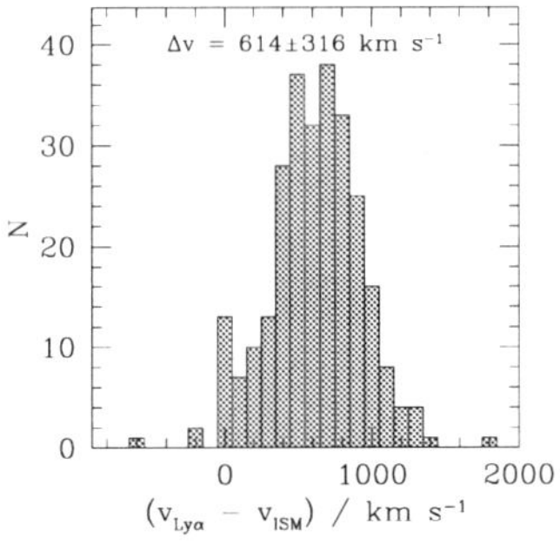

Fig. 3. Histogram of velocity differences between Lyman-α emission and interstellar absorption in the spectra of star-forming galaxies at redshift $z \sim 3$. The large velocity differences show that gas in or near these galaxies is seriously disturbed; they are generally interpreted as evidence that the combined force of numerous supernova explosions within each galaxy are leading to large outflows of gas.

way by figure 5., which shows the number density of galaxies and C IV absorption systems as a function a redshift along the sightline towards the QSO Q1422+2309. Parts of the universe that contain large numbers of galaxies tend to contain large number of metal-line absorption systems as well. The more detailed treatment of [1] shows that most C IV absorption systems lie within $\sim 2h^{-1}$ Mpc of star-forming galaxies and that a large fraction ($> 50\%$) of all CIV systems with column densities $N_{\rm CIV} > 10^{13}$ cm^{-2} lie within the $\sim 0.5h^{-1}$ Mpc radius of galaxies that outflows might be able to reach. Since winds driven by supernovae are likely to contain a reasonable fraction of the metals that the progenitor stars produced, this observations is also consistent with large scale outflows from the galaxies.

Finally, we observe a tight correlation between large scale fluctuations in galaxy and neutral hydrogen density. A few examples of the correlation are shown in figure 6. Parts of the universe that contain a large numbers of galaxies tend to contain large amounts of neutral hydrogen (top two panels), while parts of the universe that contain few galaxies tend to contain less neutral hydrogen. The sense of this correlation is not hard to understand—galaxies are formed from hydrogen—but its tightness was unexpected. The correlation coefficient is $r \sim 0.7$ between the number density of galaxies in a $10h^{-1}$ Mpc cube and the mean intergalactic Lyman-α transmissivity along the sightline segment enclosed by the cube. The large value of r shows that observations of the density of hydrogen along a single line segment allow one to predict with reasonable accuracy the galaxy density in a large surrounding three-dimensional volume. This was not

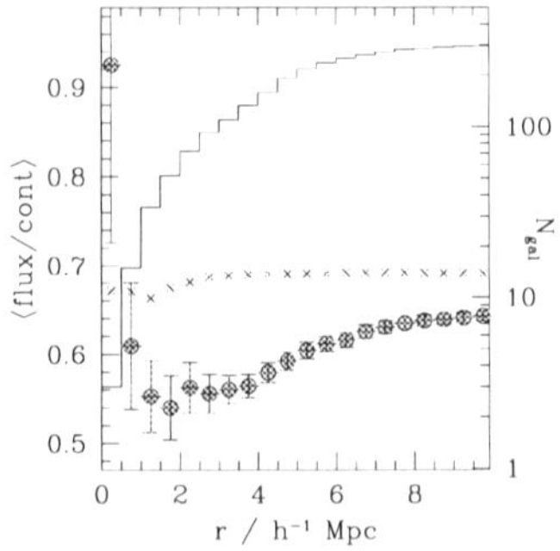

Fig. 4. Mean Lyman-α transmissivity of the intergalactic medium as a function of distance to star-forming galaxies (circles with error bars). A high average transmissivity ("⟨flux/continuum⟩") means that little of the background QSO's light was absorbed at that galactocentric radius; it corresponds to a low density of neutral hydrogen. The shape of the plot suggests first that little intergalactic neutral hydrogen is found within ~ 0.5 comoving Mpc (4×10^{23} proper cm) of star-forming galaxies, second that an excess is found at larger distances. This observation is consistent with the idea that galaxies influence their surroundings through large scale outflows. The (small) size of the current sample is indicated by the histogram, which marks the number of galaxies contributing to the measurement at each radius (refer to right axis). Crosses mark the estimated mean transmissivity when the galaxy redshifts are randomized.

expected from numerical simulations that neglect the impact of galaxies upon their surroundings; these simulations predicted $r \sim 0.5$. Both the simulations and observations are somewhat uncertain, and it would be premature to make too much of the apparent difference in correlation strength; but—as noted by [1]—an increased abundance of hydrogen in the intergalactic medium within galaxy overdensities would be one consequence of strong outflows from galaxies, since a volume of the universe that has been shocked by a galactic outflow will ultimately contain more neutral hydrogen than before it was shocked. The reason is simple. Though fast-moving outflows will at first tend to ionize hydrogen, eventually the outflow will stall, and then the compressed matter at the former shock front will support a higher hydrogen neutral fraction than before it was shocked. The increase in neutral fraction due to old winds may help to explain why simulations have difficulty reconciling the large amount of HI absorption in the spectra of QSOs with the observed ionizing background radiation at high redshift and the baryon density predicted by big-bang nucleosynthesis.

Readers are referred to [1] for a much more detailed treatment of the

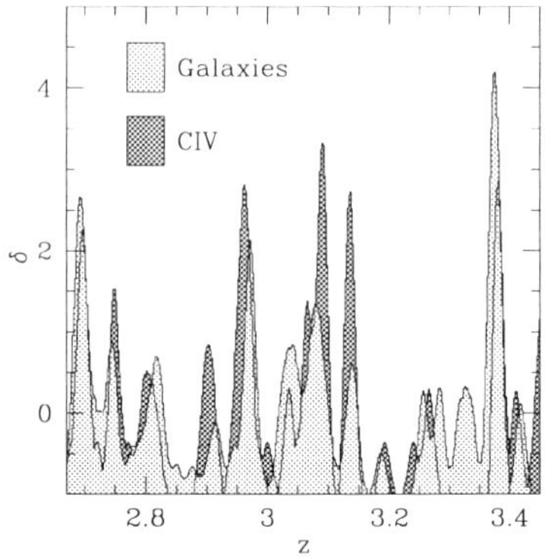

Fig. 5. The relative number density $\delta \equiv (n - \bar{n})/\bar{n}$ of galaxies and CIV absorption systems along one sightline. To reduce noise the data were smoothed on a spatial scale of $\sim$ 10 comoving Mpc ($\sim 10^{25}$cm). Metals and galaxies tend to be found in the same parts of the universe at redshift $z \sim 3$.

observations.

4. Concluding Remarks

This is a tremendous time for observational cosmology. Forty years ago our attempts to explain the present state of the universe were little more than speculation. Today we have a reasonably well motivated models for the composition of the universe soon after the big bang, for the small fluctuations in the distribution of matter at early times, and for the growth of these small fluctuations into the large structures of the present universe. Understanding the evolution of baryonic matter remains challenging, but increasingly sophisticated numerical simulations and analytic treatments are beginning to predict the properties of galaxies with some success. Although many uncertainties remain, the development of telescope and detector technology is giving us an ever improving empirical view of the history of the universe that promises to constrain and refine our theoretical understanding. The statistics are still poor, especially during the first half of history ($z > 1$), and our conclusions are still tentative. But progress has been remarkably rapid. Ten years ago only a handful of normal galaxies at $z > 1$ were known. Today there are thousands. Ten years ago theorists seldom bothered to publish predictions at $z > 1$ because they seemed unlikely to be tested. Today published predictions of galaxy and star-formation extend well beyond $z = 10$. Though the author hopes that galaxy formation will still seem to be an interesting and

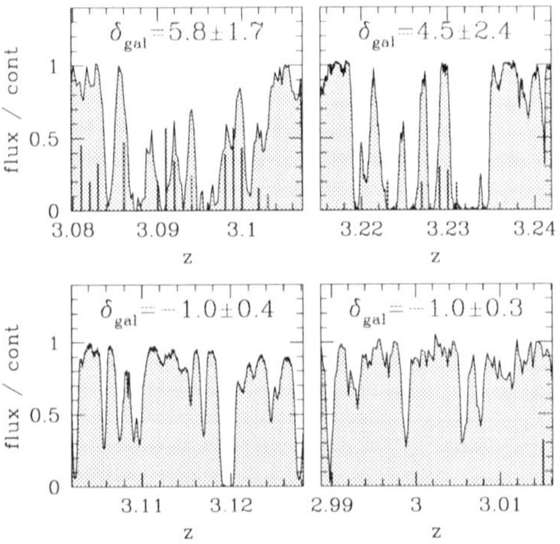

Fig. 6. The neutral hydrogen absorption within the two most significant galaxy overdensities (top panel) and underdensities (bottom panel) in our sample. The shaded curve shows the intergalactic Lyman-α transmissivity; it decreases as the density of HI increases. Vertical bars show the redshifts where galaxies have been found. The height of the bars is proportional to the galaxy's distance from the line of sight. Large amounts of neutral hydrogen are found in galaxy overdensities. The redshift length of each panel corresponds to $\sim 17.5h^{-1}$ comoving Mpc; galaxies were observed within a surrounding $\sim 10 \times 10h^{-2}$ Mpc^2 region.

unsolved problem at the 50th RESCEU Symposium, he fears that it may not.

References

[1] K. Adelberger, C. Steidel, A. Shapley, and M. Pettini, Astrophys. Jour. (2002), submitted

[2] P. de Bernardis et al. Astrophys. Jour. 564 (2002) 559.

[3] L. Cowie, A. Songaila, T.-S. Kim, and E. Hu, Astron. Jour. 109 (1995) 1522.

[4] S. Ellison, A. Songaila, J. Schaye, and M. Pettini, Astron. Jour. 120 (2000) 1775.

[5] T. Heckman, M. Lehnert, D. Strickland, and L. Armus, Astrophys. Jour. Suppl. 129 (2000) 493.

[6] H.-J. Mo and S. Mao, Mon. Not. Roy. Astron. Soc. (2002) in press.

[7] U.-L. Pen, Astrophys. Jour. Lett. 510 (1999) 1.

[8] T. Ponman, D. Cannon, and J. Navarro, Nature 397 (1999) 135.
[9] M. Rauch, W. Sargent, T. Barlow, and R. Carswell, Astrophys. Jour. 562 (2001) 76.
[10] C. Steidel, K. Adelberger, M. Giavalisco, M. Dickinson, and M. Pettini, Astrophys. Jour. 519 (1999) 1.
[11] S. White and M. Rees, Mon. Not. Roy. Astron. Soc. 183 (1978) 341.

Photometric Properties of Galaxies at $z \sim 4-5$ in the Subaru Deep Field

Kazuhiro SHIMASAKU[a,b], Masami OUCHI[a], Sadanori OKAMURA[a,b], Mamoru DOI[c], Hisanori FURUSAWA[a], Masaru HAMABE[d], Masahiko KIMURA[e,f], Yutaka KOMIYAMA[g], Masayuki MIYAZAKI[a], Satoshi MIYAZAKI[h], Fumiaki NAKATA[a], Maki SEKIGUCHI[f], Masafumi YAGI[h], Naoki YASUDA[h]

[a] *Department of Astronomy, The University of Tokyo, 7-3-1 Hongo, Bunkyo-ku, Tokyo 113-0033*; [b] *Research Center for the Early Universe, The University of Tokyo*; [c] *Institute of Astronomy, The University of Tokyo*; [d] *Department of Mathematical and Physical Sciences, Japan Women's University*; [e] *Department of Astronomy, Kyoto University*; [f] *Institute for Cosmic Ray Research, The University of Tokyo*; [g] *Subaru Telescope, National Astronomical Observatory of Japan*; [h] *National Astronomical Observatory of Japan*

Abstract

We report the detection of 1141 Lyman Break Galaxy (LBG) candidates at $z = 3.8 \pm 0.5$ and 268 candidates at $z = 4.7 \pm 0.5$ in wide-field (550 arcmin^2) multicolor data of the Subaru Deep Field. The limiting magnitudes of the $z \sim 4$ LBG sample (selected by BRi' colors) and the $z \sim 5$ sample (selected by $Vi'z'$ colors) are $i'_{AB} = 26.5$ and $z'_{AB} = 26.0$, respectively. We derive rest-frame UV (1600Å) luminosity functions (LFs) for the two samples, to find that the dust-uncorrected UV LF is roughly unchanged from $z \sim 3$ to 5, with the number density of bright galaxies being much higher than in the local universe. We also find that the dust-corrected global star formation rate is almost constant at $z \sim 1$ to 5, comparable to the estimation from number counts at 850μm.

1. Introduction

Observations of galaxies at high redshifts are crucial for understanding galaxy formation and evolution. At present, the most efficient way to select high redshift ($z > 2.5$) galaxies is the Lyman break technique (Steidel et al. 1996), and many studies have been made for Lyman Break Galaxies (LBGs) at $z \sim 2.5 - 4$ (e.g., Steidel et al. 1999, Shapley et al. 2001 [photometric properties of $z \sim 3-4$ LBGs]; Giavalisco et al. 1998, Giavalisco & Dickinson 2001 [clustering properties of $z \sim 3$ LBGs]). To extend such studies to higher redshifts, we have made wide-field multicolor imaging surveys on two blank fields with the Subaru Telescope,

Table 1. Observation Parameters for the Two Deep Fields

Field	band	time$^{(a)}$	seeing$^{(b)}$	lim. mag$^{(c)}$	obs. month
SXDF	($02^h18^m00^s.0$, $-05°12'00''$ [J2000.0])				
	B	177	0.92	27.6	Nov 2000
	V	108	0.95	26.4	Nov 2000
	R	58	0.96	26.5	Nov 2000
	i'	45	0.92	26.2	Nov 2000
SDF	($13^h24^m21^s.4$, $+27°29'23''$ [J2000.0])				
	B	151	0.86	27.7	Apr, May 2001
	V	109	0.84	27.2	Apr, May 2001
	R	73	0.83	27.0	Mar 2001
	i'	103	0.84	26.8	Apr, May, Jun 2001
	z'	68	0.82	26.0	May, Jun 2001

(a) Net exposure time in minutes. (b) FWHM seeing size in arcseconds. (c) Limiting AB-magnitude corresponding to 3σ sky brightness on a $2''\phi$ aperture.

and have constructed large samples for $z \sim 4$ and 5 LBGs. In this paper, we introduce our surveys and report initial results of studies of $z \sim 4$ and 5 LBGs in one of the two fields, focusing on their photometric properties. Our samples are based on large, contiguous areas of the sky, and thus are extremely suitable for clustering studies as well. Results on clustering analyses are reported by Ouchi et al. in this volume. We adopt $\Omega_0 = 0.3$ and $\lambda_0 = 0.7$, and use AB magnitudes.

2. Data and Selection of Lyman Break Galaxies

The data of the two fields were taken with Suprime-Cam on Subaru in November, 2000 (the Subaru/XMM-Newton Deep Field; SXDF) and in March to June, 2001 (the Subaru Deep Field; SDF) during the commissioning phase (**Table 1**). Results reported in this article are based on the data of the SDF. Suprime-Cam is an optical camera for the prime focus of Subaru, having ten $2k \times 4k$ CCDs, and it covers $34' \times 27'$ with $0.''2$ pixel^{-1} at the F/1.86 prime focus (Miyazaki et al. 1998). The SDF is the field on which three Subaru instruments, FOCAS (optical imager and spectrograph), CISCO/OHS (NIR imager and spectrograph), and Suprime-Cam, have made (and will make) extremely deep imaging and spectroscopy to study distant objects.

The data are reduced and combined using IRAF package and the mosaic-CCD data reduction software developed by us (Yagi 1998). The combined images for individual bands are aligned and smoothed with Gaussian kernels to match their seeing sizes. The final images cover a 618 arcmin^2 area with $\text{FWHM}_{\text{PSF}} = 0.''98$ for the SXDF, and a 550 arcmin^2 area with $\text{FWHM}_{\text{PSF}} = 0.''85$ for the SDF. Source detection and photometry are made using SExtractor version 2.1.6 (Bertin & Arnouts 1996). For the object detection, the i' and z' images are chosen for $z \sim 4$ and $z \sim 5$ LBGs, respectively.

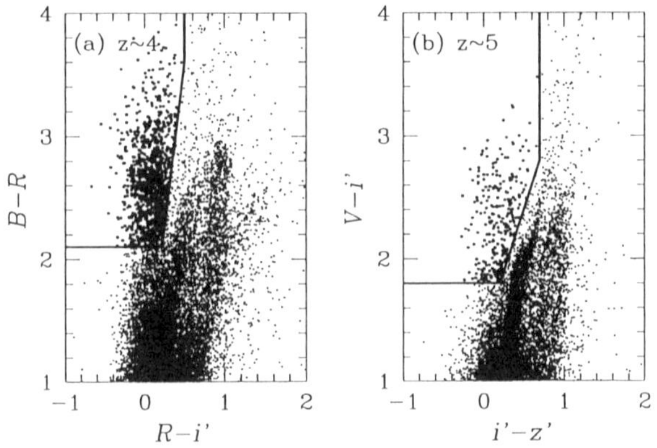

Fig. 1. Selection of $z \sim 4$ LBGs (panel [a]) and $z \sim 5$ LBGs ([b]). Objects satisfying the criteria given in the text are plotted by large dots. Small dots are detected objects. The solid line in each panel outlines the criteria.

On the basis of expectations from GISSEL96 (Bruzual & Charlot 1993, Leitherer et al. 1996) population synthesis models with dust attenuation of $E(B-V) = 0-0.5$ assuming Calzetti's (1997) extinction law, we define the photometric selection criteria for $z \sim 4$ and $z \sim 5$ LBGs as

$$z \sim 4; \quad B-R > 2.1, \quad R-i' < 0.5, \quad B-R > 5.4(R-i') + 0.9, \tag{1}$$

$$z \sim 5; \quad V-i' > 1.8, \quad i'-z' < 0.7, \quad V-i' > 2.0(i'-z') + 1.4. \tag{2}$$

We estimate from the models that the expected redshift distributions of galaxies satisfying equations (1) and (2) are, respectively, $z = 3.8 \pm 0.5$ and $z = 4.7 \pm 0.5$. There are a total of 1,141 objects down to $i'(2''\phi) = 26.5$ (for $z \sim 4$ LBGs) and 268 objects down to $z'(2''\phi) = 26.0$ ($z \sim 5$ LBGs) which meet the criteria in our catalogs (**Figure 1**). **Figure 2** presents the number counts of the objects, without correction for incompleteness of detections, against (a) i' and (b) z'. It is found from panel (a) that our counts for $z \sim 4$ LBGs are consistent with Steidel et al.'s (1999; based on *GRI*), but our data are deeper by about 1.5 mag.

We have estimated the contamination and completeness of our LBG samples by Monte Carlo simulations, generating artificial galaxies using the Hubble Deep Field North objects for which magnitudes, colors, and photometric redshifts are given in Furusawa et al. (2000) and distributing them on our original images. A detailed discussion will be given in Ouchi et al. (in preparation). Luminosity functions and star formation rate densities shown in the next section have been corrected for the contamination and completeness.

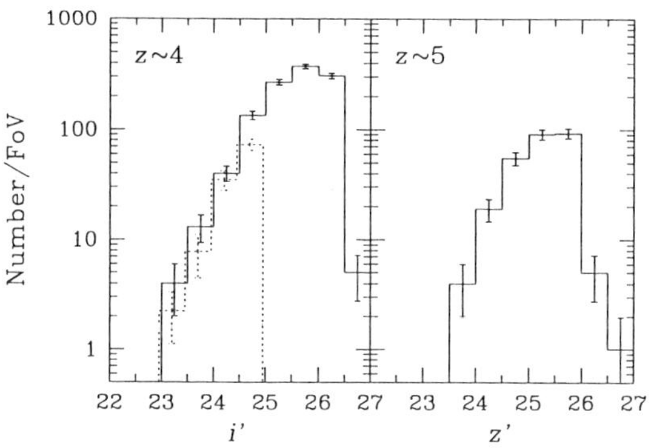

Fig. 2. Number counts of LBGs. The solid histograms indicate counts in our samples. The dotted histogram for $z \sim 4$ shows the counts given in Steidel et al. (1999). Steidel et al's (1999) counts have been normalized to an area of 550 arcmin^2.

3. Results and Discussion

Figure 3 presents histograms of $E(B-V)$ for LBGs at $z \sim 4$ and 5 in our study, together with the one for $z \sim 3$ LBGs taken from Steidel et al. (1999). We have estimated $E(B-V)$ from the slope of the rest-frame UV continuum, following Meurer et al. (1999) to convert UV color to UV absorption, and adopting Calzetti's (1997) dust extinction curve to derive $E(B-V)$. The dotted vertical line for the $z \sim 4$ and 5 panels corresponds to the boundary of our color selection; galaxies having $E(B-V)$ redder than this boundary value are not selected by our criteria. The change in the median $E(B-V)$ with redshift is found to be small, ~ 0.1 mag from $z \sim 3$ to 5. However, the fraction of dusty ($E(B-V) > 0.4$) galaxies increases with redshift. This trend is seen even if we confine ourselves to galaxies brighter than -20 mag. This seems to imply that younger LBGs are on the average dustier.

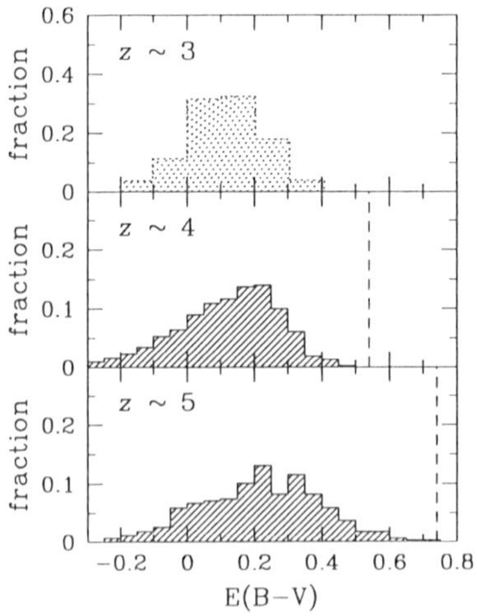

Fig. 3. $E(B-V)$ distribution of LBGs. Solid histograms show our estimations. The data for $z \sim 3$ are taken from Steidel et al. (1999).

We then derive rest-frame UV (1600 Å) luminosity functions (LFs) for $z \sim 4$ and 5 LBGs detected in our data to examine the star formation rate distribution of LBGs. The results are plotted in **Figure 4** together with measurements at $z \sim 0$ to 4 in the literature. Dust extinction has not been corrected. A comparison among redshifts shows UV-bright galaxies are much more numerous at $z \sim 3$ to 5 than at $z \sim 0$ and 1. The typical brightnesses of the LFs at $z \sim 3-5$ are about 1.5 mag brighter than those at $z \sim 0$ and 1. The number density of faint galaxies, on the other hand, does not change largely. A detailed comparison between the $z \sim 3, 4, 5$ LFs shows that the number of bright galaxies decreases slightly with redshift.

Sullivan et al. (2000) have found that the typical dust extinction of their $z \sim 0$ galaxies is $E(B-V) \simeq 0.15$, which is close to those for $z \sim 3-5$ LBGs. Thus, even if we correct the UV LFs for dust extinction, the results we obtained above for uncorrected ones do not change significantly.

By integrating UV LFs, we can obtain the global Star Formation Rate Density (SFRD) at each redshift. **Figure 5** plots the results, together with estimations based on Hα LFs. We convert UV and Hα luminosities to SFR using the formulae given in Madau et al. (1998), assuming a Salpeter IMF with

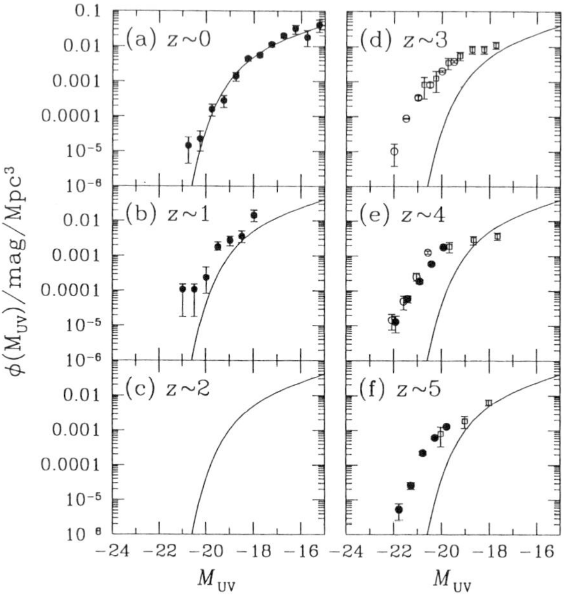

Fig. 4. UV luminosity functions. Our results are shown by solid circles in panels (e) and (f). Open circles in (d) and (e) are estimations by Steidel et al. (1999), while open squares in (e) and (f) indicate the LFs derived from the HDFN galaxies. The LFs at $z = 0$ and 1, which are for 2000Å, are taken from Sullivan et al. (2000) and Cowie et al. (1999), respectively. The solid line in each panel corresponds to the best-fit Schechter function to Sullivan et al.'s data.

$0.1 < M/M_\odot < 125$. We have corrected dust extinction for all the UV results assuming $E(B - V) = 0.15$. The dust correction lifts up the data points by a factor of 2-4, depending on the wavelength used to measure SFR. At $z \sim 1 - 5$, the SFRD is found to be roughly constant, and these values are comparable to those derived from sub-mm sources (pentagons). This result is consistent with the statement by Adelberger & Steidel (2000) based on LBGs at $z \sim 3$ that UV selected LBGs can account for most of the 850μm background light if the mean dust extinction of a factor of ~ 6 at 1600Å is assumed for LBGs.

Finally, if we integrate the dust-extinction corrected SFRD from $z = 5$ to 0, we find the density parameter of stars at present to be $\Omega_\star \simeq 0.005 \pm 0.002 h^{-1}$, where the error comes from typical uncertainties in the SFRD estimations. This value is twice as large as an estimation for galaxies in the local universe ($\Omega_\star = 0.0025^{+0.0017}_{-0.0010} h^{-1}$; Fukugita, Hogan, & Peebles 1998). However, this discrepancy

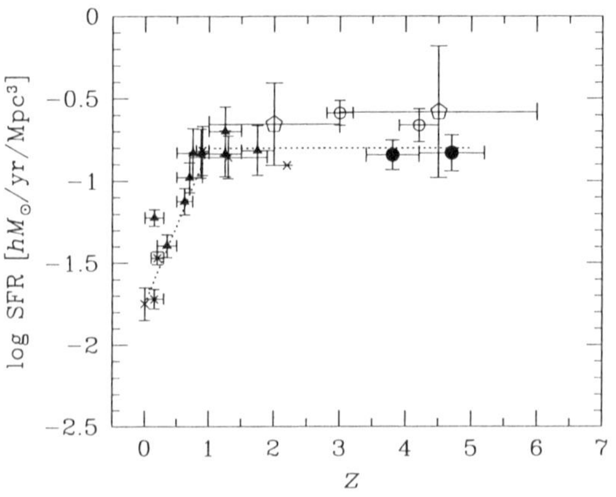

Fig. 5. Extinction-corrected global star formation rate. Estimations based on the UV LFs in our data, supplemented with the LFs in the HDFN at faint magnitudes, are shown by solid circles. Open circles are the estimations by Steidel et al. (1999) (integration is made over $M = -\infty$ and 0). Filled triangles and crosses are estimations based on UV LFs (Lilly et al. 1996; Connolly et al. 1997; Cowie et al. 1999; Sullivan et al. 2000) and Hα LFs (Gallego et al. 1995; Sullivan et al. 2000; Tresse & Maddox 1998; Yan et al. 1999; Moorwood et al. 2000; Glazebrook et al. 1999) at lower redshifts. Pentagons denote the SFRD inferred from 850μm counts by SCUBA (Barger et al. 2000). $E(B-V) = 0.15$ has been applied to all the UV-based estimates. The dotted line shows the approximation used to derive the stellar density.

is not significant, given the large uncertainties in both estimations. In any case, the crude match of the two estimations may imply that we have traced roughly the averaged star formation history in the universe since $z \sim 5$.

References

[1] Adelberger K. L. & Steidel, C. C. ApJ, 544 (2000) 218.
[2] Barger, A. J., Cowie, L. L., & Richards, E. A. AJ, 119 (2000) 2092.
[3] Bertin, E. & Arnouts, S. AAS, 117 (1996) 393.
[4] Bruzual A., G. & Charlot, S. ApJ, 405 (1993) 538.
[5] Calzetti, D. AJ, 113 (1997) 162.

[6] Connolly, A. J., Szalay, A. S., Dickinson, M. E., SubbaRao, M. U., & Brunner, R. J. ApJ, 486 (1997) L11.
[7] Cowie, L. L., Songaila, A., & Barger, A. J. AJ, 118 (1999) 603.
[8] Fukugita, M., Hogan, C. J., & Peebles, P. J. E. ApJ, 503 (1998) 518.
[9] Furusawa, H., Shimasaku, K., Doi, M., & Okamura, S. ApJ, 534 (2000) 624.
[10] Gallego, J., Zamorano, J., Aragon-Salamanca, A., & Rego, M. ApJ, 455 (1995) L1.
[11] Giavalisco, M. & Dickinson, M. ApJ, 550 (2001) 177.
[12] Giavalisco, M., Steidel, C. C., Adelberger, K. L., Dickinson, M. E., Pettini, M., & Kellogg, M. ApJ, 503 (1998) 543.
[13] Glazebrook, K., Blake, C., Economou, F., Lilly, S., & Colless, M. MN, 306 (1999) 843.
[14] Leitherer, C. et al. PASP, 108 (1996) 996.
[15] Lilly, S. J., Le Fevre, O., Hammer, F., & Crampton, D. ApJ, 460 (1996) L1.
[16] Madau, P., Pozzetti, L., & Dickinson, M. ApJ, 498 (1998) 106.
[17] Meurer, G. R., Heckman, T. M., & Calzetti, D. ApJ, 521 (1999) 64.
[18] Miyazaki, S., Sekiguchi, M., Imi, K., Okada, N., Nakata, F., & Komiyama, Y. SPIE, 3355 (1998) 363.
[19] Moorwood, A. F. M., van der Werf, P. P., Cuby, J. G., & Oliva, E. AA, 362 (2000) 9.
[20] Shapley, A. E., Steidel, C. C., Adelberger, K. L., Dickinson, M., Giavalisco, M., & Pettini, M. ApJ, 562 (2001) 95.
[21] Steidel, C. C., Giavalisco, M., Pettini, M., Dickinson, M., & Adelberger, K. L. ApJ, 462 (1996) L17.
[22] Steidel, C. C., Adelberger, K. L., Giavalisco, M., Dickinson, M., & Pettini, M. ApJ, 519 (1999) 1.
[23] Sullivan, M., Treyer, M. A., Ellis, R. S., Bridges, T. J., Milliard, B., & Donas, J. MN, 312 (2000) 442.
[24] Tresse, L. & Maddox, S. ApJ, 495 (1998) 691.
[25] Yagi, M. PhD thesis, University of Tokyo (1998)
[26] Yan, L. et al. ApJ, 519 (1999) L47.

Clustering of Galaxies at Redshift 4 and 5 in the Subaru Deep Fields

Masami OUCHI[a], Kazuhiro SHIMASAKU[a,b], Sadanori OKAMURA[a,b], Mamoru DOI[c], Hisanori FURUSAWA[a], Masaru HAMABE[d], Masahiko KIMURA[e,f], Yutaka KOMIYAMA[g], Masayuki MIYAZAKI[a], Satoshi MIYAZAKI[h], Fumiaki NAKATA[a], Maki SEKIGUCHI[f], Masafumi YAGI[h], Naoki YASUDA[h], Subaru NB Collaboration

[a] *Department of Astronomy, The University of Tokyo, 7-3-1 Hongo, Bunkyo-ku, Tokyo 113-0033*; [b] *Research Center for the Early Universe, The University of Tokyo*; [c] *Institute of Astronomy, The University of Tokyo*; [d] *Department of Mathematical and Physical Sciences, Japan Women's University*; [e] *Department of Astronomy, Kyoto University*; [f] *Institute for Cosmic Ray Research, The University of Tokyo*; [g] *Subaru Telescope, National Astronomical Observatory of Japan*; [h] *National Astronomical Observatory of Japan*

Abstract

We present preliminary results on the clustering properties of three high-z galaxy samples; 1200 Lyman Break Galaxies (LBGs) at $z = 3.8 \pm 0.5$, 270 LBGs at $z = 4.7 \pm 0.5$, and 87 Lyman α Emitters (LAEs) at $z = 4.86$. These are found in two deep and wide-fields, the Subaru/XMM Deep Field (618arcmin^2; $i' < 26$) and the Subaru Deep Field (543arcmin^2; $z' < 26$). We calculate angular correlation functions of these three populations, and find clear clustering signals in them. Assuming the redshift distributions, we estimate the correlation lengths of the $z \sim 4$ and 5 LBGs to be $r_0 = 2.7^{+0.5}_{-0.6} h^{-1}$Mpc and $4.3^{+0.5}_{-0.5} h^{-1}$Mpc, respectively, in comoving units ($\Omega_m = 0.3$ and $\Omega_\Lambda = 0.7$). We also calculate r_0 for $z \sim 4$ and 5 LBGs at the same limiting magnitude ($M_{1700} < -19.5$) to find a possible increase in r_0 from $z = 4$ to 5. This may be evidence for the biased galaxy formation. The angular correlation function of $z = 4.86$ LAEs is found to increase with decreasing angular separation, though there is a hump in the $\omega(\theta)$ at $\theta \sim 300''$, which may be caused by a large scale structure of the LAE distribution.

1. Introduction

Studies of the clustering properties of galaxies at high redshifts are essential for understanding galaxy formation and its relationship to structure formation. Giavalisco et al.(1998)[3] studied the clustering properties of Lyman Break

Galaxies (LBGs) on the basis of a large sample of bright ($R < 25.5$) LBGs lying around $z = 3$ that were selected using U_nGR colors. They found that the spatial distribution of $z \sim 3$ LBGs is strongly biased relative to the expectations for the dark matter in cold dark matter models with a linear bias of 4.5 predicted for an Einstein-de Sitter cosmology. Arnouts et al.(1999) [1] measured the correlation length of faint galaxies in the Hubble Deep Field-North (HDF-N) over the redshift range $0 < z < 4$, although errors in their measurements are large because of small statistics. In this contribution, we report our preliminary results of our studies on galaxy clustering of $z = 4$ and 5 LBGs and $z = 4.86$ Lyman α emitters (LAEs), based on unprecedentedly wide and deep field imaging data. Throughout this contribution, magnitudes are in the AB system, and all calculations assume a Λ-dominated spatially flat cosmology, $(\Omega_m, \Omega_\Lambda) = (0.3, 0.7)$.

2. Observation & Sample Selection

During the comissioning phase of Subaru Telescope, November 2000 to June 2001, we carried out multi-band, deep and wide-field optical imaging with Subaru Prime Focus Camera (Suprime-Cam [6]) in two blank fields. One is the Subaru/XMM Deep Field (SXDF: $2^h18^m00^s$,$-5°12'00''$[J2000]) covering a 618 arcmin^2 area with $i' < 26$. The other is the Subaru Deep Field (SDF: $13^h24^m21.4^s$,$-27°29'23''$[J2000]) covering a 600 arcmin^2 area with $z' < 26$. In addition to these broad-band images, we observed the SDF with a custom narrow-band, $NB711$ (λ_c=7126,$\Delta\lambda = 73$Å), to identify LAEs at $z = 4.86$ [8]. We detect 42,557 objects with $i' < 26$ for SXDF, and 56,103 objects with $z' < 26$ for SDF.

On the basis of GISSEL96 [2] population synthesis models, we define photometric selection criteria for $z \sim 4$ LBGs, $z \sim 5$ LBGs, and $z = 4.86$ LAEs [7];[8] as :

$$B - R > 2.1,\ R - i' < 0.5,\ B - R > 5.4(R - i') + 0.9\ ,\ \text{for } z \sim 4 \text{ LBGs}, \quad (1)$$

$$V - i' > 1.8,\ i' - z' < 0.7,\ V - i' > 2.0(i' - z') + 1.4\ ,\ \text{for } z \sim 5 \text{ LBGs}, \quad (2)$$

$$Ri - NB711 > 0.8,\ R - i' < 0.5,\ i' - NB711 > 0,\ \text{for } z = 4.86 \text{ LAEs}, \quad (3)$$

where $Ri \equiv (R + i')/2$. We find a total of 1192 objects satisfying the $z \sim 4$ LBG criteria (eq.[1]) from the SXDF data, 270 objects satisfying the $z \sim 5$ LBG criteria (eq.[2]) from the SDF data, and 87 objects satisfying the $z = 4.86$ LAE criteria (eq.[3]) from the SDF data. We define them as the $z \sim 4$ LBG, $z \sim 5$ LBG, and $z = 4.86$ LAE samples, respectively. We estimate the contamination and completeness of those samples with Monte Carlo simulations, generating artificial objects that mimic real high-z galaxies and distributing them on our original images (details are given in Ouchi et al. 2001 [7]). We find that the sample

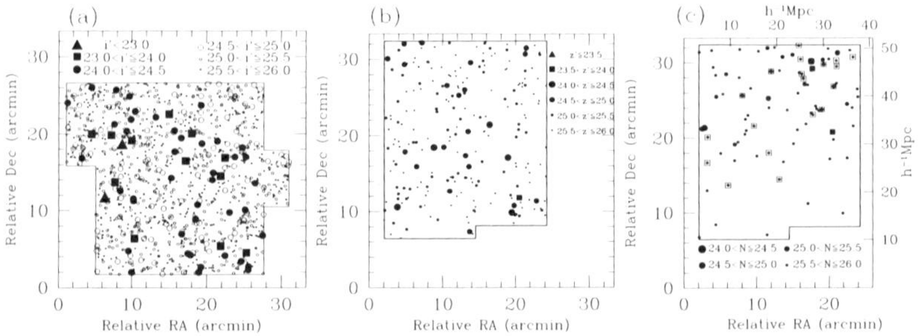

Fig. 1. Sky distribution of our three galaxy samples: (a)$z \sim 4$ Lyman break galaxies (LBGs) with $i' < 26$, (b)$z \sim 5$ LBGs with $z' < 26$, and (c)$z = 4.86$ Lyman α Emitters (LAEs) with $NB711 < 26$.

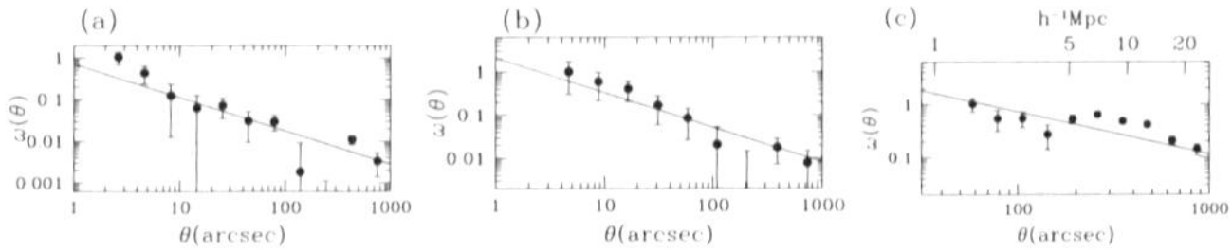

Fig. 2. Angular correlation function for (a)$z \sim 4$ Lyman break galaxies (LBGs) with $i' < 26$, (b)$z \sim 5$ LBGs with $z' < 26$, and (c)$z = 4.86$ Lyman α Emitters (LAEs) with $NB711 < 26$. The LAEs show a hump at $\theta \sim 300''$.

contamination and completeness are 17% and 45% for $z \sim 4$ LBGs (at $z = 3.8 \pm 0.5$), 29% and 40% for $z \sim 5$ LBGs (at $z = 4.7 \pm 0.5$), and 40% and 40% for $z = 4.86$ LAEs (at $z = 4.86 \pm 0.03$).

3. Clustering of Galaxies

Figure 1 shows the sky distribution of the $z \sim 4$ LBGs, $z \sim 5$ LBGs, and $z = 4.86$ LAEs. We derive the angular two-point correlation function using the estimator defined by Landy & Szalay (1993) [5]. The resulting angular correlation function for each sample is presented in Figure 2(a)-(c). Each correlation function shows a clear clustering signal with significant levels. We fit a power law, $\omega(\theta) = A_\omega \theta^{-\beta}$, to the data points, and find best-fit values of A_ω for $\beta \equiv 0.8$.

The two-point angular correlation function is related to the spatial correlation function $\xi(r) = (r/r_0)^{-\gamma}$ by an integral equation, Limber equation [9], where $\gamma = \delta + 1$. From the best-fit A_ω, we estimate the correlation lengths, r_0, of the whole $z \sim 4$ LBGs and $z \sim 5$ LBGs to be $2.7^{+0.5}_{-0.6} h^{-1}$Mpc and $4.3^{+0.5}_{-0.5} h^{-1}$Mpc, respectively, using the predicted redshift distribution of the sample galaxies, as-

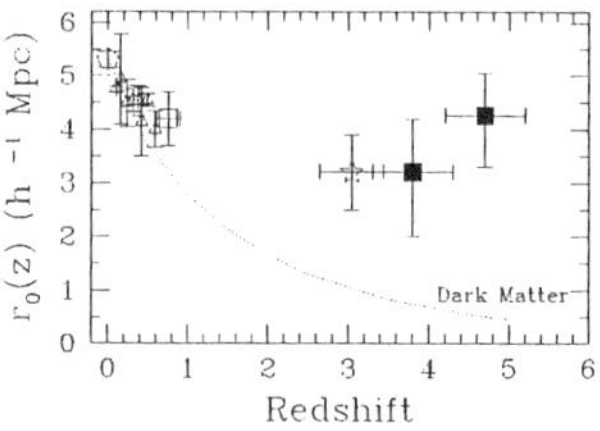

Fig. 3. Correlation length at various redshifts in comoving units. The filled squares indicate our measurements for $z \sim 4$ and 5 LBGs. The star mark shows the value for $z \sim 3$ LBGs[4]. These correlation lengths are derived at a common brightness limit ($M_{1700} < -19.5$) for $z = 3 - 5$ LBGs. We show the correlation length of $> L_B^*$ galaxies at $z < 1$ in the literature (references in Ouchi et al. 2001[7]).

suming that the clustering pattern is fixed in comoving coordinates in the redshift ranges of our samples. Since the amplitude of the galaxy-galaxy clustering is dependent on luminosity and type of galaxies, we estimate correlation lengths of $z \sim 4$ and 5 LBGs for the same limiting magnitudes ($M_{1700} < -19.5$), so as to make a fair comparison of the clustering amplitudes at $z \sim 4$ and 5. Figure 3 plots r_0 of $z \sim 4$ and 5 LBGs with $M_{1700} < -19.5$ against redshift, together with the one for $z \sim 3$ LBGs with a similar magnitude limit [4]. A slight increase in r_0 from $z = 3$ to 5 is found. This may support the biased galaxy formation scenario which predicts that the galaxy-galaxy clustering amplitude increases with redshift at high-z. In Figure 2(c), we find that the $\omega(\theta)$ of $z = 4.86$ LAEs has a hump at $\sim 300''$ corresponding to $8h^{-1}$Mpc (in comoving units). This hump may be due to a large-scale distribution of $z = 4.86$ LAEs, though spectroscopic follow-up is needed to confirm the structure.

References

[1] Arnouts, S. et al., MNRAS 310 (1999) 540.
[2] Bruzual A., G. & Charlot, S., ApJ 405 (1993) 538
[3] Giavalisco, M. et al., ApJ 503 (1998) 543
[4] Giavalisco, M. & Dickinson, M., ApJ 550 (2001) 177
[5] Landy, S. D. & Szalay, A. S., ApJ 412 (1993) 64
[6] Miyazaki, S. et al., SPIE 3355 (1998) 363
[7] Ouchi, M. et al., ApJ 558 (2001) L83
[8] Ouchi, M. et al., Submitted to ApJ (astro-ph/0202204)
[9] Peebles, P. J. E. 1980, *The Large-Scale Structure of the Universe.* (1980) Princeton Univ. Press.

Constraining Cosmological Parameters from Clusterings of Intermediate-redshift Galaxies

Takahiko MATSUBARA
Department of Physics and Astrophysics, Nagoya University, Chikusa, Nagoya 464-8602, Japan
Alexander S. SZALAY
Department of Physics and Astronomy, The Johns Hopkins University, 3400 North Charles Street, Baltimore, MD 21218, USA

Abstract

We estimate how the data of apparent redshift-space clusterings in future redshift surveys can constrain various cosmological parameters. In modern redshift surveys, both depths and sky coverage of the surveys are greater than ever, so traditional techniques used in constraining cosmological models does not suffice. We take advantage of a maximum likelihood method which is appropriate for deep redshift surveys of wide sky coverage. Using this method, the Fisher information matrix is evaluated to show the effectiveness of constraining cosmological models by any given sample of redshift survey. As a result, we show intermediate-redshift galaxies, such as Luminous Red Galaxies in the Sloan Digital Sky Survey, can constrain cosmological parameters, including cosmological constant, unexpectedly well.

1. Introduction

In deep, wide redshift surveys like the Sloan Digital Sky Survey (SDSS), the apparent clustering properties in observable redshift space are complicated because of the redshift-evolution and redshift-distortion effects. However, these two effects also have invaluable information on cosmology. The redshift-evolution of the density fluctuations depends on the growth factor which is a function of the density parameter $\Omega_{\rm M}$ and the normalized cosmological constant parameter Ω_Λ. It is also a function of the galaxy bias b which contains the information on the unknown galaxy formation process. The redshift-distortion depends on the density-velocity relation which is a function of ${\Omega_{\rm M}}^{0.6}/b$ [1], and also on $\Omega_{\rm M}$ and Ω_Λ through geometrical distortions [2, 3, 4].

When our main purpose is not determining the power spectrum itself, it is desirable to directly analyze the data to obtain cosmological information without any intermediate statistical quantities like the power spectrum. It is especially true when the apparent clustering in redshift space is quite different from that in real space. Therefore, we take advantage of the maximum-likelihood method to directly constrain cosmological models from the data.

2. Maximum-likelihood Method for the Apparent Redshift-space Clustering

In maximum-likelihood methods, we directly deal with the density fluctuations in redshift space. In order to do so, first we place cells in redshift space and consider the number of objects n_i in a cell i. Each count represents the smoothed density field contaminated by shot noise. For given mean number densities, $\bar{n}_i$, the density fluctuations $d_i = n_i/\bar{n}_i - 1$ are the fundamental quantities we will analyze. In the following, the most important statistical quantity is the correlation matrix, $C_{ij} = \langle d_i d_j \rangle$. This matrix is related to a continuous two-point correlation function in redshift space $\xi(\boldsymbol{s}, \boldsymbol{s}') = \langle \delta(\boldsymbol{s})\delta(\boldsymbol{s}') \rangle$, where $\delta(\boldsymbol{s})$ is the underlying continuous density contrast which survey objects trace.

According to the Bayesian statistics, the likelihood function for a set of cosmological parameters (θ_α) is evaluated by an identity $L(\theta_\alpha|\delta_i) \propto L(\delta_i|\theta_\alpha)L(\theta_\alpha)$. In the following, we assume we do not have any prier information for the parameters so that $L(\theta_\alpha) = \text{constant}$. In linear regime, the function $L(\delta_i|\theta_\alpha)$ is given by a multi-variate Gaussian distribution function, $\propto \exp(-\frac{1}{2}\boldsymbol{\delta}^{\mathrm{T}}\boldsymbol{C}^{-1}\boldsymbol{\delta})/\sqrt{\det \boldsymbol{C}}$, where $C_{ij}(\boldsymbol{\theta})$ is the correlation matrix for a fixed set of cosmological parameters. While it is straightforward to evaluate the likelihood function in linear regime, the correlation matrix is a huge matrix. We are now developing a novel, fast algorithm to calculate the correlation matrix for arbitrary sample, which will be presented elsewhere in near future. In the following, we focus on how much information can be extracted using the maximum-likelihood analysis of the redshift-space clusterings.

3. Fisher Information Matrix Evaluations

The Cramér-Rao bound $\langle \Delta\theta_\alpha \Delta\theta_\beta \rangle \geq (F^{-1})_{\alpha\beta}$ gives the minimum error variances in the likelihood analysis for a given set of data. Thus the Fisher information matrix $F_{\alpha\beta} = -\langle \partial^2 \ln L/\partial\theta_\alpha\partial\theta_\beta \rangle$ is a key quantity in estimation theory. When the number of independent data is large enough, the Cramér-Rao inequality becomes equality.

In our analysis, the model parameters are the mass density parameter Ω_{M}, the normalized cosmological constant Ω_Λ, the baryon fraction $\Omega_{\mathrm{B}}/\Omega_{\mathrm{M}}$, the Hubble constant divided by 100 km/s/Mpc, h, the primordial spectral index n, the normalization of the power spectrum, σ_8, and the bias parameter b. We assume a

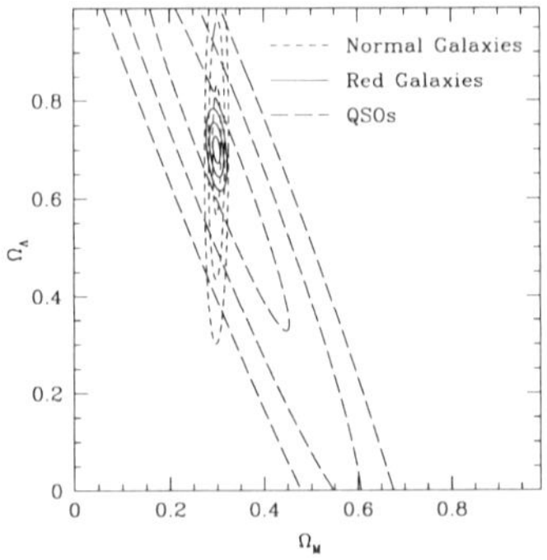

Fig. 1. The concentration ellipses for the cosmological parameters $\Omega_{\rm M}$ and Ω_Λ.

fiducial model which is given by $\Omega_{\rm M} = 0.3$, $\Omega_\Lambda = 0.7$, $\Omega_{\rm B} = 0.04$, $h = 0.7$, $n = 1$, $\sigma_8 = 1$. The fiducial bias parameters are $b = 1$ for main galaxies and $b = 2$ for LRGs and QSOs.

We assume the SDSS sample in this analysis. There are three kinds of samples in the SDSS survey, i.e., about 1,000,000 main galaxies, 120,000 luminous red galaxies (LRGs), and 100,000 quasars (QSOs). The largest-ever redshift-space distribution map for each sample will be obtained in next four years. For the details of those sample, see [5]. The mean redshift for each sample is roughly given by 0.1, 0.35, and 1.0 for main galaxies, LRGs, and QSOs, respectively.

In Fig. 1 plotted the expected errors in simultaneously evaluating the mass density parameter and the cosmological constant. Three ellipses for each sample give expected 1σ, 2σ, and 3σ errors. One can notice that the LRGs (which is referred to as Red Galaxies in the figure) impose the best constraint on the $\Omega_{\rm M}$-Ω_Λ plane. The main galaxies can only probe shallow redshift region, so that the cosmological constant does not strongly contribute the redshift-space clustering. Although the QSOs can probe deep redshift region, the QSO sample is too sparse to detect the redshift-space clustering itself. The LRGs have nicely intermediate properties for the deepness and the density of the objects (see also [6]).

In Fig. 2 plotted the expected errors in simultaneously evaluating the four parameters, $\Omega_{\rm M}$, Ω_Λ, $\Omega_{\rm B}/\Omega_{\rm B}$, and b. In each panel, marginalized errors are plotted for each set of two parameters. In this case, the QSOs does not constrain each parameter well, again because of their sparseness. Besides the cosmological constant, main galaxies and LRGs can impose similar constraints, but the LRGs are basically better in each case. This is because we only use the linear regime of the clustering so that the largeness of the probed region in LRG sample, rather

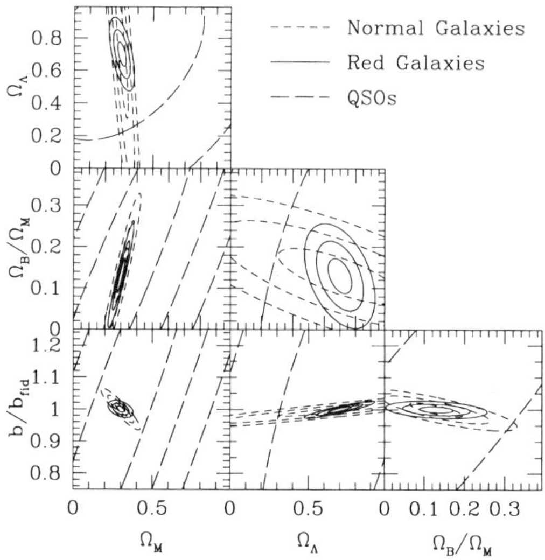

Fig. 2. The marginalized concentration ellipses for four cosmological parameters.

than the density of the objects in main galaxies, is critical for the parameter estimation.

4. Conclusions

In this work, we showed: 1) Direct determinations of cosmological parameters using maximum-likelihood method is promising for the data of future redshift surveys. 2) A novel, fast algorithm to calculate correlation matrix is being developed. 3) The samples of normal galaxies are dense, but have low-redshift, small volume. The sample of QSOs have high-redshift, large volume, but are sparse. The samples of LRGs have appropriately intermediate properties and can impose the best constraints on cosmological parameters.

References

[1] N. Kaiser, MNRAS 227 (1987) 1.
[2] C. Alcock and B. Paczyński, Nature 281 (1979) 358.
[3] W. E. Ballinger, J. A. Peacock and A. F. Heavens, MNRAS 282 (1996) 877.
[4] T. Matsubara and Y. Suto, ApJL 470 (1996) L1.
[5] D. G. York et al. (SDSS collaboration), AJ 120 (2000) 1579.
[6] T. Matsubara and A. S. Szalay, ApJL 556 (2001) L67.

How Can We Study the Local Universe with CMB Photons?

Asantha COORAY
California Institute of Technology, Pasadena, California 91125, USA

Abstract

We show how observations of temperature fluctuations in cosmic microwave background (CMB) can be used to extract information related to the large scale structure, including dark matter distribution, pressure and halo velocities involving the line of sight, transverse and rotational components. The frequency spectrum can be used to separate the SZ effect from thermal fluctuations. The measurement of higher order correlations in CMB can be used to extract effects involving lensing and kinetic SZ effects. A high signal-to-noise arcminute scale CMB experiment can be used to construct the weak lensing convergence, which is useful given that the background source, CMB anisotropies at the last scattering surface, is well understood.

1. Introduction

The cosmic microwave background (CMB) is now well known as a probe of the early universe. The temperature fluctuations in the CMB, especially the acoustic peaks in the angular power spectrum of CMB anisotropies, capture the physics of primordial baryon-photon fluid undergoing oscillations in the potential wells of dark matter [1]. In transit to us, CMB photons also encounter the large scale structure that defines the local universe; thus, several aspects of photon properties, such as the frequency or direction of propagation, are affected. In the reionized epoch, variations are imprinted when photons are Compton-scattered via electrons, moving with respect to the CMB.

These gravitational and scattering effects, though some times insignificant compared to primary fluctuations, leave certain imprints in the anisotropy structure while inducing higher order correlations. These signatures can then be used to extract the properties of large scale structure that led to changes in the CMB temperature. Here, we will summarize what these signatures are, and, how they can be used to study the local universe with CMB photons.

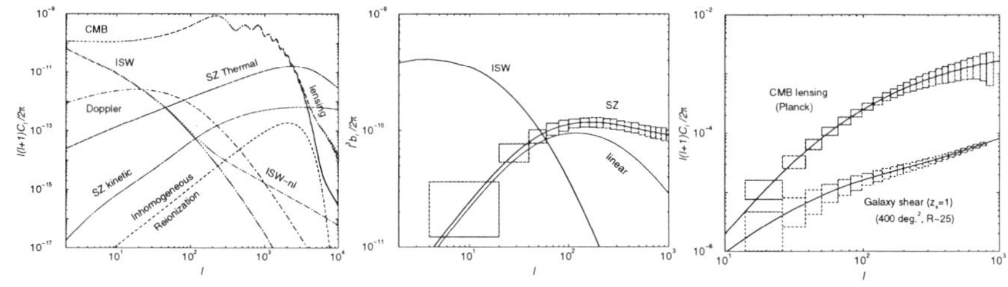

Fig. 1. *Left*: Power spectrum for the temperature anisotropies in the fiducial ΛCDM model with $\tau = 0.1$. The curves show the local universe contributions to CMB due to gravity (ISW and lensing) and scattering (Doppler, SZ effects, patchy reionization). *Middle*: The lensing-secondary correlations and error bars show how well lensing-SZ correlation can be measured with Planck. *Left*: CMB as a weak lensing experiment and error bars show the reconstruction of convergence with Planck.

2. Local Universe Contributions to CMB

Integrated Sachs-Wolfe Effect: The differential redshift effect from photons climbing in and out of a time-varying gravitational potential along the line of sight is called the integrated Sachs-Wolfe (ISW; [2]) effect. The ISW effect is important for low matter density universes $\Omega_m < 1$, where the gravitational potentials decay at low redshift, and contributes anisotropies on and above the scale of the horizon at the time of decay. The non-linear contribution to the ISW effect comes from large scale structure momentum density field. In [3], we presented a model for the non-linear contribution; one aspect of this non-linear effect is that it results from the transverse velocity of halos across the line of sight and results in a dipolar temperature change aligned with the direction of motion. This contribution, however, produces fluctuations which are of order 0.5 μK when transverse velocities are of order 100 km sec^{-1}. The possibility to measure transverse velocities challenges improvements in the experimental front to reach such low temperatures.

Weak Gravitational Lensing: Gravitational lensing of the photons by the intervening large-scale structure both redistributes power in multipole space and enhances it due power in the density perturbations. The most effective structures for lensing lie half way between the surface of recombination and the observer in comoving angular diameter distance. In the fiducial ΛCDM cosmology, this is at $z \sim 3.3$, but the growth of structure skews this to somewhat lower redshifts. In general, the efficiency of lensing is described by a broad bell shaped function between the source and the observer, and thus, correlates well with a large number of tracers of the large scale structure from low to high redshifts.

Since the lensing effect involves the angular gradient of CMB photons and leaves the surface brightness unaffected, its signatures are at the second order in temperature. A quadratic statistic can be devised to probe the gradient structure of CMB temperature and to extract the projected line of sight dark matter density field. This can be best achieved with the divergence of the temperature weighted gradient statistic of [4] and we show errors for a construction of convergence in figure 1. Since lensing has a distinct non-linear mode coupling behavior, it can produce a non-Gaussian signature in CMB data. The idea behind here is that the potentials which lensed CMB also contribute to first order temperature fluctuations from effects such as ISW. These correlations are discussed in detail in [5].

Sunyaev-Zel'dovich Thermal Effect: At small angular scales, probably, the best known effect is the thermal Sunyaev-Zel'dovich (SZ; [6]) effect. The SZ effect arises from the inverse-Compton scattering of CMB photons by hot electrons along the line of sight. This effect has now been directly imaged toward massive galaxy clusters, where temperature of the scattering medium can reach as high as 10 keV producing temperature changes in the CMB of order 1 mK. Note that the effect is proportional to the temperature weighted electron density, or pressure in a cluster. Using a distribution of halos with gas in hydrostatic equilibrium, we can calculate the pressure clustering and project it along the line of sight to obtain the resulting angular power spectrum of temperature fluctuations [7].

Note that the SZ effect also bears a spectral signature that differs from the other effects. The upscattering of photons moves them from low to high frequencies, with no effect at a frequency of $\sim$ 217 GHz. This leads to decrements at low frequencies and increments at high frequencies. An experiment such as Planck with sensitivity beyond the peak of the spectrum can separate out the SZ contribution based on the spectral signature [8]. This is important since statistics related to the SZ effect can be studied separately uncontaminated by primary anisotropies and confusing foregrounds. For example, in [9], we introduced the CMB^2-SZ power spectrum as a probe of the lensing-SZ correlation which involves the CMB and frequency cleaned SZ maps. This statistic can be used to directly estimate how pressure correlates with dark matter; in figure 1., we show the expected errors for the Planck mission.

Linear Doppler Effect: The bulk flow of the electrons that scatter the CMB photons leads to a Doppler effect. Its effect on the power spectrum peaks around the horizon at the scattering event projected on the sky today (see Fig. 1.). On scales smaller than the horizon at scattering, the contributions are significantly canceled as photons scatter against the crests and troughs of the perturbation.

As a result, the Doppler effect is moderately sensitive to how rapidly the universe reionizes since contributions from a sharp surface of reionization do not cancel.

Kinetic Sunyaev-Zel'dovich Effect: The Ostriker-Vishniac/kinetic SZ (kSZ) effect arises from the second-order modulation of the Doppler effect by density fluctuations [10]. Due to the density weighting, kSZ effect peaks at small scales and avoids cancellations associated with the linear effect; the linear effect can also be modulated via the fraction of electrons ionized [11]. Due to density modulation, the kSZ effect has a distinct non-Gaussian behavior [5]. The bulk velocities can be extracted with higher order correlations such as the SZ^2-kSZ^2 statistic of [7]. For a given cluster, the temperature fluctuation is proportional to the line of sight peculiar velocity and when the SZ thermal contribution is removed with frequency information, the kinetic SZ effect should dominate. In addition to the peculiar motion, any coherent rotation of the cluster will produce a dipole-like temperature distribution toward clusters, especially when the rotational axes is aligned perpendicular to the line of sight [12]. This allows an extremely useful probe of the cluster rotations and the angular momentum of gas distribution within galaxy clusters.

Acknowledgements: I thank Wayne Hu for collaborative work and thank the organizers of RESCEU 2001 for their hospitality and opportunity to present this work. This study was partially supported by the Sherman Fairchild foundation.

References

[1] Hu W., Sugiyama N., Silk J. 1997, Nature 386, 37 and references therein
[2] Sachs, R. K., & Wolfe, A. M., 1967, ApJ, 147, 73.
[3] Cooray, A. 2001, Phys. Rev. D., submitted (astro-ph/0109162). also, Seljak, U. 1996a, ApJ, 460, 549.
[4] Hu, W. ApJ, 557, L79.
[5] Goldberg, D. M. & Spergel, D. N. 1999, Phys. Rev. D., 59, 103002; Cooray, A., Hu, W. 2000, ApJ, 534, 533;
[6] Sunyaev, R.A. & Zel'dovich, Ya. B. 1980, MNRAS, 190, 413.
[7] Cooray, A. 2001, Phys. Rev. D., 64, 063514. Cooray, A. 2000, Phys. Rev. D., 62, 103506.
[8] Cooray, A., Hu, W., Tegmark, M. 2000, ApJ, 544, 1;
[9] Cooray, A. 2001, Phys. Rev. D., 64, 043516;
[10] Ostriker, J.P., & Vishniac, E.T. 1986, Nature, 322, 804.
[11] Aghanim, N., Desert, F. X., Puget, J.L., Gispert, R. 1996, A&A, 311, 1.
[12] Cooray, A., Chen, X. 2001, PRD submitted (astro-ph/0107544).

H_2, D/H and the CMBR Temperature at $z = 3.025$ Toward QSO 0347–3819

Sergei A. Levshakov
National Astronomical Observatory, Mitaka, Tokyo 181-8588, Japan and Physico-Technical Institute, St. Petersburg 194021, Russia
Paolo Molaro
Osservatorio Astronomico di Trieste, Via G.B. Tiepolo 11, I-34131, Trieste, Italy
Miroslava Dessauges-Zavadsky
European Southern Observatory, Karl-Schwarzschild-Str. 2, D-85748, Garching, Gernany and Observatoire de Genève, CH-1290 Sauverny, Switzerland
Sandro D'Odorico
European Southern Observatory, Karl-Schwarzschild-Str. 2, D-85748, Garching, Gernany

Abstract

A new molecular hydrogen cloud has been identified at $z = 3.025$ in the absorption-line spectrum of the quasar 0347–3819 observed with the UVES spectrograph at the VLT/Kueyen telescope. At the same redshift numerous metal lines and the D I Ly-5, Ly-8, Ly-10, and Ly-12 lines were detected. The simultaneous analysis of metal and hydrogen lines yielded D/H $= (3.75 \pm 0.25) \times 10^{-5}$. This value is consistent with SBBN if the baryon-to-photon ratio, η, lies within the range $4.4 \times 10^{-10} \leq \eta \leq 5.3 \times 10^{-10}$, implying $\Omega_b h^2_{100} = 0.018 \pm 0.002$ (1 σ c.l.). The population of the ground state rotational levels of H_2 from $J = 0$ to 5 revealed a Galactic-type UV radiation field in the H_2-bearing cloud ruling out UV pumping as an important mechanism for C^+. Accounting for two other important excitation mechanisms such as collisions and FIR photon absorption, we were able to measure for the first time the temperature of the cosmic background radiation at $z > 3$, $T_{\rm CMBR} = 12.1^{+1.7}_{-3.2}$ K, from the analysis of the C^+ fine-structure lines. This result supports the value of $T_{\rm CMBR} = 10.968 \pm 0.004$ K predicted at $z = 3.025$ by the hot Big Bang cosmology.

1. Introduction

In the standard Big Bang model (SBB) the temperature of the relic radiation from the hot phase of the Universe is predicted to increase linearly with redshift z: $T_{\rm CMBR}(z) = T_{\rm CMBR}(0)\,(1 + z)$. At the present epoch direct measurements show that $T_{\rm CMBR}(0) = 2.725 \pm 0.001$ K ($1\,\sigma$ c.l.), and that the relic radiation follows a Planck spectrum with very high precision [1]. However, at earlier cosmological epochs $T_{\rm CMBR}$ cannot be measured directly. It was suggested that this scaling in proportion to $(1 + z)$ can be tested by observing the population of excited fine-structure lines in the QSO absorption spectra [2].

The relative population of the fine-structure levels may not, however, be caused by photo-absorption of the CMBR only. Non-cosmological sources such as particle collisions, pumping by UV radiation or by IR dust emission may compete with the CMBR to populate the excited fine-structure levels. Only independent knowledge of the ambient radiation field, particle densities and the kinetic temperature of the gas allow to disentangle the contribution of the background radiation from that of other mechanisms. For these reasons previous studies set only upper limits to $T_{\rm CMBR}$ [3], [4], [5], [6], [7], [8].

The physical parameters in question can be accurately estimated if the absorber is a diffuse molecular cloud. Molecules allow to measure the volumetric gas density $n_{\rm H}$, the kinetic temperature $T_{\rm kin}$ and the intensity of the UV radiation field through the analysis of their distribution on the low rotational levels. In particular, intervening molecular clouds showing H_2 and the fine-structure lines of C I and C II provides a unique opportunity to measure the cosmic microwave background radiation temperature in early cosmological epochs and to test the SBB predictions.

Another observational test of the SBB model is the measurement of the hydrogen isotopic ratio at high z. The standard (homogeneous) Big Bang nucleosynthesis (SBBN) predicts the same D/H abundance ratio for any direction in the early Universe since "no realistic astrophysical process other than the Big Bang could produce significant D" [9]. Deuterium is created exclusively in BBN and therefore we can expect that the D/H ratio decreases with cosmic time due to conversion of D into ^{3}He and heavier elements in stars. It is clear that the precise measurements of the D/H values at high redshift are extremely important to probe whether BBN was homogeneous. The choice of the appropriate BBN model may in turn place constrains on different models of structure formation.

In this contribution we discuss the role of the H_2-bearing cloud with respect to the measurement of D/H and $T_{\rm CMBR}$ at $z_{\rm abs} = 3.025$ toward the quasar 0347–3819.

2. Observations and results

The spectroscopic observations of Q0347–3819 ($z_{\rm em} = 3.23$, V = 17.3) obtained with the VLT 8.2 m telescope are described in [10] and [11]. The spectrum was obtained with the resolutions FWHM $\simeq$ 7.0 km s^{-1} and $\simeq$ 5.7 km s^{-1} in the UV and near-IR ranges, respectively. A signal-to-noise ratio of S/N $\simeq 10-40$ per resolution element was achieved in these observations.

The analysis of the H_2, deuterium and metal absorption-lines was performed in [11]. We found that the fractional abundance of H_2 is $f_{H_2} = (3.3 \pm 0.2) \times 10^{-6}$. This value is similar to that observed in the Galactic diffuse clouds with low color excesses, $E(B-V) < 0.1$. The kinetic gas temperature is found to be less than 430 K. The population of the low rotational levels is represented by a single excitation temperature of $T_{ex} = 825 \pm 110$ K. The observed population ratios of H_2 in different rotational states and the C II*/C II ratio give the number density $n_H \simeq 6$ cm^{-3}, and the cloud dimension along the line of sight $D \simeq 14$ pc. The relative populations of H_2 in the $J = 4$ and 5 rotational levels correspond to the rate of photo-destruction $I \simeq 2 \times 10^{-10}$ s^{-1}. The photo-absorption rate, $\beta_0 \simeq I/0.11$, is thus equal to $\beta_0 \simeq 2 \times 10^{-9}$ s^{-1}. Taking into account that the Galactic interstellar radiation field value ranges in 5×10^{-10} s$^{-1} \leq \beta_0 \leq 1.6 \times 10^{-8}$ s^{-1} [12], we conclude that the UV radiation fields in the $z_{abs} = 3.025$ absorbing cloud and in the Galactic ISM are very much alike. We also found that the formation rate of H_2 upon grain surfaces, $R\,n_H \simeq 3 \times 10^{-16}$ s^{-1}, is similar to that measured in the Galaxy [12].

The simultaneous analysis of metal profiles and the H+D Lyman series lines yielded a new estimation of the hydrogen isotopic ratio D/H $= (3.75 \pm 0.25) \times 10^{-5}$ in this DLA. It implies that the present-day baryon density is $\Omega_b\,h^2 = 0.018 \pm 0.002$. The obtained D/H value is in good agreement with the hydrogen isotopic ratio found in the Lyman Limit Systems in [13], [14], [15]. We thus summarize that when all most accurate measurements of D/H in high redshift QSO absorbers are considered, one finds the *same* D/H which is about 2.5 times the mean ISM value of D/H $= (1.50 \pm 0.10) \times 10^{-5}$ [16].

Recently, an attempt was made to measure T_{CMBR} in the H_2 molecular cloud at $z = 2.34$ toward Q1232+0815 [17]. However, further analysis of this system [18] revealed inconsistency of the H_2 and HD column density measurements [19] which places the value of T_{CMBR} found in [17] to an upper limit only.

In our analysis of the DLA system at $z = 3.025$ [19] we obtained a self-consistent solution for both molecular and metal absorption lines. We have considered collisions and excluded fluorescence and dust emission as significant processes in the population of the C^+ excited level $J = 3/2$. Our measurement of $N(\text{C II}^*)/N(\text{C II}) = (3.8 \pm 0.3) \times 10^{-3}$ leads to the most probable value of $T_{CMBR} = 12.1^{+1.7}_{-3.2}$ K which is fully consistent with the predicted temperature $T_{CMBR} = 10.968 \pm 0.004$ K within 1 σ confidence intervals.

We conclude that our result, together with upper limits measured in the above mentioned papers support the linear evolution of the CMBR within the framework of the hot Big Bang cosmology. Being considered together, these D/H and T_{CMBR} measurements show again that the fundamental concepts of cosmology and of Big Bang nucleosynthesis are consistent with observations.

References

[1] J. C. Mather, D. J. Fixsen, R. A. Shafer, C. Mosier, and D. T. Wilkinson, Astrophys. J. 512 (1999) 511.

[2] J. N. Bahcall, and R. A. Wolf, Astrophys. J. 152 (1968) 701.

[3] D. M. Meyer, J. H. Black, F. H. Chaffee, C. B. Foltz, and D. G. York, Astrophys. J. Lett. 308 (1986) L37.

[4] A. Songaila, L. L. Cowie, S. Vogt, M. Keane, A. M. Wolfe, E.M. Hu, A. L. Oren, D. R. Tytler, and K. M. Lanzetta, Nature 371 (1994) 43.

[5] L. Lu, W. L. W. Sargent, D. S. Womble, and T. A. Barlow, Astrophys. J. Supl. Ser. 107 (1996) 475.

[6] J. Ge, J. Bechtold, and J. H. Black, Astrophys. J. 474 (1997) 67.

[7] K. C. Roth, and J. M. Bauer, Astrophys. J. Lett. 515 (1999) L57.

[8] J. Ge, J. Bechtold, and V. P. Kulkarni, Astrophys. J. Lett. 547 (2001) L1.

[9] D. N. Schramm, in *Primordial Nuclei and Their Galactic Evolution*, eds. N. Prantzos, M. Tosi, and R. von Steiger (1998) Kluwer Academic Publ., Dordrecht, The Netherlands, p. 3.

[10] S. D'Odorico, M. Dessauges-Zavadsky, and P. Molaro, Astron. Astrophys. 368 (2001) L1.

[11] S. A. Levshakov, M. Dessauges-Zavadsky, S. D'Odorico, and P. Molaro, Astrophys. J. 565 (2002) in press (astro-ph/01055529).

[12] M. Jura, Astrophys. J. 197 (1975) 575.

[13] D. Tytler,, J. M. O'Meara, N. Suzuki, and D. Lubin, Phys. Rev. 333 (2000) 409.

[14] S. A. Levshakov, W. H. Kegel, and F. Takahara, Astrophys. J. 499 (1998) L1.

[15] S. A. Levshakov, D. Tytler, and S. Burles, Astron. Astrophys. Trans. 19 (2000) 385.

[16] J. L. Linsky, 1998, Space Sci. Rev. 84 (1998) 285.

[17] R. Srianand, P. Petitjean, and C. Ledoux, Nature 408 (2000) 931.

[18] D. A. Varshalovich, A. V. Ivanchik, P. Petitjean, R. Srianand, and C. Ledoux, Astron. Lett. 27 (2001) 683.

[19] P. Molaro, S. A. Levshakov, M. Dessauges-Zavadsky, and S. D'Odorico, Astron. Astrophys. (2002) in press (astro-ph/0111589).

Large Scale Structure in the SDSS Galaxy Survey

Andrei Doroshkevich
Theoretical Astrophysics Center, Juliane Marie vej. 30, DK 2100, Copenhagen Ø, Denmark
Douglas Tucker
Fermi National Accelerator Laboratory, MS 127, P.O. Box 500, Batavia, IL 60510 USA
Sahar Allam
National Research Institute for Astronomy & Geophysics, Helwan Observatory, Cairo, Egypt

Abstract

We make use of the Minimal Spanning Tree technique [1, 2] to investigate the large scale galaxies distribution in the SDSS Early Data Release (SDSS EDR) [3] and to quantify some of its statistical characteristics .

1. Introduction

With the advent of the Durham/UKST Galaxy Redshift Survey (DURS, [4]) and the Las Campanas Redshift Survey (LCRS, [5]), the galaxy distribution on scales of up to ~300 h^{-1} Mpc could be studied. Now these investigations can be extended with the public data sets from the SDSS EDR which contain in two slices ≈ 25 000 galaxies for distances $D \leq 600h^{-1}$Mpc.

The MST technique described in [1, 2] allows us to discriminate filamentary and wall-like structure elements located presumably within low and high density regions (LDRs and HDRs) and to estimate their parameters for different threshold overdensities. In the SDSS EDR the subsample of wall-like HDRs were identified with clusters extracted using threshold density contour equal to the mean density of galaxies and a threshold cluster richness of $N_{thr} = 50$. The morphology of the HDRs is found to be sheet-like. The complementary subsample of LDRs, containing those galaxies from the full samples *not* included in the HDR subsample, is populated by filaments and poorer groups of galaxies.

In this paper we present the mass function of structure elements (which we call galaxy 'clouds') and the main characteristics of the walls.

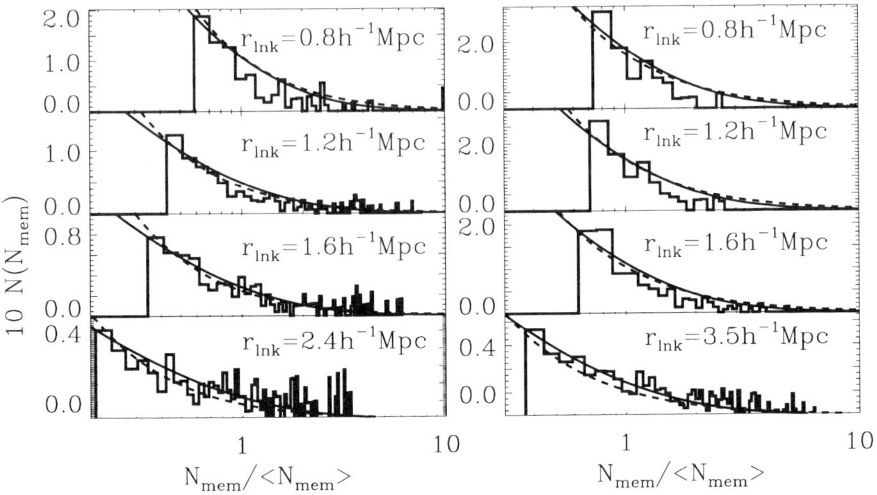

Fig. 1. Mass functions galaxy clouds, $N(N_{mem}/\langle N_{mem}\rangle)$, selected in HDRs (left panels) and LDRs (right panels) for four threshold linking lengths. Fits (1) and (2) are plotted by solid and dashed lines, respectively.

2. Mass function of the structure elements

Three samples of high density galaxy groups and one sample of unrelaxed structure elements – walls and filaments – were selected within HDRs and LDRs separately for the threshold richness $N_{mem} \geq 5$. The mass functions for each of these samples are plotted in Fig. 1. The main properties of the selected clouds are listed in Table 1, where r_{lnk} and δ_{thr} are the threshold linking length and overdensity bounded by the clouds, f_{gal} is the fraction of galaxies within the selected clouds, and $\langle N_{mem}\rangle$ is the mean richness of individual clouds. Differences in $\langle N_{mem}\rangle$ and f_{gal} for clouds selected with the same threshold overdensity within HDRs and LDRs illustrate the impact of environment on the cloud properties.

As was shown in [6, 7], in Zel'dovich theory the mass function of structure elements at small redshifts can be approximated by a one-parameter function,

$$N(x)dx = 8.6\kappa_{ZA}\exp(-3.1x^{1/3})\mathrm{erf}(0.6x^{0.75})\frac{dx}{x^{0.75}}, \quad x = \frac{M}{\langle M\rangle}, \tag{1}$$

which can be compared with mass function from the Press-Schechter formalism,

$$N_{PS}(x)dx = \frac{5\kappa_{PS}}{8\sqrt{\pi}}\exp(-\xi^{1/3})\frac{dx}{\xi^{5/6}}, \quad \xi = \frac{15}{8}x = \frac{15}{8}\frac{M}{\langle M\rangle}. \tag{2}$$

Here the fitting parameters κ_{ZA} and κ_{PS} take into account two cutoffs of measured mass function at low and larger masses.

Table 1. Parameters of groups of galaxies selected in HDRs and LDRs.

$r_{lnk}h^{-1}$Mpc	δ_{thr}	f_{gal}	$\langle N_{mem}\rangle$	$r_{lnk}h^{-1}$Mpc	δ_{thr}	f_{gal}	$\langle N_{mem}\rangle$
	HDR				LDR		
0.8	45.	0.1	8.5	0.8	45.	0.02	6.8
1.2	13.4	0.2	11.6	1.2	13.4	0.06	7.1
1.6	5.6	0.3	14.5	1.6	5.6	0.13	8.1
2.4	2.9	0.4	28.5	3.5	0.8	0.34	13.2

Relations (1) & (2) characterize the mass distribution of dark matter clouds associated with the observed galaxy groups. Both functions are quite similar to each other despite the fact that they are based on different assumptions about of the process of cloud formation. Both functions plotted in Fig. 1 fit reasonably the observed mass distribution for all the very different samples of clouds. For the largest r_{lnk} a significant excess of very massive clouds demonstrates the impact of percolation process which results in the formation of network of filaments and the largest walls. This process is not described by theory. For clouds selected within HDRs some excess of larger clouds demonstrates the acceleration of merging of clouds within HDRs. More detailed analysis of the process of cloud formation within HDRs and LDRs can be performed for richer observed samples.

3. Parameters of the wall-like structure elements

The rich sample of walls extracted from the SDSS EDR also permits more refined estimates of wall properties, such as the mean velocity dispersion of galaxies within walls, $\langle w_w\rangle$, and the mean wall separation, $\langle D_{sep}\rangle$. Comparison of the mean overdensity $\langle\delta\rangle$ and the mean thickness of walls, $\langle h\rangle$, measured both in the radial (r) and in the transverse (t) directions illustrates the influence of the velocity dispersion of galaxies on the observed wall thickness. The mean dimensionless surface density of walls, $\langle q_w\rangle$, measured by the number of galaxies per 1 Mpc^2 normalized by the mean density of galaxies and a typical scale of primordial perturbations [6, 8], is closely linked with the dimensionless amplitude of perturbations, $\tau_m = \sqrt{q_w/6.55}$, which is found to be close to that measured by the CMB anisotropy and σ_8 ($\tau_m \approx 0.27\sqrt{\Gamma/0.2}$, [8]). These parameters are listed in Table 2, where the values from the SDSS EDR are compared with those from the DURS and LCRS catalogues and with those from mock catalogues simulating the SDSS EDR [8]. The method used is the same as was described in [6, 7, 8].

Results obtained for the SDSS EDR are close to those found for the mock

Table 2. Wall properties in observed and simulated catalogues

sample	$\langle q_w \rangle / \Gamma$	$\tau_m / \sqrt{\Gamma}$	$\langle \delta_r \rangle$	$\langle \delta_t \rangle$	$\langle h_r \rangle$	$\langle h_t \rangle$	$\langle w_w \rangle$	$\langle D_{sep} \rangle$
observed samples								
SDSS	2.4 ± 0.7	0.60 ± 0.09	1.5	3.4	11.5	5.0	333 ± 71	67 ± 13
LCRS	2.5 ± 0.9	0.62 ± 0.10	3.0	7.4	8.6	2.8	247 ± 48	60 ± 10
DURS	2.2 ± 0.6	0.58 ± 0.08	1.7	6.5	9.7	4.9	280 ± 52	44 ± 10
mock catalogues in real and redshift spaces for the model with $\Gamma = 0.2$								
redshift	2.7 ± 0.5	0.63 ± 0.06	1.8	3.8	11.8	6.5	338 ± 65	50 ± 10
real	2.1 ± 0.4	0.57 ± 0.06	4.3	4.6	4.8	4.2	305 ± 47	50 ± 10

Here $\langle h \rangle$ and $\langle D_{sep} \rangle$ are given in h^{-1}Mpc, $\langle w_w \rangle$ – in km/s .

catalogues, whereas those for the LCRS and DURS are shifted due to poorer wall statistics and, for the LCRS, the impact of the thinness and nonhomogeneity of the observed slices. These results verify that, indeed, the observed walls represent recently formed Zel'dovich pancakes.

Acknowledgments

Funding for the SDSS Archive has been provided by the Alfred P. Sloan Foundation, the National Aeronautics and Space Administration, the National Science Foundation, the U.S. Department of Energy, the Japanese Monbukagakusho, the Max Planck Society, the University of Chicago, Fermilab, the Institute for Advanced Study, the Japan Participation Group, the Johns Hopkins University, the Max-Planck-Institute for Astronomy (MPIA), the Max-Planck-Institute for Astrophysics (MPA), New Mexico State University, Princeton University, the United States Naval Observatory, and the University of Washington. This paper was supported in part by Denmark's Grundforskningsfond through its support for an establishment of the Theoretical Astrophysics Center.

References

[1] A. Doroshkevich et al., MNRAS 315 (2000) 767.
[2] A. Doroshkevich et al., MNRAS 322 (2000) 369.
[3] S. Stoughton et al. 2001, AJ, submitted
[4] A. Ratcliffe et al. MNRAS 281 (1996) L47
[5] S.A. Shectman et al. ApJ 470 (1996) 172
[6] M. Demiański and A. Doroshkevich MNRAS, 306 (1999) 779.
[7] M. Demiański and A. Doroshkevich 2002, MNRAS, submitted.
[8] A. Doroshkevich et al. 2002, MNRAS, submitted.

Numerical Study of Weak Lensing

Takashi HAMANA
National Astronomical Observatory, Mitaka 181-8588, Japan

Abstract

We performed a series of numerical simulations of weak lensing that combines very large N-body simulations for the evolution of the dark matter distribution in the universe with the ray-tracing simulation for examining the weak lensing effects. We present some selected result from the numerical experiments; the skewness of the lensing convergence, the source-lens clustering effect on the cosmic shear statistics and statistical properties of fluctuations in angular separations of nearby light ray pairs (the so-called lensing excursion angle) induced by weak lensing by large-scale structures. We also briefly present our project of very wide/deep ray-tracing simulations.

1. Introduction

The cosmic shear—coherent distortions in distant galaxy images due to the weak lensing by large-scale structures—is now recognized as a powerful tool to measure the mass distribution in the universe as well as a promising way to measure the cosmological parameters [6] [7]. Although its signal is very weak, recent reports on the detections demonstrate that a well developed data analysis algorithm has been established [8].

So far, the detections were obtained from relatively small fields, which limit the statistical analysis of the surveys to second order statistics (the variance or two-point correlation function of the cosmic shear). Their amplitude reflects the amplitude of the density contrast, and thus provides a constraint on the combination of the values of $\Omega_{\rm m}$ and σ_8. The skewness of the lensing convergence is, on the other hand, known to be sensitive to $\Omega_{\rm m}$ almost independently on σ_8 [3], and thus it may break the degeneracy among the cosmological parameters constrained from the second order statistics. A precise measurement of the skewness is, therefore, one of main goals of on-going wide field cosmic shear surveys such as the DESCART project (for more information about DESCART project, see http://terapix.iap.fr/Descart/.).

Such wide field cosmic shear surveys have a potential for probing the cosmological models as accurately as the cosmic microwave background. In order to get their ability to the fullest, there are, however, some issues which should be developed/examined in details: (i) making an accurate theoretical prediction of the cosmic shear statistics at intermediate scales (0.5-10 arcmin). On such scales the signals will be detected easily, however neither perturbation theory nor the hierarchical ansatz apply [9] [3]. (ii) examining possible corrections which arise from higher order correction terms in calculations of the theoretical predictions, for example, Born approximation and lens-lens couplings [3], the lensing magnification effect [2], and the source-lens clustering [4]. They are especially important for the convergence skewness because it is (in the perturbation theory sense) a quantity of 4th order of $\delta^{(1)}$. (iii) examining the impact of the intrinsic shape correlation of source galaxies. (iv) developing a robust way to correct defects in instruments, in particular the point spread function anisotropy.

We present selected results from ray-tracing simulations combined with large N-body simulations [1], [5]. Experiments were performed to explore source-lens clustering effects on the convergence skewness [4], the impact of Born approximation and lens-lens couplings on cosmic shear statistics [3], statistical properties of fluctuations in angular separations of nearby light ray pairs (the so-called lensing excursion angle) induced by weak lensing by large-scale structures [1]. We also briefly describe new project of very large/deep ray-tracing simulations.

2. Results and discussion

2.1. Weak Lensing Predictions at Intermediate Scales

As pointed out in previous studies, the measurement of the skewness of the convergence field κ will be useful in breaking the degeneracy among the cosmological parameters constrained from weak lensing observations. The combination of shot noise and finite survey volume implies that such a measurement is likely to be done in a range of intermediate scales ($0.5'$ to $20'$) where neither perturbation theory nor the hierarchical ansatz apply. We explore the behavior of the skewness of κ at these intermediate scales, based on results for the non-linear evolution of the mass bispectrum [3]. We combined different ray-tracing simulations to test our predictions, and we find that our calculations describe accurately the transition from the weakly non-linear to the strongly non-linear regime. We also found that the single lens-plane approximation remains accurate even in the non-linear regime, and we explicitly calculate the corrections to this approximation. We also discuss the prospects of measuring the skewness in upcoming weak lensing surveys. See [3] for details.

2.2. Source-lens clustering effects on the skewness of the lensing convergence

The correlation between source galaxies and lensing potentials causes a systematic effect on measurements of cosmic shear statistics, known as the source-lens clustering (SLC) effect. The SLC effect on the skewness of lensing convergence, S_3, is examined using a nonlinear semi-analytic approach and is checked against numerical simulations. The semi-analytic calculations have been performed in a wide variety of generic models for the redshift distribution of source galaxies and power-law models for the bias parameter between the galaxy and dark matter distributions. The semi-analytic predictions are tested successfully against numerical simulations. We find the relative amplitude of the SLC effect on S_3 to be of the order of five to forty per cent. It depends significantly on the redshift distribution of sources and on the way the bias parameter evolves. We discuss possible measurement strategies to minimize the SLC effects. See [4] for details.

2.3. Numerical study of statistical properties of the lensing excursion angles

We explore, for the first time, statistical properties of fluctuations in angular separations of nearby light ray pairs (the so-called lensing excursion angle) induced by weak lensing by large-scale structures using the ray-tracing simulations. We found that the probability distribution function (PDF) of the lensing excursion angles is not simply Gaussian but has an exponential tail. It is, however, found that the tail, or more generally non-Gaussian nature in the PDF has no significant impact on the weak lensing of the CMB. Moreover, we found that the variance in the lensing excursion angles predicted by the power spectrum approach is in good agreement with our numerical results. These results demonstrate a validity of using the power spectrum approach to compute lensing effects on the CMB. See [1] for details.

2.4. Very wide/deep ray-tracing simulations

We performed very wide/deep ray-tracing simulations combined with N-body simulations carried out by the Virgo Consortium (for more information about the Virgo project see http://www.mpa-garching.mpg.de/NumCos.) The ΛCDM Hubble Volume simulation (http://www.mpa-garching.mpg.de/Virgo/hubble.html) data is used for the very wide ray-tracing simulation. We generate the lensing convergence and shear map of a continuous $0.4\times\pi$ ster (a fixed source redshift of z=1). This lensing data set is used to study non-Gaussian contribution to the cosmic variance in a cosmic shear survey. The ΛCDM Very Large Simulation (http://www.mpa-garching.mpg.de/Virgo/VLS.html) is used for a very deep ray-tracing simulation which is used to study the lensing effect on the CMB and to study the lensing magnification effects on a deep survey.

Acknowledgments

The author would like to thank Y. Mellier, S. Colombi, A. Thion, F. Bernardeau, L. Van Waerbeke R. Scoccimarro, J. E. G. T. Devriendt, N. Yoshida, M. Bartelmann, B. Menard, C. Pfrommer S. D. M. White, and Y. Suto for many discussions, this paper is based on the works done in collaboration with them. He acknowledges support from Research Fellowships of the Japan Society for the Promotion of Science.

References

[1] T. Hamana, and Y. Mellier, MNRAS, 327 (2001) 169.
[2] T. Hamana, MNRAS, 326 (2001) 326.
[3] L. Van Waerbeke, T. Hamana, R. Scoccimarro, S. Colombi, and F. Bernardeau, MNRAS, 322 (2001) 918.
[4] T. Hamana, S. T. Colombi, A. Thion, J. E. G. T. Devriendt, Y. Mellier, and F. Bernardeau, MNRAS in press (astro-ph/0012200).
[5] T. Hamana, S. Colombi and Y. Suto, A&A, 367, (2001) 18.
[6] Y. Mellier, ARA&A, 37, (1999) 127.
[7] Bartelmann M., Schneider P., Physics Report, 340, (2001) 291.
[8] L. Van Waerbeke, et al., A&A, 358, (2000) 30.
[9] B. Jain, U. Seljak, S. D. M. White, ApJ, 530, (2000) 547.

Weak Lensing Observations using Suprime-Cam

Satoshi MIYAZAKI
National Astronomical Observatory of Japan, 2-21-1 Osawa, Mitaka-shi, Tokyo 181-0015, Japan

Abstract

Weak gravitational lensing provides promising measures to probe mass distribution in the universe. Precise measurements of faint galaxies over wide field of view (FOV) are crucial to obtain reliable results. Subaru prime camera (Suprime-Cam) has the widest FOV among the world largest telescopes and therefore it can be one of the most suitable instruments for weak lensing. It is our primary goal to realize natural seeing limited image at all times in developing the Suprime-Cam. After the optical adjustments at the telescope, we routinely obtain the image size better than 0.5 arcsec over 34×27 arcmin^2 entire FOV. Analysis of both clusters and 2.2 deg^2 blank field images taken during the commissioning phase prove that the Suprime-Cam is actually a powerful tool for weak lensing observations.

1. Suprime-Cam

Suprime-Cam is a mosaic CCD camera with 80 mega pixels (10240×8192) built for the wide field prime focus of 8.2 m Subaru telescope The focal plane consists of ten high-ρ type CCD developed by MIT Lincoln Laboratory, which provides 34×27 arcmin^2 FOV with the pixel sampling of 0.2 arcsec.

Image quality is crucial for weak lensing analysis. There are three main factors which determine the optical performance of the camera; the primary mirror, the corrector and telescope tracking. Image degradation due to the primary mirror is contolled well below 0.2 arcsec FWHM judging from results of the Shack-Haltman tests. The contribution of the wide field corrector is smaller than 0.22 arcsec FWHM according tests at a lab. The telescope tracking error is proved to be less than 0.07 arcsec for typical ten minutes exposure if assisted by the auto guider. Therefore, the image size of the camera is expected to be natural seeing limited under ordinary conditions.

The focal depth is $\pm 30\mu$m and the co-planarity below the depth is required all over the focal plane whose physical size is 15cm×12cm. Since some CCDs have slope larger than 200 μm from one corner diagonally to the opposite, special

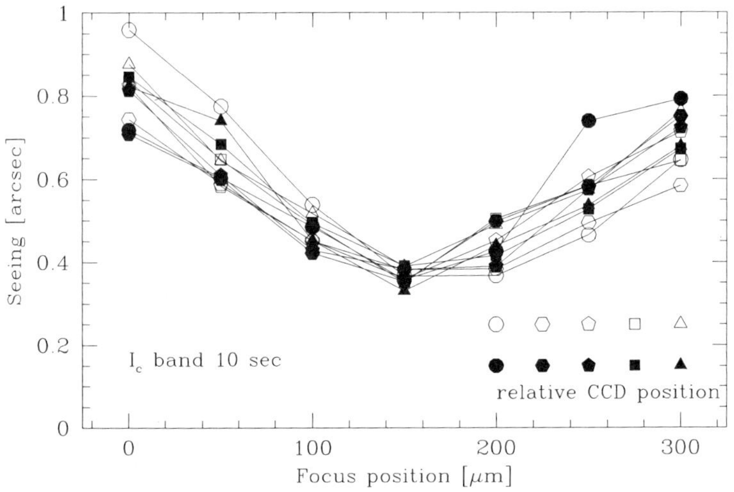

Fig. 1. Stellar image sizes averaged over each ten CCDs as a function of focus position. The expsoure is in I_c band and the integration time is ten seconds at each position.

cares should have been taken to meet the specifications[1, 2]. Precise and stable support of a dewar which has the focal plane array is another technical challenge. We employ low CTE (coefficient of thermal expansion) metal, Invar, to build a support truss structure inside a camera frame.

We here show results of performance verifications at the telescope. These were taken after the adjustment the optics and mechanics. Fig.1 shows stellar image sizes averaged over each ten CCDs and plotted as a function of focus position. I band filter is used and the integration time is ten seconds at each position. Since the best focus position is the same for all the chips, the co-planarity of the focal plane seems accomplished at satisfying level even under ~0.4 arcsec seeing. The median seeing in I band, which is one of the most frequently used filter, measured over one year and a half is 0.61 arcsec.

We employ ellipticities of stars, $\vec{e}$, as a index of PSF anisotropy. Left panel of Fig.2 shows the stellar ellipticities measured on a image of 30 min (6min $\times$ 5) exposure in R band, which is created by precise stacking procedure described later in this section. The seeing size is a typical 0.65 arcsec. The variance of the ellipticity, σ_e, is about 2% and mean value of the ellipticity is consistent with zero. These values are quite small compared with the existing system[3], which demonstrates that we succeeded in the optical and mechanical adjustments of the telescope/camera system. We examined the position dependence of the anisotropy of PSF, which is shown in Fig.2 right. The ellipticity map is not axis-symmetric and cannot be explained based on simple optical calculation. However, the variation of $\vec{e}$ is continuous and can be modeled by polynomial functions of field positions and can be corrected. After the correction, σ_e decreases down to 1% level.

Because of the brightness of the sky especially in red band, the exposure time is limited typically shorter than 5~10 minutes. Multiple exposure is inevitable to have longer exposure time. Position error in aligning these images can be another cause of PSF anisotropy. Simple simulations tell us that one pixel error induces 3 % ellipticity under 0.6 arcsec seeing. This is comparable with cluster induced shear. Therefore precise alignment well below 1 pixel error is required to obtain reliable results from weak lensing.

We first adopt a simple geometrical model as the following. Optical distortion is modeled by 4 order polynomial function of the distance from the optical axis. Displacement and rotation of each CCD from its nominal position are set as free parameters. Offset and rotation of the telescope pointings between exposures are set free as well. All these parameters can be determined by minimizing the distance of stars identified common between the exposures (control stars). Alignment error is estimated by the residual of positions of the control stars and the typical RMS value is $\sim$ 0.5 pixel. This is sufficiently small for most of the applications but not negligible for weak lensing based on the discussion above. We then parameterize the residual $(\Delta x, \Delta y)$ from a reference frame as polynomial function of field position (x, y) and the parameters are obtained by minimizing the χ^2. Sixth order polynomial is usually sufficient from our experiences. Each CCD image is warped by using so obtained polynomials before stacked. After this procedure, the stack error become small enough; 0.05~ 0.07 pixel. Appropriate external catalog should be used to obtain accurate astrometric solutions but no catalog is available to match the faint magnitude considering here (R $>$ 22). We therefore employ the first frame of the exposure, which is corrected by the simple geometrical model, as a reference frame.

2. Weak Lensing Analysis

Armed with the newly developed camera and the precise mosaic-stacking software we have carried out weak lensing observations. Here we show two examples; one is a mass reconstruction of the cluster A2390 (z=0.23) and the other is cosmic shear estimate of a 2.2 sq deg blank field.

Fig.3 (left) presents the result of mass reconstruction of A2390 based on Kaiser and Squires algorithm[4]. The exposure time is 30 minutes in R-band

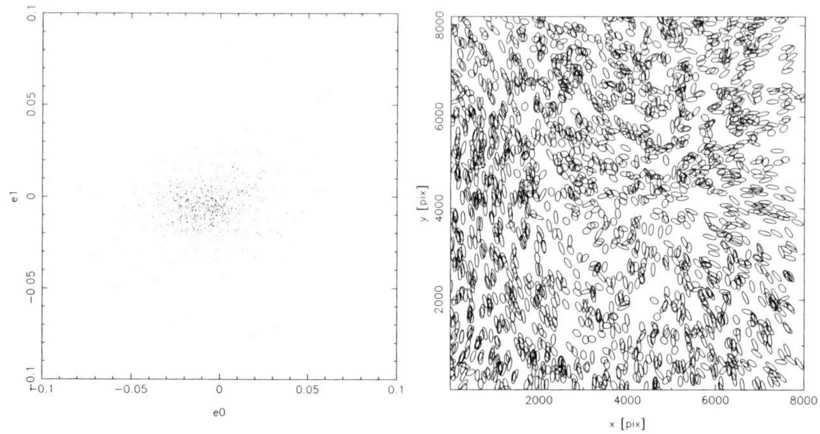

Fig. 2. Ellipticities of stars evaluated on a median image composed of five 6 minutes exposures (Left). The ellipticity map is shown on the right panel.

and the galaxies fainter than 23.5 magnitude are used for the analysis. Mass concentration is clearly detected up to the cluster outer region of $> 1h_{50}^{-1}$Mpc. The radial profile of the mass concentration can be estimated from tangential shear and the result is shown in Fig.3 (right). The error is estimated by using the orthogonal shear components. Since the cluster is a nearby one, the lensing strength is almost independent of the background galaxy redshift. We then assumed that the source galaxies were all located at redshift of 1.

Allen et al.[5] present that the radial mass profile estimated from the X-ray observation by Chandra can be well modeled by NFW density profile[6], which is shown as a dashed-line in the Fig.3 (right). Our result agrees quite well with it.

Squires et al.[7] shows the radial profile based on the weak lensing analysis, which is also shown in the figure. Compared with their result, the error bar of our data is smaller and wider FOV is covered. This demonstrates that the Suprime-Cam has clear advantage for weak lensing observation compared with the 2k×2k pixels camera which was typical on 4 m class telescope in 90's .

We notice the discrepancy of factor of about two between our results and Squires et al's. We feel that the discrepancy is due to the difference of the way to evaluate the relationship between shears ($\vec{\gamma}$) and observed change of galaxy ellipticities ($\vec{\Delta e}$). We adopt *shear polarisability* (P_γ) tensor defined as $\vec{\Delta e} = P_\gamma \vec{\gamma}$ which is worked out by Luppino and Kaiser[8]. P_γ depends on several factors including seeing and the MTF of the system. It is evaluated from images of stars and galaxies and can be noisy especially if faint objects are considered.

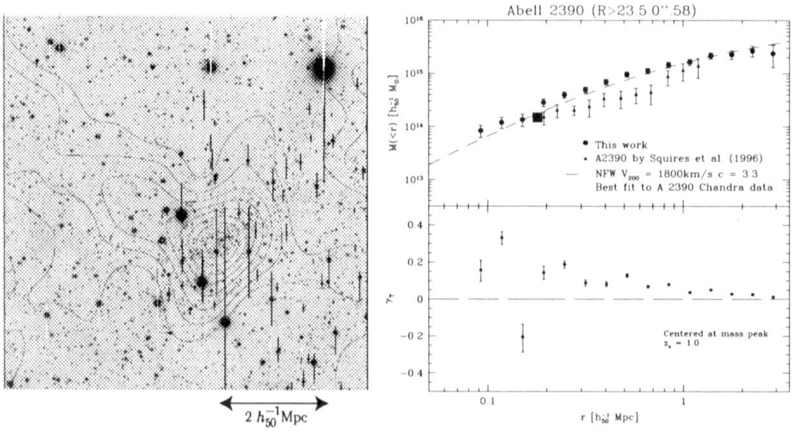

Fig. 3. Reconstructed mass map of Abell 2390 (Left) and the radial mass profile (Right). The tangential shear (γ_T) used to obtain the profile are shown in the lower right panel.

Calibration of the procedure is quite important. Hopefully, we establish some standard clusters catalog by which all the weak lens searchers can calibrate their instruments and analysis procedures. Nearby and well studied clusters such as A2390 would be a good candidate as the member of the catalog.

Second example of our weak lensing study is cosmic shear statistics measured on the blank field where no known cluster exists. We have carried out 3×3 contiguous fields observations which cover 2.2 deg^2 in total. The exposure time in each field is 30 minutes in R band. We calculate RMS shear in aperture defined as

$$\gamma_{rms}^2(\Theta) = \sum_{\alpha=0}^{1} (\frac{1}{N} \sum_{k=1}^{N} \gamma_\alpha)^2, \tag{1}$$

where N is a number of galaxies in an aperture and (γ_0, γ_1) are components of the shear $\vec{\gamma}$. The contribution from intrinsic shape of galaxies into the statistics is estimated from the randomized catalog and subtracted in each aperture. We then obtain the ensemble average of the $\gamma_{rms}^2(\Theta)$ and the results are shown in Fig.4. Four curves in the figure are different realizations of the CDM cosmology (Table 1.) estimated using power spectrum P_δ prescribed by Peacock and Dodds[9]. Several groups announced (Maoli et al.[10] and references therein) the detection of the cosmic shear and claim that SCDM model is not acceptable. They are

Table 1. Cosmological models considered in Fig.4.

Model	Ω_0	Λ	σ_8
SCDM	1.0	0.0	1.00
sCDM	1.0	0.0	0.60
OCDM	0.3	0.0	1.02
ΛCDM	0.3	0.7	1.02

consistent with other models although it is not possible to discriminate among them due to the lack of statistics. Our data is consistent with others and has slightly smaller error bars.

It is, however, risky to make conclusive arguments from this data alone. Because the cosmic shear signal is quite small, evaluation of systematic error is essentially important. One of the major possible sources which causes the error are the procedure to correct shapes of galaxies. As is shown in Fig.2, the PSF anisotropy is 2% under median seeing of about 0.6 arcsec, the correction itself is small, which implies systematic error should be small even if it exists. We calculate Eqn.(1) for 8000 corrected stars over the FOV and then obtain the ensemble average and the variance. The result is $2.4\times10^{-6}\pm4.0\times10^{-6}$ for $\Theta = 10$ arcmin. This is approximately $\frac{1}{10}$ of one tick of vertical axis of Fig.4, which is quite small. In order to compare our results with others which are based on relatively shallower galaxy catalogs, we only use galaxies of $R < 24.5$ which is 1.5 magnitude brighter than S/N=5 limiting magnitude with the aperture of 2 arcsec ϕ. Therefore, it is not suprising if precision of the shape measurent/correction is high.

The error bar shown in the Fig.4 is estimated from variance of 10 times randomization of the galaxy orientation. We have worse S/N at smaller angular scale because each aperture has fewer number of galaxies. The signal itself decreases on large scale. Therefore, we can expect that there is some appropriate angular scale in-between to evaluate cosmic shear signal. Judging from our result, the data around 20 arcmin shows the highest S/N ratio.

Cosmic variance is another major source of systematic error. It is necessary to observe several independent fields to argue this issue. We then hope to obtain at least another field of comparable size to estimate the cosmic variance.

3. Conclusion

We have built a prime focus camera for Subaru (Suprime-Cam) and finished installation on the telescope. We have observed A2390 and the mass distri-

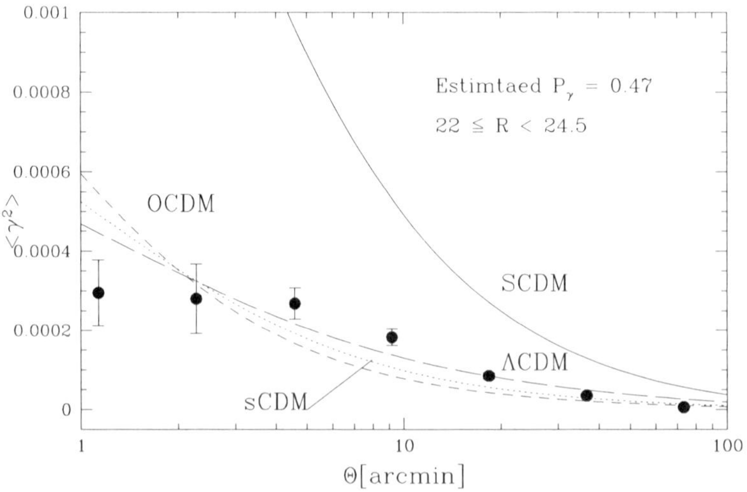

Fig. 4. RMS shear in aperture estimated in the Suprime-Cam's 2.2 deg^2 field

bution estimated from weak lensing analysis is consistent with the one estimated by X-ray observation by Chandra. The radial mass profile turned out to be well described by a NFW density profile from radius 0.1 - 3 h_{50}^{-1} Mpc. Cosmic shear signal estimated on 2.2 deg^2 blank field is consistent with previously published results and our signal-to-noise ratio is highest. We have to consider the cosmic variance next to argue the universality of the result. These results demonstrate that the Suprime-Cam is a powerful tool for weak lensing observations.

References

[1] S. Miyazaki, M. Sekiguchi, K. Imi et al., SPIE 3355 (1998) 363

[2] F. Nakata, S. Miyazaki, N. Okada et al., *in Proc. of 4th ESO CCD Workshop* (2000) 133 Kluwer Academic Publisher

[3] L. Van Waerbeke, Y. Mellier, T. Erben, et al. A&A 358 (2000) 30.

[4] N. Kaiser and G. Squires, Astrophys. J. 404 (1993) 441.

[5] S.W. Allen, S. Ettori and A.C. Fabian, MNRAS 324 (2001) 877.

[6] J.F. Navarro, C.S. Frenk and S.D.M. White, Astrophys. J. 490 (1997) 493

[7] G. Squires, N. Kaiser, G. Fahlman, et al. Astrophys. J. 469 (1996) 73.

[8] G. A. Luppino and N. Kaiser, Astrophys. J. 475 (1997) 20L.

[9] J.A. Peacock and S.J. Dodds, MNRAS 280 (1996) 19L.

[10] R. Maoli et al., A&A 368 (2001) 766

Gravitational Lensing of the Most Distant Known Supernova, SN1997ff

Christofer GUNNARSSON
Fysikum, Stockholm University, SCFAB, SE-106 91 Stockholm, Sweden

Abstract

We investigate the effect of gravitational lensing on the farthest known supernova, SN1997ff. The SN was found at $z \sim 1.7$ in the Hubble Deep Field North and is most likely of Type Ia. Due to our poor knowledge of the properties of the lensing foreground galaxies, we conclude that large magnification effects are possible for reasonable lens parameter values implying that this single SN does not put any strong constraints on the cosmological parameters, grey dust obscuration or luminosity evolution of SNIa until we can model the lensing with high accuracy.

1. Introduction

A primary goal for cosmology is to determine the total energy density of the Universe, Ω and its constituents. It has been long recognised that this goal is achievable through the study of the redshift-distance relation of Type Ia supernovae (SNe). The relation is sensitive to different values of the cosmological parameters and this difference is more prominent at high redshifts. Therefore, the recently discovered SN at $z \sim 1.7$ [1] might prove to be invaluable in this respect. However, different systematic effects such as obscuration by grey dust, luminosity evolution of Type Ia SNe and gravitational lensing are also possibly more severe at high redshifts. In Ref. [2, 3], the systematic effects of gravitational lensing on a large sample of SNe have been investigated. Here, (with more details in Ref. [4]), we investigate the effects of gravitational lensing due to galaxies lying close to the line-of-sight to SN1997ff, generalising the work in Refs. [5] and [1] by investigating the combined effects from a larger number of galaxies and estimating the masses and velocity dispersions of the lensing galaxies from the measured luminosities. Riess et al. [1] argued that the observed brightness of SN1997ff suggests that there cannot be a sizeable luminosity evolution for Type Ia's nor significant extinction by dust. Our work shows that the possible lensing magnification effects are large enough that the data is also consistent with an intrinsically dimmer supernova, or with significant dust density along the line-of-sight.

2. Method

From the Hubble Deep Field North (HDF-N; [6]), we obtain the relative positions and redshifts of galaxies lying in the proximity of the line-of-sight to SN1997ff. All (11) galaxies within $0 \leq z \leq 1.7$ and which are lying closer than 10 arc-seconds to the line-of-sight to SN1997ff and its host galaxy have been included in the study. We model the matter distribution of the galaxies as truncated isothermal spheres characterised by their velocity dispersion, v, which can be estimated from the Faber-Jackson relation

$$\frac{v}{v_*} = \left(\frac{L}{L_*}\right)^{0.25} = 10^{0.1(M-M_*)}. \tag{1}$$

Here, M is the absolute magnitude of the galaxy as measured in the b_J magnitude system and L is the luminosity. The star indicates typical galaxy values on v, L and M (and the mass, m, below). We calculate M by performing cross-filter K-corrections (assuming early-type galaxy spectra) on the observed magnitudes in the r and i bands (ST magnitude system) into rest-frame b_J.

To estimate the masses of the lensing galaxies, we combine the observed luminosities with the mass-to-luminosity ratio [7]

$$\frac{m}{m_*} = \left(\frac{L}{L_*}\right)^{1.25} = 10^{0.5(M_*-M)}, \tag{2}$$

valid for early-type galaxies lying in the fundamental plane. From the velocity dispersion and mass of the galaxies, we can compute the truncation radii of the halos. The magnification and deflection of the light from SN1997ff are calculated by using the multiple lens-plane method [8, Ch. 9] where the mass of each of the lensing galaxies is projected onto a plane at the galaxy redshift. Next, the deflection angle and magnification in each plane is computed and by tracing the light-rays backwards from the observer to the source, we can determine the magnification and position of the ray in the source plane. Our calculations have been made using $\Omega_M = 0.3$, $\Omega_\Lambda = 0.7$, $h = 0.7$ and the filled beam approximation when calculating cosmological distances. A typical value of $v_* = 238$ km/s was obtained in Ref. [9] for this cosmology.

3. Results and discussion

3.1. *Magnification*

In Fig. 1., the magnification of SN1997ff is given as a function of the mass normalisation in the left panel and of the velocity dispersion normalisation in the

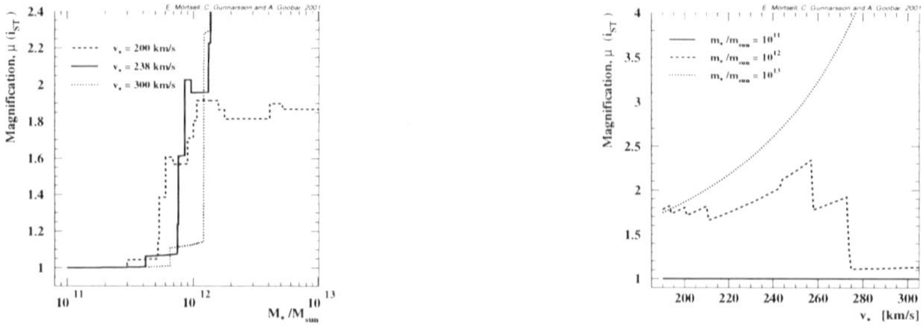

Fig. 1. The magnification, μ, as a function of the typical galaxy mass (left panel) and μ as a function of typical galaxy velocity dispersion (right panel). Both as obtained from the i_{ST} band.

right panel. The masses and velocity dispersions are calculated from the observed luminosities i_{ST}-band but the r_{ST}-band give very similar results. For a circularly symmetric lens, the magnification is determined by the mass within the radius defined by the impact parameter of the light-ray and the surface mass density (or *convergence*) at this radius. The effect of the convergence here is to increase the magnification. Therefore, a light-ray that is passing a plane outside a halo for some specific values of m_* and v_* might pass inside the same halo for some other values, thereby gaining the convergence component of the magnification or vice versa. This is what causes the discontinuities in the magnification in the plots. Of course, these effects are unphysical in the sense that they are very sensitive to the specific modeling of the halos, in this case the steepness of the density profile and the cut-off radii. It is also an indication that lensing effects are very model dependent, and thus very detailed, individual modeling of the lensing galaxies is necessary to make robust predictions of the magnification.

3.2. Clues from host galaxy

The appearance of the host galaxy might offer some clues to the magnitude of the lensing effect. Riess et al. [1] studied the lensing effects of the foreground galaxy closest to the line-of-sight and concluded that the probability of a large magnification is small with such a small ellipticity as is observed (axis ratio $\sim$ 0.85). For a single lens, this is true but when including all foreground galaxies close to the line-of-sight, the situation proves to be much more complicated. The results in Ref. [4] shows that there is no simple relation between magnification and ellipticity when more than one lens is included in the study.

4. Conclusions

Our lensing analysis show that a large range of magnifications of SN1997ff is possible for reasonable values of the galaxy masses and velocity dispersions. The value of the magnification is very sensitive to details in the modelling of the matter distribution in the lensing galaxies. The apparent (lack of) ellipticity of the host galaxy can be shown not to put any strong constraints on the magnitude of the magnification effect. Thus we conclude, that in order to use the apparent magnitude of a single high redshift SN to infer the values of any cosmological quantities, or even to place meaningful limits on the possible dimming of Type Ia SNe by intergalactic grey dust or luminosity evolution, very careful modelling of the galaxies along the line-of-sight is needed in order to control the systematic effects from lensing.

This work has been carried out together with Edvard Mörtsell and Ariel Goobar. The author would like to thank Lars Bergström, Joakim Edsjö and Peter Nugent for helpful discussions and Tomas Dahlén for providing galaxy spectral templates. I would also like to thank the Swedish Research Council for financial support.

References

[1] A.G. Riess et al., ApJ 560 (2001) 49, astro-ph/0104455.

[2] E. Mörtsell, A. Goobar, & L. Bergström, ApJ 559 (2001) 53, astro-ph/0103489.

[3] R. Amanullah et al., *in preparation.*

[4] E. Mörtsell, C. Gunnarsson and A. Goobar, ApJ 561 (2001) 106

[5] G.F. Lewis & R.A. Ibata, MNRAS 324 (2001) L25, astro-ph/0104254.

[6] S.D.J. Gwyn, & F.D.A. Hartwick, ApJ 468 (1996) L77
(see also `http://astrowww.phys.uvic.ca/grads/gwyn/pz/hdfn/spindex.html`)

[7] P.J.E. Peebles 1993, *Principles of physical cosmology* (1993) Princeton University Press, Princeton.

[8] P. Schneider, J. Ehlers, & E.E. Falco, *Gravitational Lenses* (1992) Springer Verlag, Berlin.

[9] G. Wilson, N. Kaiser, G.A. Luppino, & L.L. Cowie, ApJ 555 (2001) 572, astro-ph/0008504.

Is the Luminosity Function of Galaxies Universal?

Sadanori OKAMURA[1)], Masafumi YAGI[2)], Nobunari KASHIKAWA[2)], Maki SEKIGUCHI[3)], Mamoru DOI[4)], Kazuhiro SHIMASAKU[1)], and Naoki YASUDA[2)]
[1)] *Department of Astronomy and Research Center for the Early Universe (RESCEU), School of Science, The University of Tokyo, 7-3-1 Hongo, Bunkyo-ku, Tokyo 113-0033, Japan okamura@astron.s.u-tokyo.ac.jp;*
[2)] *National Astronomical Observatory of Japan, Mitaka, Tokyo 181-8588 Japan;*
[3)] *Institute for Cosmic Ray Research, The University of Tokyo, Kashiwa, Chiba, 277-8582 Japan;*
[4)] *Institute of Astronomy, School of Science, The University of Tokyo, Mitaka, 181-0015 Japan*

Abstract

The 'universal luminosity function hypothesis' is investigated on the basis of a new dataset for 10 nearby Abell clusters. It is shown clearly that the total luminosity function is different between field and cluster environments. The type specific luminosity function of $r^{1/4}$-like galaxies show the signature suggesting the presence of significant dwarf population in this type. While the type-specific luminosity function of exp.-like galaxies is quite similar among the 10 clusters, that of $r^{1/4}$-like galaxies exhibits diversity, which is characterized by the dip at $M_{\rm R} \sim -18.5$ mag, the boundary between giant-dominated and dwarf-dominated domains. The depth of the dip is found to be correlated with the velocity dispersion of the cluster. The hypothesis is shown to be valid neither for the total luminosity function nor for the type-specific luminosity function.

1. Introduction

The luminosity function (hereafter LF) is an important tool to investigate the galaxy population. The LF of galaxies of all types is referred to as the total (or general) LF while that of galaxies of specific type is referred to as the type-specific LF.

Schechter (1976) [1] proposed an analytic function,

$$\phi(L)dL = \phi^* \left(\frac{L}{L^*}\right)^\alpha e^{-\left(\frac{L}{L^*}\right)} d\left(\frac{L}{L^*}\right), \tag{1}$$

which represents the total LF well. He suggested that the shape of the total LF of cluster galaxies is universal and only the multiplicative constant ϕ^* differs among clusters. This 'universal (total) LF hypothesis' has been examined by many authors since then. It is, however, not yet very clear whether the total LF is universal *among clusters* or *between field and cluster* environments.

A direction to clarify the situation was suggested by Binggeli et al (1988) [2], who studied the LF of galaxies in the Virgo cluster and in the local field. They constructed the type-specific LF (hereafter TSLF) for each of the different morphological types as well as the total LF. They proposed the following hypothesis: the TSLF is universal and it is the different type mix that is responsible for the difference, if any, in the total LF between field galaxies and cluster galaxies as well as the diversity seen among different clusters. This 'universal TSLF hypothesis' has not been examined extensively yet due to the observational difficulty that we need both luminosity and morphological type.

We investigate the 'universal LF hypothesis' on the basis of a new dataset for 10 nearby Abell clusters ($z \sim 0.01 - 0.08$). This work is based on a series of papers by Yagi et al. (2002a, b) [3] [4].

2. Observational Material

Since the details of the observation and data reduction are given in [3] and [4], we give only a very brief summary here.

Imaging observations were made in the R band during 4 periods, May 1994, October-November 1994, May 1995, and October 1995, using our mosaic CCD camera mounted at the F/8 Cassegrain focus of the 1-m Swope telescope at the Las Campanas Observatory. Four blank fields were also observed for statistical field subtraction. Star/galaxy discrimination and galaxy extraction are performed with five criteria, which are complementary to one another. We set $R = 20$ mag as the limiting magnitude of our galaxy catalog, which is set by the secure galaxy extraction procedure. The detection limit is in fact much fainter ($R \sim 22.5$).

We apply a single classification scheme based on the bulge-to-total luminosity ratio, B/T, to both giant and dwarf galaxies. We estimate the B/T ratio based on a kind of profile shape parameter. Dressler (1980) [5] classified bright galaxies in 55 clusters down to $m_V \sim 16$ mag by eye inspection of photographic images. Eight of our ten clusters are included in the Dressler's sample. We classify galaxies into two broad types, '$r^{1/4}$-like' and 'exp.-like'. We set the boundary at B/T=0.35 using the conventional classification by Dressler as the fiducial. This boundary gives 70% completeness to both our $r^{1/4}$-like sample (for Dressler's E/S0 galaxies) and our exp.-like sample (for Dressler's Sa-Irr galaxies).

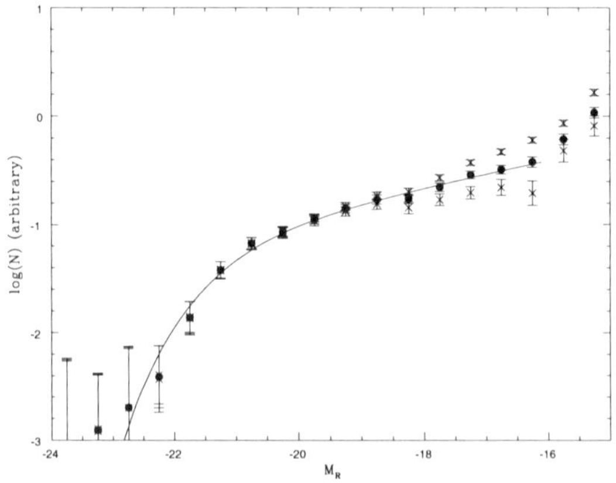

Fig. 1. Composite total luminosity function of the 10 clusters (filled circles). The two series of crosses represent the results of field subtraction changed by ±40%.

Field correction is made by subtracting the mean number count of the blank fields, which are subjected to the same analysis as the cluster fields. On the average, about 1620/1020 galaxies included in a cluster/blank field. Accordingly, about 6000 galaxies are available to the present LF study. The uncertainty of the statistical field subtraction (±40%) is the largest source of the systematic error.

3. Total Luminosity Function

The composite total luminosity function of the 10 clusters shown in Fig. 1 is approximated well by the Schechter function at $M_{\rm R} < -16$. The best-fit Schechter parameters are $M_* = -2.5\log L_* = -21.3$ and $\alpha = -1.31$. As shown in Fig. 2, our result (error ellipses) is consistent with those of previous studies (circles with error bars).

Figure 3 shows the error ellipses overlaid on the results of previous studies for the total LF of *field galaxies*. The systematic offset is evident. In particular, the LFs in the red band from two modern surveys, Las Campanas Redshift Survey (LCRS) [6] and the Sloan Digital Sky Survey (SDSS) [7] have significantly different parameters *. The total LF of cluster galaxies has brighter M_* and smaller α (steeper faint end slope) than the total LF of field galaxies.

*The parameter value in [7] is based on the Petrosian magnitude. The data point in Fig. 2 indicated by the arrow and marked as SDSS shows the parameter values translated to the case of isophotal magnitude on which all other studies are based.

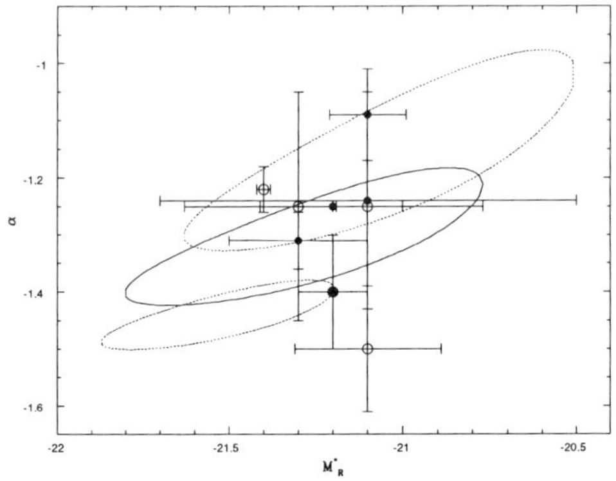

Fig. 2. Schechter parameters of *cluster galaxies*. Our result is shown by the ellipses of 1σ error. The solid, upper broken, and lower broken ellipses correspond to $\pm 0\%$, $+40\%$, and -40% change in field subtraction, respectively. Filled and open circles represent the results from red bands and blue bands, respectively. See [4] for details.

4. Type-Specific Luminosity Function

Figure 4 shows the composite TSLF of the 10 clusters. The composite TSLF of $r^{1/4}$-like galaxies shows the deviation from the Schechter function at $M_{\rm R} > -18$, which indicates the presence of significant dwarf population. It is noted that the sign of the dwarf population is also seen in the TSLF of exp.-like galaxies, but it becomes noticeable at $M_{\rm R} \sim -16$, two magnitudes fainter than the case of $r^{1/4}$-like galaxies. The number of $r^{1/4}$-like dwarfs in $-18 < M_R < -16$ is much larger than the simple extrapolation of the Schechter function fitted to $r^{1/4}$-like giants ($M_R < -18$), but it is nearly the same as the number of exp.-like galaxies in the same magnitude range. This is why the $r^{1/4}$-like dwarfs do not manifest themselves in the composite total LF shown in Fig. 1.

A similar signature which suggests the presence of significant dwarf population is recently found in the TSLF of *quiescent field* galaxies in the 2dF survey (Madgwick et al. 2001) [9] and in the TSLF of *red field* galaxies in the Subaru Deep Field (Furusawa et al. 2002) [10].

TSLFs of $r^{1/4}$-like galaxies in individual clusters exhibit diversity while those of exp.-like galaxies are quite similar among all the 10 clusters. The diversity is characterised by the dip at $M_R \sim -18.5$, the boundary between the giant-dominated and dwarf-dominated domains. The dip is only barely visible in the composite LF (left panel of Fig. 4). We found that the depth of the dip is correlated with the velocity dispersion of the cluster in the sense that the dip is

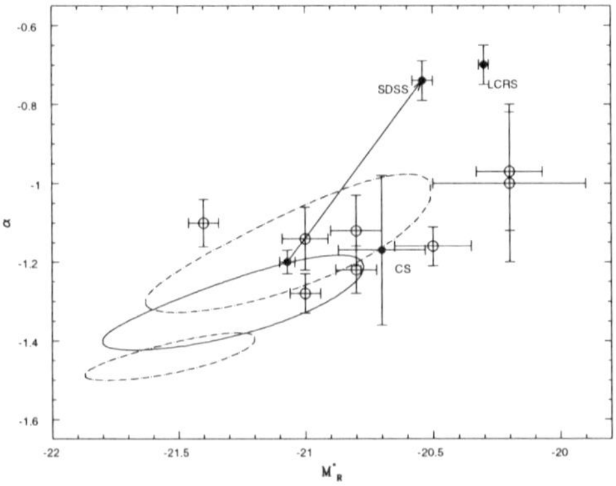

Fig. 3. Our result (ellipses) is compared with Schechter parameters of *field galaxies*. Most of the previous studies are based on blue band magnitude and their M_* are simply transformed to R band (open circles). The LCRS, SDSS, and CS (Century Survey; Geller et al. 1997 [8]) shown by the filled circles are based on the red bands. See [4] for details.

deeper for clusters with larger velocity dispersions (Okamura et al. 2001) [11].

5. Conclusions

We investigate the 'universal LF hypothesis' on the basis of a new imaging dataset for 10 nearby Abell clusters. The composite total LF of the 10 clusters is fitted by the Schechter function well. Our Schechter parameters are consistent with previous studies. We show that the total LF of cluster galaxies is different from that of field galaxies. The cluster LF has brighter M_* and smaller α (steeper faint end slope) than the field LF.

The composite TSLF of $r^{1/4}$-like galaxies shows the deviation from the Schechter function at $M_R > -18$, which indicates the presence of significant dwarf population in this type.

TSLFs of $r^{1/4}$-like galaxies in individual clusters exhibit diversity while those of exp.-like galaxies are quite similar among all the 10 clusters. The diversity is characterised by the dip at $M_R \sim -18.5$, the boundary between the giant-dominated and dwarf-dominated domains. We found that the depth of the dip is correlated with the velocity dispersion of the cluster.

In summary, the 'universal luminosity function hypothesis' is shown to be valid neither for the total luminosity function nor for the type-specific luminosity function.

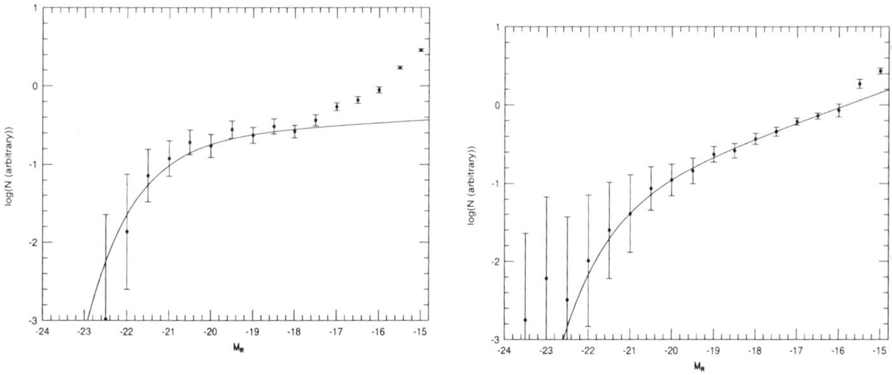

Fig. 4. Composite TSLF of the 10 clusters. The left and the right panels are for $r^{1/4}$-like and exp.-like galaxies, respectively. The curves represent the Schechter function fitted to the data in $M_R < -18$ for $r^{1/4}$-like galaxies and $M_R < -16$ for exp.-like galaxies.

References

[1] P. Schechter, Astrophys. J. 203 (1976) 297.

[2] B. Binggeli, A. Sandage, and G.A. Tammann Ann. Rev. Astron. Astrophys. 26 (1988) 509.

[3] M. Yagi, N. Kashikawa, M. Sekiguchi, M. Doi, N. Yasuda, K. Shimasaku, and S.Okamura Astron. J. 123 (2002) 66.

[4] M. Yagi, N. Kashikawa, M. Sekiguchi, M. Doi, N. Yasuda, K. Shimasaku, and S.Okamura Astron. J. 123 (2002) 87.

[5] A. Dressler Astrophys. J. Suppl. 42 (1980) 565.

[6] H. Lin, R.P. Kirshner, S.A. Shectman, S.D. Landy, A. Oemler, D.L.Tucker, and P.L. Schecter Astrophy. J. 464 (1996) 60.

[7] M.B. Blanton et al. Atron. J. 121 (2001) 235.

[8] M.J. Geller et al. Astron. J. 114 (1997) 2205.

[9] D.S. Magdvick et al. (2001) astro-ph/0107197

[10] H. Furusawa et al. (2002) in preparation.

[11] S. Okamura, M.Yagi, M. Sekiguchi, M. Doi, K. Shimasaku, N. Kashikawa, and N.Yasuda, *The New Era of Wide Field Astronomy* (2001) ASP Conf. Ser. Vol. 232, p.158.

The MAGNUM Project: AGN Variability as a New Technique for Distance Determination

Yuzuru Yoshii
Institute of Astronomy, Graduate School of Science, The University of Tokyo, Mitaka, Tokyo 181-8588, Japan

Abstract

The MAGNUM (Multi-color Active Galactic Nuclei Monitoring) Project is designed to carry out multi-color monitoring observations of hundreds of active galactic nuclei (AGNs) over several years in order to measure the distance to these far away objects using a simple physical principle and thereby to determine the cosmological parameters in models of the expanding Universe. The Project started in 1995 and has progressed through the efforts of scientists at the University of Tokyo, the National Astronomical Observatory of Japan, the University of Hawaii, and the Australian National University. We have constructed a new 2m telescope which we are now commissioning at the University of Hawaii's Haleakala Observatory on the Island of Maui. Engineering first light was obtained in August 2000. Here we present an overview of the MAGNUM Project and report the current status of the Project.

1. Introduction – Principle and Project Purpose

The MAGNUM Project conducts long term monitoring of many AGNs and quasars from ultraviolet to near infrared wavelengths. We aim to measure the distances of AGNs and quasars from those nearby out to those at redshift of $z = 1$ by using photometric observations obtained with the MAGNUM 2m telescope. Our ultimate goal is to determine the destiny of the Universe by deciding whether it will continue to expand forever or it will begin to contract at some time in the future.

It is expected that there are hot dust grains near an AGN as illustrated in Figure 1, that the highest temperature these dust grains can reach is limited by the evaporation temperature of dust grains about 1500K, and that the location near the AGN where this temperature is reached is dependent upon the absolute luminosity of the AGN. The dust grains absorb radiation emitted by the AGN at ultraviolet and optical wavelengths and they re-emit this energy in the near infrared wavelength region. Typically, emission from hot dust grains dominates the AGN spectrum at near infrared wavelengths (Rees et al. 1969; Lebofsky & Rieke 1980; Barvainis 1987, 1990; Kobayashi et al. 1993).

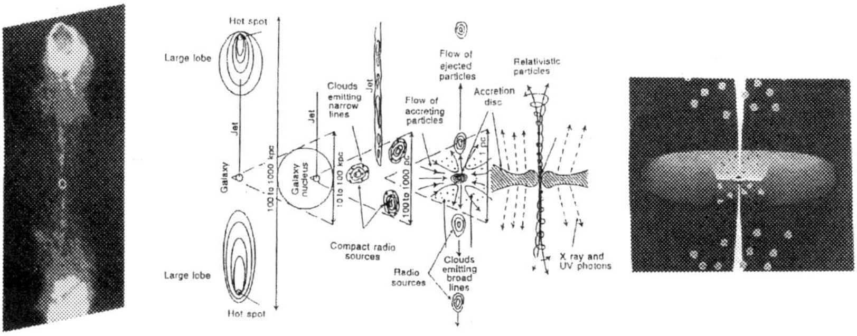

Figure 1. The schematic structure of an active galaxy on various scales, taken from Collin (1987). The dust torus is considered to surround the nucleus as illustrated on the right.

Most AGNs are variable at ultraviolet and optical wavelengths (Neugebauer et al. 1989). The brightness variation of the AGN produces a brightness variation of the dust grains, delayed by the light travel time from the AGN to the surrounding dust torus (Barvainis 1987). By measuring delay time between optical variations and near infrared variations, we measure the radius of dust torus, or the distance from the active center to the hottest dust grains. This radius is proportional to the square root of the absolute luminosity of the central source, and the mechanism that establishes this proportionality is the evaporation of the dust at a constant temperature about 1500K. Figure 2 shows that the data of several AGNs based on monitoring observations in the literature fit the relation between the delay time and absolute luminosity predicted by this principle.

The luminosity distance from an observer to the AGN is then obtained from the directly observed mean apparent luminosity and the absolute luminosity estimated from the delay time. By simple calculation, the delay time ranges from several tens of days for an AGN in a low luminosity Seyfert 1 galaxy to a few months to several years for quasars. In order to measure these time delays from multi-color monitoring observations for a sufficient number of AGNs and quasars, we need exclusive use of a relatively large aperture telescope. Monitoring observations of this kind have not been done before, at least partly because of the difficulty in getting enough observing time on such a telescope.

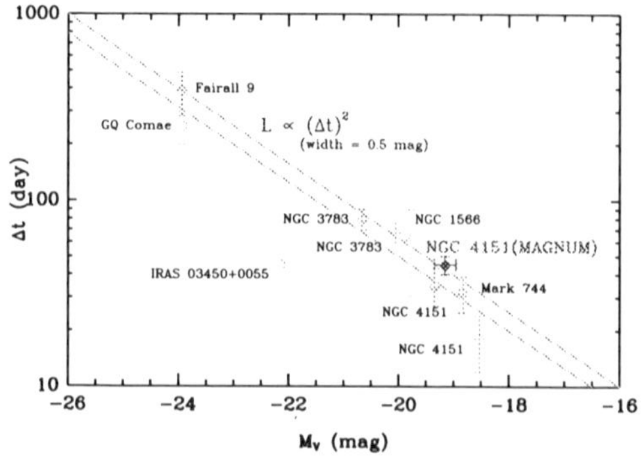

Figure 2. The delay time between optical and infrared brightness variations plotted against the absolute V-luminosity of AGNs. Thick lines are those expected from the MAGNUM principle. Open circles show the data of several AGNs in the literature (Clavel et al. 1989; Baribaud et al. 1992; Glass 1992; Oknyanskii 1993; Sitko et al 1993; Nelson 1996). A closed circle is for NGC 4151 observed by T. Minezaki with the MAGNUM telescope.

Measuring the time variability of faint objects over a wide wavelength range with our new 2m telescope will open unknown aspects in universe, and improve our understanding of the inner structure of AGNs. Since these monitoring observations are relatively routine, we plan for all the observations to be done either in an automatic unmanned mode, or by a remote observer who could be located in Japan.

2. Telescope

The MAGNUM 2m telescope has a thin (10cm) meniscus primary mirror of Corning ultra-low expansion material. This thin mirror reduces the total telescope weight and improves the thermal performance. The mirror is supported by a wiffle tree type passive support mechanism.

The telescope has f/9 Ritchey-Chretien optics with a field view of 33 arc minutes (Figure 3). The composite focal ratio of f/9 was specified after taking into account the expected seeing size and the pixel size of the adopted infrared array. We use an alt-azimuth mount with a tertiary mirror mounted on the Cassegrain instrument rotator. The tertiary mirror itself rotates, allowing us to quickly and automatically change instruments by selecting any of four instruments mounted at one of bent Cassegrain foci.

The secondary mirror was designed to be under-sized and act as the system pupil. It has a central hole in order to reduce the infrared background from the telescope. The telescope mount carries He-gas hoses from a He compressor to supply the mechanical coolers incorporated into the instruments at the bent Cassegrain foci.

Table 1. The specification of the MAGNUM telescope.

primary diameter	2020cm
primary f-ratio	1.75
composite f-ratio	9
optics type	Ritchey-Chretien
field of view	33.3 arc minutes
mount type	alt-azimuth
drive	friction drive
absolute pointing	6 arc seconds
image quality	80% in ϕ 0.6 arc second
focus	bent Cassegrain

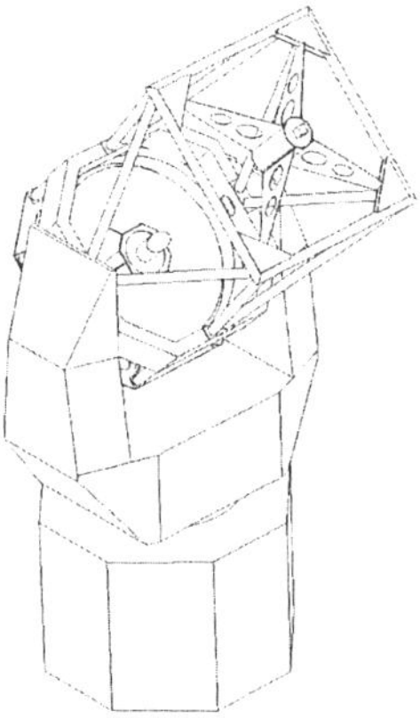

Figure 3. The MAGNUM 2m telescope

3. Observatory

The University of Hawaii allows the MAGNUM Project to install the telescope in the 10m diameter northern dome of the LURE lunar ranging complex (Figure 4) at their Haleakala Observatory which is located near the Haleakala summit at a height of 3050m at 140 degrees west longitude, 20 degrees 43 minutes north latitude. This observatory site is one of two good astronomical observing sites in the Hawaiian Islands and is easily accessible from Japan.

Figure 4. The northern LURE dome with the observatory control rooms on the right.

MAGNUM observations are planned to be made remotely, with the observatory unmanned. Therefore it is very important that the observatory be equipped with automated systems which provide real time status information about the weather, telescope, dome, and instruments. For example, the cloud cover is monitored with a camera working at 8-12μm. The temperature, humidity, rain, wind speed and wind direction are monitored with a weather system. Another system monitors the telescope truss and mirror temperatures, the electric power quality, and the pressure of the He-gas used by the mechanical coolers on the instruments. All monitoring data are obtained in real time and logged.

Table 2. The specification of the MAGNUM cloud monitor

wavelength	8-12μm
optics	aspherical reflective mirror system and camera lens
sensitivity	0C cloud of 0.01% thickness
sensor	micro thermal bolometer array 320 × 240
pixel scale	0.6 deg/pixel
monitor interval	2 min

Figure 5. The MAGNUM cloud monitor.

Our newly developed cloud monitor measures the cloud cover in real time (Figure 5). The Sloan Digital Sky Survey (SDSS) cloud camera is well known, and demonstrates the effectiveness of a cloud camera for remote observations. However, the SDSS cloud camera uses liquid nitrogen for cooling, and a mechanical scanning mechanism, both of which are not suitable for unmanned operation. Therefore, we have built a new type of cloud monitor, which utilizes a thermal bolometer array in conjunction with aspherical reflective mirrors to make an unvignetted image of the cloud cover at 8-12μm. The camera operates at ambient temperature, so the system is free from daily maintenance. For this wavelength region, the aspherical mirrors were easily made using a diamond turning machine. The cloud camera can detect thin cirrus as well as thick cloud, and is useful not only for observatory operation but also for data quality control.

During the daytime, the dome temperature is air-conditioned to the expected night time temperature in order to reduce the effect of dome and mirror seeing. During observations, the dome is vented in order to equalize the inside and outside temperatures. We have not installed adaptive optics since such systems are expensive and require high maintenance. The mean temperature at Haleakala is a little bit higher than that at Mauna Kea and this makes it easier to operate unmanned. However, the seeing conditions may be a little bit worse at Haleakala than those at Mauna Kea because the jet stream goes nearer Haleakala.

4. Observing Instruments

The observing instruments are mounted at bent Cassegrain foci. Four ports are available, and a rotating tertiary mirror selects which pre-installed instrument will be brought into use. Two instruments are being built for the MAGNUM Project. The first is a multi-color imaging photometer (MIP) and another is a wide field camera (WFC). The remaining two instrument ports can be used for other instruments.

The MIP is a camera which has the capability of imaging over a very wide wavelength region, from the ultraviolet to the near infrared, and is used for multi-color monitoring observations of AGNs. The camera is equipped with both a 1024 × 1024 CCD for use at optical wavelengths ($UBVRI$) and a 256 × 256 InSb array for near infrared wavelengths shorter than 4.6μm ($ZJHK'KL'$).

Table 3. The major characteristics of the MAGNUM MIP.

Optical	detector	SITE thin CCD
	format	1024 × 1024 partially used
	pixel size	24μm
	pixel scale	0.277 arc sec/pixel
	field of view	89 arc seconds
	photometry band	U, B, V, R, I
Infrared	detector	SBRC InSb
	format	256 × 256
	pixel size	30μm
	pixel scale	0.346 arc sec/pixel
	field of view	89 arc seconds
	photometry band	Z, J, H, K', K, L'

The WFC is a mosaiced optical CCD camera, which has eight 2k × 4k thick CCDs manufactured by Hamamatsu Photonics Co. The camera has a VRI broad band imaging capability producing an 8k × 8k image with a pixel scale of 0.16 arc seconds. The CCDs are cooled using a GM mechanical cooler and require a large shutter and a field-flattener and corrector. This low cost camera of a very simple design will be used to investigate the mass distributions of clusters of galaxies at $z \sim 0.5$, by observing the gravitational lensing effects produced by the cluster mass distribution on the images of galaxies that lie behind the cluster. By imaging many clusters of galaxies we expect to discover supernovae which can be used as standard candles to measure the distance to the clusters in which they are found.

Table 4. The major characteristics of the MAGNUM WFC.

detector	Hamamatsu thick 2k × 4k CCD
format	8192 × 8192
pixel size	15μm
pixel scale	0.16 arc sec/pixel
field of view	23 × 23 arc minutes
photometry band	V, R, I

5. Automated Operation

We have designed the MAGNUM observatory to be operated unmanned, both automatically and remotely. Most of the MAGNUM observatory systems have a link to the observatory communication network where they are controlled by the MAGNUM scheduler according to a parameter file, which describes all the tasks and related parameters, and the objects with their required observations. We can modify the parameter file by placing a new file over the network from a remote site. The parameter file can describe repeated observations so that the MAGNUM observatory could run forever without modification of the parameter file.

As each observation completes, the scheduler uses the parameter file, taking into account observing history, moon, object magnitudes, cloud cover and so on, to decide which observation to undertake next. This dynamic rescheduling of ground based observations is somewhat challenging as compared to the HST scheduling operation, because the seeing and weather on the Earth change, and do not allow advance prediction of what can be observed.

We intend to minimize the daily maintenance work and schedule major maintenance twice a year, at which time instrument changes may also be made. Routine checks will be made by UH staff when required. Using the network, we can monitor the observatory status and obtain a quick look at the data in real time.

6. Data Acquisition and Automated Reduction

The amount of observational data will easily exceed a few Gbytes for each night of observations. Because it is expensive to transfer such a large amount of data over the network from Maui to Japan, we transfer the observing data and various monitoring data by air mail. A weekly posting of the data tapes is acceptable. Status reports of the observations and fault reports will be transfered via the network.

Automatic data reduction is important for automatic observations. We are building an automated analysis facility for MAGNUM images. The proto-type of the MAGNUM analysis facility is IRAF based, and is already working and being used for reducing data obtained with the PICNIC infrared camera developed by the National Astronomical Observatory, Japan.

We have adopted a CCD controller and an infrared array controller developed by Dr. Leach's CCD labs at San Diego State University. His controllers have the capability to control either an infrared array or a CCD. We use a controller at 1MHz sampling 4 channels for the infrared array, and a second controller with 1 channel for the CCD. Digital data are transferred from the controllers via fiber optics to an instrument control work station. The MAGNUM observations will be made automatically using the MAGNUM scheduling system, with instrument control working under the direction of the scheduler. All the data are written first to hard disks on the instrument computer, then recorded to DATs, with one copy archived in the dome and the other copy sent by mail to Japan.

7. Summary

We have built a new remote controlled observatory with a 2m automated telescope as well as new infrared and optical instruments for the multi-color monitoring of many AGNs over several years, in order to measure AGN luminosity distances.

The telescope has been installed on Haleakala, and engineering first light was obtained in August 2000. In early 2001, we started optical and near infrared monitoring of several AGNs in parallel with the commissioning of the telescope, instrumentation and software. After the air-conditioning system in the dome was installed, and the primary mirror support system was adjusted, the telescope achieved 0.5 arcsec FWHM PSFs which include the effects of seeing, and optics. Figure 6 demonstrates this performance of 0.5 arcsec.

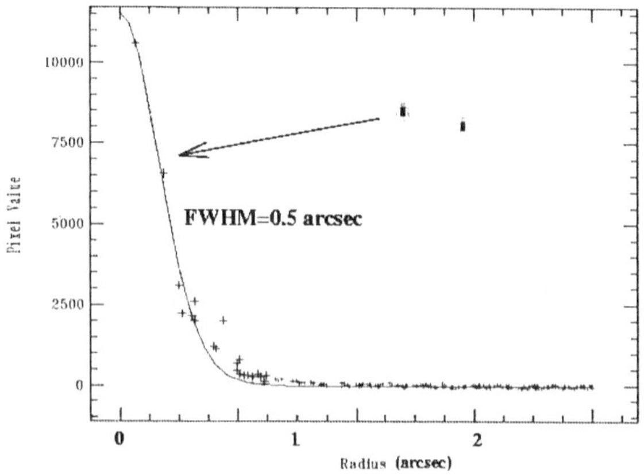

Figure 6. A stellar image taken by the multi-color imaging photometer(MIP) attached to the MAGNUM telescope. The exposure time was 0.365 sec and the filter was $K(2.15\mu\text{m})$. The 0.5 arcsec measured FWHM of the PSF includes the effects of telescope and camera optics, dome and atmospheric seeing.

A multi-color imaging photometer (MIP) is now in operation with an imaging capability over a wide wavelength range from 0.3μm to 4μm. For test observations in the commisioning phase we have monitored several AGNs at least once a week in the V and K bands, and obtained their light curves with much smaller magnitude errors and much shorter time intervals than ever. A cross correlation analysis gives their delay times which well fit the relation of $L \propto (\Delta t)^2$, confirming that the principle described in §1 can actually works (see Figure 2).

In a few years the MAGNUM Project will produce a large database of high precision monitoring observations which can be used to estimate the absolute luminosities of AGNs and quasars independently of their redshifts. By comparing many of estimated luminosity distances with those predicted by models of the expanding Universe, the MAGNUM Project is expected to eventually determine the geometry of the Universe and resolve fundamental questions as to whether the Universe will expand forever, and whether the cosmological constant is necessary.

Y.Y. thanks the MAGNUM personnel including Y. Kobayashi, T. Minezaki, T. Aoki, K. Enya, M. Suganuma, H. Tomita, and B. A. Peterson for their unlimited contribution to the Project, the UH for allowing our use of the Haleakala Observatory site, and the EOS/EOST for manufacturing a telescope of high performance. Y.Y. particularly thanks K. Sato and all other RESCEU members for their continuous support and understanding. This Project has been funded by the Grant-in-Aid for the Center of Excellence (COE) Research (07CE2002) of the Ministry of Education, Science, and Culture in Japan.

References

[1] S. Collin, *L'Active dance les Galaxies* (1987) Paris.
[2] M. J. Rees, J. I. Silk, M. W. Werner, and N. C. Wickramasinghe, Nature 223 (1969) 788
[3] M. J. Lebofski, and G. H. Rieke, Nature 284 (1980) 410
[4] R. Barvainis, Astrophys. J. 320 (1987) 537
[5] R. Barvainis, Astrophys. J. 353 (1990) 419
[6] Y. Kobayashi, S. Sato, T. Yamashita, H. Shiba, and H. Takami, Astrophys. J. 404 (1993) 94
[7] G. Neugebauer, B. T. Soifer, K. Matthews, and J. H. Elias, Astron. J. 97 (1989) 957
[8] J. Clavel, W. Wamstecker, and I. S. Glass, Astrophys. J. 337 (1989) 236
[9] T. Baribaud, D. Alloin, I. Glass, and D. Pelat, Astron. Astrophys. 256 (1992) 375
[10] I. S. Glass, MNRAS 256 (1992) 23p
[11] V. L. Oknyanskii, Astron. Lett. 19 (1993) 416
[12] M. L. Sitko, A. K. Sitko, A. Siemiginowska, and R. Szczebra, Astrophys. J. 409 (1993) 139
[13] B. O. Nelson, Astrophys. J. 465 (1996) L87

Explosive Nucleosynthesis in The First Stars

Ken'ichi Nomoto,[1,2] Keiichi Maeda,[1] Hideyuki Umeda,[1,2] and Marii Shirouzu[1]
[1]*Department of Astronomy, School of Science, The University of Tokyo, 7-3-1 Hongo, Bunkyo-ku, Tokyo 113-0033, Japan*
[2]*Research Center for the Early Universe (RESCEU), School of Science, The University of Tokyo, 7-3-1 Hongo, Bunkyo-ku, Tokyo 113-0033, Japan*

Abstract

We calculate stability, evolution and explosion of massive Pop III stars. We find the upper limit mass of the pulsationally stable Pop III ZAMS stars is about $130M_\odot$, and the mass loss rate of unstable stars may be low. The nucleosynthesis results are compared with abundances of metal-poor halo stars to constrain the IMF of Pop III stars. The interesting trends of the observed ratios [Zn, Co, Mn, Cr/Fe] of the very metal-poor stars can be related to the mass ratio between the complete Si burning region and the incomplete Si burning region. We find that yields of Fe core collapse Hypernovae can be consistent with the abundances of extremely metal-poor stars, if significant amount of complete Si-burning products are mixed out in the ejecta.

1. Introduction

Massive Pop III stars are important since they make the first metal enrichment of the universe, which affects significantly the following chemical and dynamical evolution of galaxies. There are several suggestions that the IMF of Pop III is different from Pop I and II, and the typical mass range of the first stars is likely to be $100 - 200M_\odot$ (e.g., [1][2][3]). If this is the case, the abundance pattern of the early galaxy should be similar to the yield of these stars.

Recent observations have shown that the abundance pattern of halo stars are different for $[Fe/H] < -2$. Specifically, the Co/Fe and Zn/Fe ratios increase significantly for smaller metallicity while Mn/Fe and Cr/Fe decreases ([4][5][6]). These interesting trends challenge the previous supernova models.

Recently, Type Ic supernova (SN Ic) 1998bw, which was associated with GRB980425 ([7][8]), and SN Ic 1997ef ([9][10]) have been found to have such a large kinetic explosion energy as $E \gtrsim 10^{52}$ erg. This is more than one order of magnitude larger than typical SNe, so that these objects may be called "Hy-

Table 1. The results of the stability analysis for Pop III and Pop I stars. ○ and × represent that the star is stable and unstable, respectively. The e-folding time for the fundamental mode is shown after × in units of 10^4yr.

mass ($M_\odot$)	80	100	120	150	180	300
Pop III	○	○	○	× (9.03)	× (4.83)	× (2.15)
Pop I	○	× (7.02)	× (2.35)	× (1.43)	× (1.21)	× (1.71)

pernovae". These SNe produced more ^{56}Ni than the average core collapse SN. Their progenitors' masses are estimated to be $M \gtrsim 25M_\odot$. These massive stars are likely to form black holes, while less massive stars form neutron stars (see, however, [11]).

We investigate the characteristics of nucleosynthesis in such energetic core-collapse hypernovae, the systematic study of which has not yet been done. We examine both spherical and aspherical explosion models and discuss their contributions to the Galactic chemical evolution.

2. Stability and Mass Loss of Massive Pop III stars

To determine the upper limit mass of the Zero Age Main Sequence (ZAMS), we analize a linear non-adiabatic stability of massive ($80M_\odot$ - $300M_\odot$) Pop III stars using a radial pulsation code ([12]). Stellar structures are calculated by the classical fitting method. We adopt e^- scattering and Kramers formula for the opacity, and the main nuclear reactions are pp-chain, hot-CNO cycle, and 3α reaction. Because the CNO elements are absent during the early stage of their evolution, the CNO cycle does not operate and the star contracts until temperature rises sufficiently high for the 3α reaction to produce ^{12}C. We calculate that these stars have $X_{\rm cno} \sim 1.6 - 4.0 \times 10^{-10}$, and the central temperature $T_c \sim 1.4 \times 10^8 K$ on thier ZAMS. We also examine the models of Pop I stars for comparison.

Table 1 shows the results for our analysis. The critical mass of ZAMS Pop III star is $128M_\odot$ while that of Pop I star is $94M_\odot$. This difference comes from very compact structures (with high T_c) of Pop III stars.

Stars more massive than the critical mass will undergo pulsation and mass loss. We note that the e-folding time of instability is much longer for Pop III stars than Pop I stars with the same mass, and thus the mass loss rate is much lower. These results are consistent with [13] and [14]. We thus assume that the mass loss is negligible in Pop III stars in the following calculations.

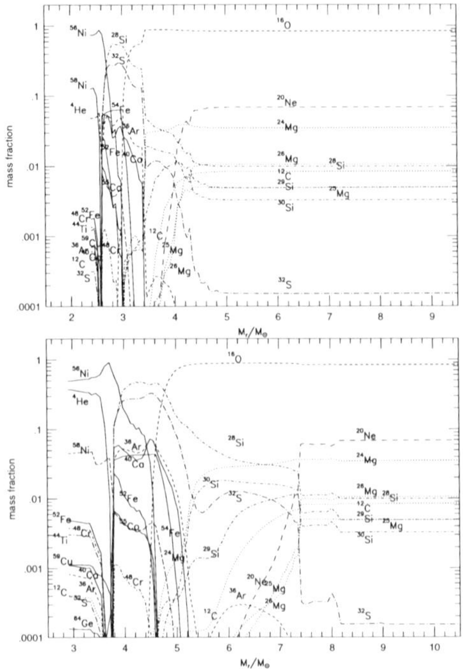

Fig. 1. The isotopic composition of the ejecta with $E = 1 \times 10^{51}$erg (upper panel) and $E = 3 \times 10^{52}$erg (lower panel) for $M = 40M_\odot$ ([29]).

3. Nucleosynthesis in Hypernova Explosions

In core-collapse supernovae/hypernovae, stellar material undergoes shock heating and subsequent explosive nucleosynthesis (Figure. 1). Iron-peak elements are produced in two distinct regions, which are characterized by the peak temperature, $T_{\rm peak}$, of the shocked material. For $T_{\rm peak} > 5 \times 10^9$K, material undergoes complete Si burning whose products include Co, Zn, V, and some Cr after radioactive decays. For $4 \times 10^9{\rm K} < T_{\rm peak} < 5 \times 10^9$K, incomplete Si burning takes place and its after decay products include Cr and Mn (e.g., [15][16][17]).

We note the following characteristics of nucleosynthesis with very large explosion energies ([18][19]):

1) Both complete and incomplete Si-burning regions shift outward in mass compared with normal supernovae, so that the mass ratio between the complete and incomplete Si-burning regions becomes larger. As a result, higher energy explosions tend to produce larger [(Zn, Co)/Fe], small [(Mn, Cr)/Fe], and larger

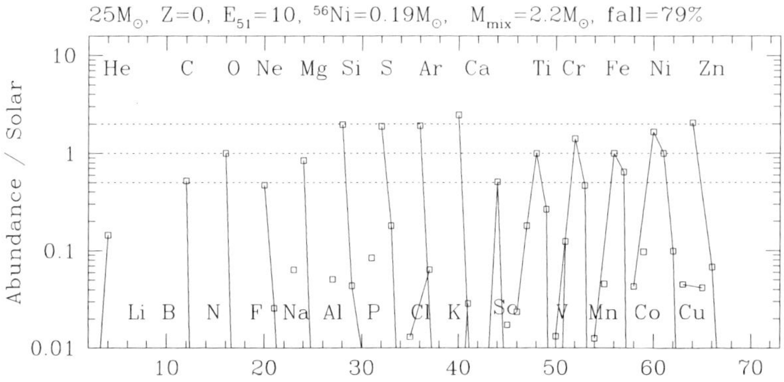

Fig. 2. The abundance pattern for $M = 25M_{\odot}$, $E_{\rm exp} = 10^{52}$erg normalized by the solar abundance. See [30] for details.

[Fe/O]. The elements synthesized in this region such as ^{56}Ni, ^{59}Cu, ^{63}Zn, and ^{64}Ge (which decay into ^{56}Co, ^{59}Co, ^{63}Cu, and ^{64}Zn, respectively) are ejected more abundantly than in normal supernovae.

2) In the complete Si-burning region of hypernovae, elements produced by α-rich freezeout are enhanced because nucleosynthesis proceeds at lower densities (i.e., higher entropy) and thus a larger amount of ^{4}He is left. Hence, elements synthesized through capturing of α-particles, such as ^{44}Ti, ^{48}Cr, and ^{64}Ge (decaying into ^{44}Ca, ^{48}Ti, and ^{64}Zn, respectively) are more abundant.

3) Oxygen burning takes place in more extended, lower density regions for the larger explosion energy. Therefore, more O, C, Al are burned to produce a larger amount of burning products such as Si, S, and Ar. Therefore, hypernova nucleosynthesis is characterized by large abundance ratios of [Si/O], [S/O], [Ti/O], and [Ca/O]. The overall abundance pattern normalized by the solar abundances for the massive energetic SNe II is shown in Figure 2.

4. Signatures of Hypernova Nucleosynthesis in Chemical Evolution

The abundance pattern of metal-poor stars with [Fe/H] < -2 provides us with very important information on the formation, evolution, and explosions of massive stars in the early evolution of the galaxy.

In the early galactic epoch when the galaxy is not yet chemically well-mixed, [Fe/H] may well be determined by mostly a single SN event ([24]). The formation of metal-poor stars is supposed to be driven by a supernova shock, so that [Fe/H] is determined by the ejected Fe mass and the amount of circumstellar hydrogen swept-up by the shock wave ([5]). Then, hypernovae with larger E are likely to induce the formation of stars with smaller [Fe/H], because the mass of interstellar hydrogen swept up by a hypernova is roughly proportional to E ([5][25]) and the ratio of the ejected iron mass to E is smaller for hypernovae than for canonical supernovae.

4.1. Zn, Co, Mn, Cr

The observed abundances of metal-poor halo stars show quite interesting pattern. There are significant differences between the abundance patterns in the iron-peak elements below and above [Fe/H]~ -2.5 - -3.

1) For [Fe/H]$\lesssim -2.5$, the mean values of [Cr/Fe] and [Mn/Fe] decrease toward smaller metallicity, while [Co/Fe] increases ([4][5]).

2) [Zn/Fe]~ 0 for [Fe/H] $\simeq -3$ to 0 ([26]), while at [Fe/H] < -3.3, [Zn/Fe] increases toward smaller metallicity (Figure 3; [6][27]).

These trends cannot be explained with the conventional chemical evolution model that uses previous nucleosynthesis yields.

The larger [(Zn, Co)/Fe] and smaller [(Mn, Cr)/Fe] in the supernova ejecta can be realized if the mass ratio between the complete Si burning region and the incomplete Si burning region is larger, or equivalently if deep material from complete Si-burning region is ejected by mixing or aspherical effects. This can be realized if (1) the mass cut between the ejecta and the collapsed star is located at smaller M_r ([28]), (2) E is larger to move the outer edge of the complete Si burning region to larger M_r ([29]), or (3) asphericity in the explosion is larger.

Also a large explosion energy E results in the enhancement of the local mass fractions of Zn and Co, while Cr and Mn are not enhanced ([30]). Therefore, if hypernovae made significant contributions to the early Galactic chemical evolution, it could explain the large Zn and Co abundances and the small Mn and Cr abundances observed in very metal-poor stars.

The dependence of [Zn/Fe] on M and E is summarized in Figure 3. Models with $E_{51} = E/10^{51}$ergs do not produce sufficiently large [Zn/Fe]. To be compatible with the observations of [Zn/Fe] ~ 0.5, the explosion energy must be much larger, i.e., $E_{51} \gtrsim 20$ for $M \gtrsim 20M_\odot$, i.e., hypernova-like explosions of massive stars ($M \gtrsim 25M_\odot$) with $E_{51} > 10$ are responsible for the production of Zn.

4.2. Ni

The observed trend of another iron-peak element, Ni, is also interesting. Unlike the elements we have focused on, [Ni/Fe] of metal-poor stars shows no clear trend (see e.g., [5][28]). Theoretically, this is understood as the fact that Ni

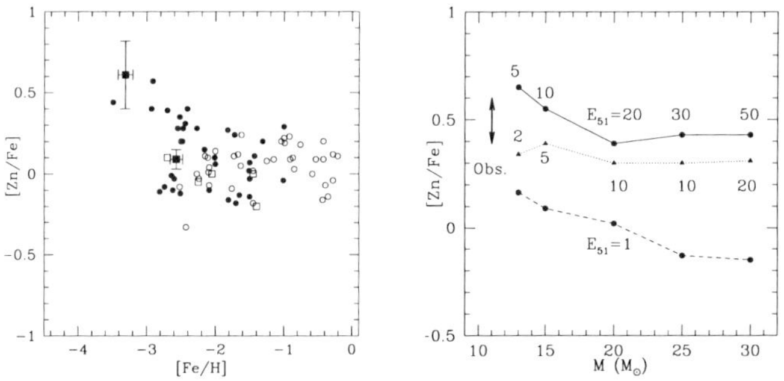

Fig. 3. Left: Observed abundance ratios of [Zn/Fe]. These data are taken from Primas et al. (2000; [6]) (filled circles), Blake et al. (2001: [27]) (filled square) and from Sneden et al. (1991; [26]) (others) ([30]).
Right: The maximum [Zn/Fe] ratios as a function of M and $E_{51} = E/10^{51}$ergs ([30]). The observed large [Zn/Fe] ratio in very low-metal stars ([Fe/H] < -2.6) found in Primas et al. (2000; [6]) and Blake et al. (2001; [27]) are represented by a thick arrow.

is produced abundantly by both complete and incomplete Si-burning. Recently, Ellison et al. (2001; [31]) observed DLA abundance and found [Co/Fe] > 0, which is similar to the metal-poor halo stars. They, on the other hand, did not find oversolar [Ni/Fe]. Similar results have also been found by Norris et al. (2001; [32]), who observed abundances of five halo stars with [Fe/H] $\lesssim -3.5$. They discussed that the results are inconsistent with the predictions of Nakamura et al. (1999; [28]), where the enhancement of [Co/Fe] appears to be accompanied by the enhancement of [Ni/Fe].

We note that the increase in [Ni/Fe] along with the increase in [Co/Fe] is not significant in the hypernova models by Umeda & Nomoto (2002; [30]); for example, Co/Fe in the hypernova is larger by a factor of 17 than a normal SN II, while Ni is larger only by a factor of 1.8.

The apparent difference from the results in Nakamura et al. (1999; [28]) can be understood as follows. In Nakamura et al. (1999; [28]), the explosion energy was fixed to be $E_{51} = 1$. They obtained the larger Co/Fe ratio for more massive SNe II by assuming "deeper" mass-cuts so that Y_e in the explosive burning region is smaller ($Y_e \simeq 0.495$). For smaller Y_e, the Co abundance is larger, but the abundance of Ni (especially ^{58}Ni) is enhanced by a larger factor than Co. Therefore, the increase in [Ni/Fe] with [Co/Fe] was unavoidable, unless neutrinos substantially enhance Y_e in the deep complete Si burning region.

In Umeda & Nomoto models, mass-cuts of the larger Co (and Zn) models are not deeper in M_r and Y_e in the complete Si burning region is not small. This is because, they assume larger explosion energy for more massive stars, which shifts the mass-cut outwards in M_r. As a result, the dominant Ni isotope is ^{60}Ni. In our model, with increasing E, the abundance of Co, Zn, and ^{58}Ni increase more than ^{60}Ni. Therefore, Co and Zn abundances can be enhanced without appreciable increase in the Ni abundance. In this sense, the abundance trends of very metal-poor stars is better explained with hypernova models rather than the simple "deep" mass-cut models ([28]).

4.3. Pair Instability Supernovae ?

It is quite likely that the IMF of Pop III stars is different from that of Pop I and II stars, and that more massive stars are abundant for Pop III (e.g., [1]). If $M \lesssim 130M_\odot$, then these stars are likely to form black holes either without explosion or with energetic explosions. Nucleosynthesis of the latter case may not be so different from the models considered here. This might favor the scenario that invokes the hypernova-like explosions for large [Zn/Fe].

Stars as massive as $\sim 130 - 300M_\odot$ enter into the $e^\pm$ pair instability region during the central oxygen burning stage. When these stars enter this region, the core contracts rapidly and temperature rises quickly. This leads to explosive oxgen burning, and the thermal energy produced by this rapid oxygen burning exceeds the gravitational binding energy. Hence the stars disrupt completely. These SNe are called Pair Instability SNe (PISNe) ([33]). During the explosion, explosive burning takes place to produce ^{56}Ni and other heavy elements. Their nucleosynthesis is different from core-collapse SNe. In particular, PISNe produce [Zn/Fe] < -1.5, because in PISNe, iron peak elements are mostly produced by incomplete Si burning so that the mass fraction of complete Si burning elements is much smaller than SNe II ([30]). We thus conclude that PISNe are unlikely to produce a large enough Zn/Fe ratio to explain the observations.

A star of $300M_\odot \lesssim M \lesssim 10^4 M_\odot$ also enters into the $e^\pm$ pair instability region during or before the central oxygen burning stage. If the star is not rotating sufficiently fast, it collapses into a black hole with no heavy element ejection. However, if the rotation is fast enough, an accretion disk would be formed around the central black hole. In this case, high energy jets along the rotational axis may escape deep gravitational potential (e.g., [34]). It is expected such jets induce heavy element synthesis, although no detailed calcations have been performed yet.

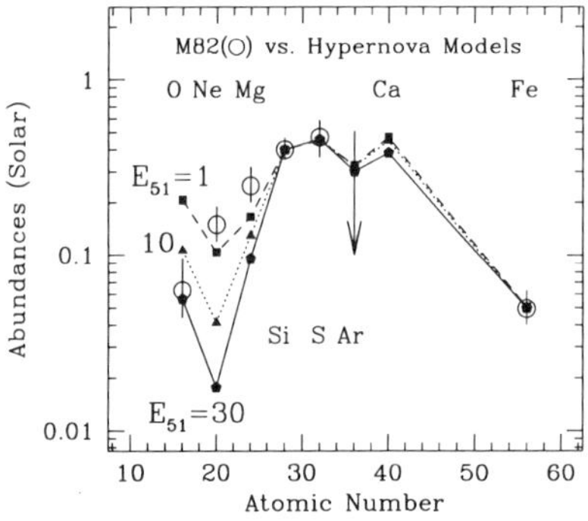

Fig. 4. Abundance patterns in the ejecta of $25M_\odot$ metal-free SN II and hypernova models compared with abundances (relative to the solar values) of M82 observed with ASCA (open circles: [35]). The filled square, triangle, and pentagons represent E_{51}=1, 10, and 30 models, respectively.

5. Starburst Galaxy M82 and Hypernovae

X-ray emissions from the starburst galaxy M82 were observed with ASCA and the abundances of several heavy elements were obtained [35]. Tsuru et al. [35] found that the overall metallicity of M82 is quite low, i.e., O/H and Fe/H are only 0.06 - 0.05 times solar, while Si/H and S/H are $\sim$ 0.40 - 0.47 times solar. This implies that the abundance ratios are peculiar, i.e., the ratio O/Fe is about solar, while the ratios of Si and S relative to O and Fe are as high as $\sim$ 6 - 8. These ratios are very different from those ratios in SNe II. Compared with normal SNe II, the important characteristic of hypernova nucleosynthesis is the large Si/O, S/O, and Fe/O ratios. Figure 4 shows the good agreement between the hypernova model ($E_{51} = 30$) and the observed abundances in M82.

Hypernovae could also produce larger E per oxygen mass than normal SNe II, as required for M82. We therefore suggest that hypernova explosions may make important contributions to the metal enrichment and energy input to the interstellar matter in M82. The age of starburst activity is estimated to be $\lesssim 10^7$ years [36], which is so young that only massive stars ($M > 25\ M_\odot$) contributed to nucleosynthesis in M82.

6. Hypernovae vs. Type Ia Supernovae

The large [Fe/O] observed in some metal-poor stars and galaxies might possibly be the indication of Hypernovae rather than Type Ia supernovae (SNe

Ia).

To distinguish between Hypernova ([29][30]) and SN Ia ([37][38]) nucleosynthesis, following abundance ratios are useful:

[Ti/Fe] $\sim$ 0.3 - 0.5 in Hypernovae; $<$ 0 in SNe Ia.
[Mn/Fe] $<$ 0 in Hypernovae; $>$ 0 in SNe Ia.
[Ni/Fe] $\sim$ 0 - 0.3 in Hypernovae; $\sim$ 0 - 0.5 in SNe Ia.
[Zn/Fe] $\sim$ 0 - 0.5 in Hypernovae; $<$ 0 in SNe Ia.
[Si/Fe] $\sim$ 0.3 - 0.5 in Hypernovae; ~ -0.5 - 0 in SNe Ia.

To reproduce the solar abundance ratios with the combination of SN II and SNe I products, there exist certain constraints on the model abundances of SNe Ia. For example, [Si/Fe] $\sim$ 0.3 in SNe II ([39]) so that [Si/Fe] < -0.2 for SNe Ia ([37]). For a system with a non-solar abundance pattern, this constraint would be weaker.

7. Concluding Remarks

We have shown that signatures of hypernova nucleosynthesis are seen in the large [(Ti, Zn)/Fe] ratios in very metal poor stars and the large Si/O and Fe/O ratios in the starburst galaxy M82. (See also the abundance pattern in X-ray Nova Sco; [40][41]). These properties of hypernova nucleosynthesis suggest that hypernovae of massive stars may make important contributions to the Galactic (and cosmic) chemical evolution, especially in the early low metallicity phase. This may be consistent with the suggestion that the IMF of Pop III stars is different from that of Pop I and II stars, and that more massive stars are abundant for Pop III.

Acknowledgments

This work has been supported in part by the grant-in-Aid for Scientific Research (07CE2002, 12640233) of the Ministry of Education, Science, Culture, and Sports in Japan.

References

[1] Nakamura, F., & Umemura, M. 1999, ApJ, 515, 239
[2] Ferrara, A. 2000, in ASP Conf. Ser. 222, *The Physics of Galaxy Formation*, ed. M. Umemura & H. Susa, (ASP), 91

[3] Coppi, P. et al. 2000, in ASP Conf. Ser. 222, *The Physics of Galaxy Formation*, ed. M. Umemura & H. Susa, (ASP), 15
[4] McWilliam, A., Preston, G.W., Sneden, C., & Searle, L. 1995, AJ, 109, 2757
[5] Ryan, S.G., Norris, J.E. & Beers, T.C. 1996, ApJ, 471, 254
[6] Primas, F., Reimers, D., Wisotzki, L., Reetz, J., Gehren, T., & Beers, T.C. 2000, in *The First Stars*, ed. A. Weiss, et al. (Springer), 51
[7] Iwamoto, K., Mazzali, P.A., Nomoto, K., et al. 1998, Nature, 395, 672
[8] Woosley, S.E., Eastman, R.G., & Schmidt, B.P. 1999, ApJ, 516, 788
[9] Iwamoto, K., Nakamura, T., Nomoto, K., et al. 2000, ApJ, 534, 660
[10] Mazzali, P.A., Iwamoto, K., & Nomoto, K. 2000, ApJ, 545, 407
[11] Wheeler, J. C., Yi, I., Höflich, P. A., & Wang, L. 2000, ApJ, 537, 810
[12] Shibahashi,H. (2000) private communication
[13] Ibrahim, A., Boury, A., & Noels A. 1981, A&A, 103, 390
[14] Baraffe, I., Heger, A., & Woosley, S. E. 2001, ApJ, 550, 890
[15] Hashimoto, M., Nomoto, K., & Shigeyama, T. 1989, A&A, 210, L5
[16] Woosley, S. E., & Weaver, T. A. 1995, ApJS, 101, 181
[17] Thielemann, F.-K., Nomoto, K., & Hashimoto, M. 1996, ApJ, 460, 408
[18] Nomoto, K., et al. 2001a, in *Supernovae and Gamma Ray Bursts*, eds. M. Livio, et al. (Cambridge Univ. Press) 144 (astro-ph/0003077)
[19] Nomoto, K., Maeda, K., Umeda, H., & Nakamura, T. 2001b, in *The Influence of Binaries on Stellar Populations Studies*, ed. D. Vanbeveren (Kluwer) 507 (astro-ph/0105127)
[20] Nakamura, T., Mazzali, P. A., Nomoto, K., & Iwamoto, K. 2001a, ApJ, 550, 991
[21] Mazzali, P.A., Nomoto, K., Patat, F., & Maeda, K. 2001, ApJ, 559, 1047
[22] Maeda, K., Nakamura, T., Nomoto, K., Mazzali, P.A., & Hachisu, I. 2002, ApJ, 565, 405
[23] Patat, F., et al. 2001, ApJ, 555, 900
[24] Audouze, J., & Silk, J. 1995, ApJ, 451, L49
[25] Shigeyama. T., & Tsujimoto, T. 1998, ApJ, 507, L135
[26] Sneden, C., Gratton, R.G., & Crocker, D.A. 1991, A&A, 246, 354
[27] Blake, L.A.J., Ryan, S.G., Norris, J.E., & Beers, T.C. 2001, Nucl.Phys.A.
[28] Nakamura, T., Umeda, H., Nomoto, K., Thielemann, F.-K., & Burrows, A. 1999, ApJ, 517, 193
[29] Nakamura, T., Umeda, H., Iwamoto, K., Nomoto, K., Hashimoto, M., Hix, R.W., Thielemann, F.-K. 2001b, ApJ, 555, 880
[30] Umeda, H., & Nomoto, K. 2002, ApJ, 565, 385
[31] Ellison, S. L., Ryan, S. G., & Prochaska, J. X. 2001, MNRAS, in press (astro-ph/0104466)

[32] Norris, J.E., Ryan S.G., & Beers, T.C. 2001, ApJ, 561, 1034
[33] Barkat, Z, Rakavy, G, & Sack, N. 1967, Phys.Rev.Lett., 18, 379
[34] Fryer, C. L., Woosley, S. E., & Heger, A. 2001, ApJ, 550, 372
[35] Tsuru, T. G., Awaki, H., Koyama K., Ptak, A. 1997, PASJ, 49, 619
[36] Stickland, D. 2001, in *Chemical Enrichment of Intracluster and Intergalactic Medium*, ASP Conference Series, in press (astrp-ph/0107116)
[37] Iwamoto, K., Brachwitz, F., Nomoto, K., et al. 1999, ApJS, 125, 439
[38] Nomoto, K., et al. 1997b, Nucl. Phys. A621, 467c
[39] Nomoto, K., et al. 1997a, Nucl. Phys. A616, 79c
[40] Israelian, G., Rebolo, R., Basri, G., Casares, J., & Martin, E.L., 1999, Nat, 401, 142
[41] Podsiadlowski, Ph., Nomoto, K., Maeda, K., Nakamura, T., Mazzali, P.A., & Schmidt, B. 2002, ApJ, 567, 491

Environmental Variation of Photometric Properties of Galaxies along the Large Scale Structure around a z=0.4 Cluster

T. Kodama[1], Ian Smail[2], F. Nakata[1], S. Okamura[1], R. G. Bower[2]
[1]*Dept of Astronomy, Univ of Tokyo, Bunkyo-ku, Tokyo 113-0033, Japan*
[2]*Dept of Physics, Univ of Durham, South Road, Durham DH1 3LE, UK*

Abstract

We present deep, panoramic multi-color imaging of the distant rich cluster A 851 ($z = 0.41$) using Suprime-Cam on Subaru. These images cover a 27′ field of view ($\sim 11 h_{50}^{-1}$ Mpc), and by exploiting photometric redshifts, we can isolate galaxies in a narrow redshift slice at the cluster redshift. Using a sample of ~ 2700 probable cluster members ($< M_V^* + 4$), we trace the network of filaments and subclumps around the cluster core. The depth of our observations, combined with the identification of filamentary structure, gives us an unprecedented opportunity to test the influence of the environment on the properties of low luminosity galaxies. We find an abrupt change in the colors of faint galaxies ($> M_V^* + 1$) at a local density of 100 gal. Mpc^{-2}. The transition in the color-local density behavior occurs at densities corresponding to subclumps within the filaments surrounding the cluster. Identifying the sites where the transition occurs brings us much closer to understanding the mechanisms which are responsible for establishing the present-day relationship between environment and galaxy characteristics.

1. Introduction

Clusters of galaxies are continuously growing through the accretion of galaxies and groups from the field. The star formation activity in the accreting galaxies must be quenched during the assimilation of the galaxies into the cluster. This transformation is a key process in creating the environmental dependence of galaxy properties, and may also underpin the observed evolution of galaxy properties in distant clusters [e.g. 1.2] However, the physical mechanism which is responsible for these changes has not yet been identified [3,4,5].

The advent of Suprime-Cam, a revolutionary wide-field camera on the Subaru telescope, has opened a new window in this field. Its 27′ field of view can trace the variation of galaxy properties from the cluster cores out to the surrounding field in an attempt to identify the environment where the decline in the star formation in accreted galaxies begins.

2. Large scale structure

As a first step towards a systematic study of distant clusters with Subaru and Suprime-Cam, we obtained deep ($\sim M^* + 4$) $BVRI$ imaging of the rich cluster A 851 at z=0.4. We constructed an I-band selected sample which contains 15,055 galaxies brighter than $I = 23.4$.

In order to assign cluster membership across our large field and to faint magnitudes, we applied photometric redshift technique [6] to isolate the galaxies in a narrow redshift slice around the cluster redshift. By applying the redshift cut of $0.32 < z_{\rm phot} < 0.48$, we can reduce the field contamination by factor 10 at $I = 23.4$, while keeping most of the cluster members (80%, from the estimate using the spectroscopic members [7]).

After this photometric redshift selection, it is straightforward to map out the distribution of galaxies which are likely to be associated with the cluster. As shown in Fig. 1, several large scale structures are visible around the cluster core, comprising many clumps and filamentary extensions coming out directly from the core. Importantly, most of these extensions from the core are aligned with the surrounding subclumps.

The structures identified in this region are qualitatively similar to those seen in cosmological simulations of the growth of clusters which exhibit the filamentary/clumpy substructures on similar scales [e.g. 8] It appears therefore that we are witnessing A 851 as it assembles through the accretion of galaxies and groups along the filaments onto the cluster core from the surrounding field.

2.1. *Environmental dependence of galaxy properties*

By exploiting the striking large scale structure around this cluster, we can investigate the influence of environment on the photometric properties of galaxies. We define the environment for each galaxy using the local surface number density, Σ, of members, calculated from the 10 nearest neighbors. We statistically subtract residual field contamination using a Subaru blank field data [9].

We divide the density distribution into three regimes as indicated in Fig. 2. There is a close correspondence between the local density and structure: the high density region corresponds to the cluster core within $\lesssim 1$ Mpc; the medium density region includes the structures defining the filaments surrounding the cluster; and the low density region comprising the rest of the volume (Fig. 1).

As shown in Fig. 2, the color distribution in the high density region is strongly peaked at $(V - I)_c \sim 2$, the color of an early-type cluster member.

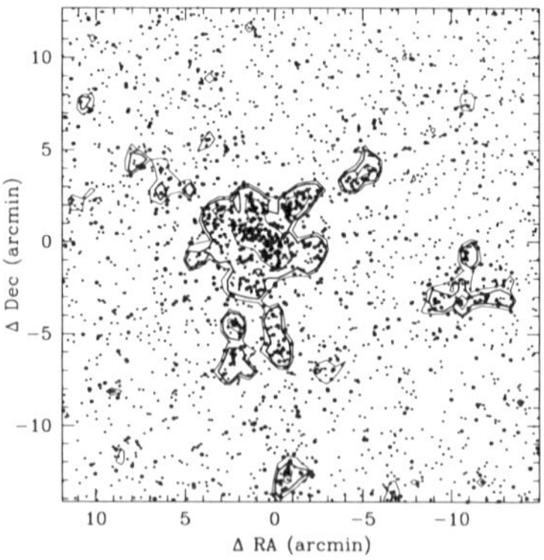

Fig. 1. A wide-field map of A 851 from our Suprime-Cam *BVRI* imaging. The field is 11 Mpc on a side. The points identify cluster members based on photometric redshifts. The remaining field contamination is not subtracted. The large and small circles separate the colors of galaxies at the color 0.5 mag bluer than the color-magnitude sequence in $B - I$, where large ones are in red. The contours indicate the local surface density of the member galaxies, corresponding to $\log_{10}\Sigma = 1.8$, 2.0 (thick) and 2.5 Mpc^{-2} (after correcting for field contamination), respectively. The thick contour traces the boundary where the color distribution of galaxies changes dramatically. Large scale structure is clearly visible. The overall shape resembles an 'octopus', with a round head and many legs.

However, as we move to lower densities, the distribution becomes dramatically bluer. Notably, this color transition with local density occurs quite abruptly at $\log_{10}\Sigma \sim 2$, indicating a threshold effect in transforming galaxy properties. The boundary corresponding to this critical density is highlighted in Fig. 1. Surprisingly, this transition occurs in subclumps well outside the core. It is important to note that the changes in galaxy properties as a function of local density are most prominent in faint galaxies ($> M_V^* + 1$).

Our ability to pin-point this environment is a fundamental step towards identifying the dominant mechanism behind the environmental dependence of galaxy properties. Since the ram-pressure stripping is largely supressed in groups [4], the effective mechanism at this hierarchy is either galaxy–galaxy collisions (which cause cold disk gas to be driven to the galaxy center creating a star burst [3]); and 'suffocation' (where *warm* gas in the galaxy's halo is shock heated by the intra-cluster medium so that it can no longer cool and replenish the cold gas in the disk [10]). Our data do not distinguish directly between these possibilities, but they can be distinguished by tracing the variation in the individual components of galaxies, bulges and disks, between the cluster and the field.

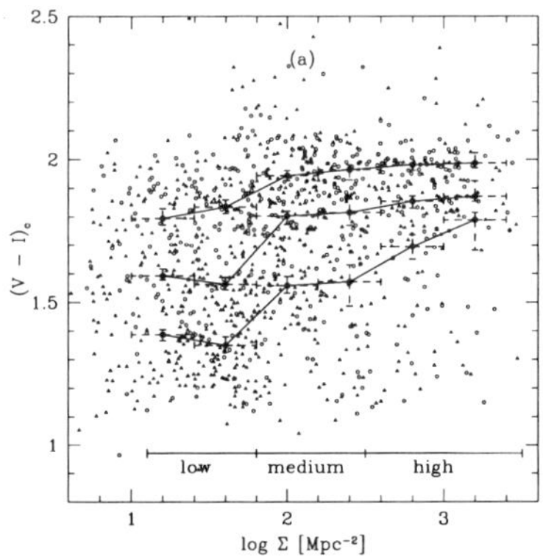

Fig. 2. The variation in color versus local galaxy density, Σ, for cluster members brighter than $I = 23.4$. The open circles and filled triangles show the galaxies brighter or fainter than $I = 21.4$ (M_V^*+2), respectively. The three red lines represent the loci of the 25, 50, and 75th percentile colors. The colors are corrected for the slope of the color-magnitude relation. We also correct for the residual field contamination using the blank field data.

Acknowledgments: We are grateful to Dr Y. Komiyama for his assistance during our observations. We acknowledge the Suprime-Cam team for allowing us to use the blank field data. TK and FN acknowledge the Japan Society for the Promotion of Science for support through its Research Fellowships for Young Scientists. IRS acknowledges support from the Royal Society and the Leverhulme Trust. We acknowledge the Daiwa Anglo-Japanese Foundations for the Daiwa-Adrian Prize 2001.

References

[1] H. Butcher, A. Oemler, 1984, ApJ, 285, 426.
[2] T. Kodama, R. G., Bower, 2001, MNRAS, 285, 426.
[3] B. Moore, et al. 1996, Nature, 379, 613.
[4] M. G. Adabi, R. G. Bower, J. F. Navarro, 2000, MNRAS, 314, 759.
[5] M. L. Balogh, J. F. Navarro, S. L. Morris, 2000, ApJ, 540, 113.
[6] T. Kodama, E. F. Bell, R. G. Bower, 1999, MNRAS, 302, 152.
[7] A. Dressler, et al., 1999, ApJS, 122, 51.
[8] S. Ghigna, et al., 1998, MNRAS, 300, 146.
[9] M. Ouchi, et al., 2001, ApJ, 558, L83.
[10] R. B. Larson, B. M. Tinsley, C. N. Caldwell, 1980, ApJ, 237, 692.

Determining the Supernova Direction by its Neutrinos

Shin'ichiro ANDO* and Katsuhiko SATO*,**
*Department of Physics, School of Science, The University of Tokyo, 7-3-1 Hongo, Bunkyo-ku, Tokyo 113-0033, Japan
**Research Center for the Early Universe (RESCEU), School of Science, The University of Tokyo, 7-3-1 Hongo, Bunkyo-ku, Tokyo 113-0033, Japan

Abstract

We simulate supernova neutrino events at SuperKamiokande detector, using a realistic supernova model and several realistic neutrino oscillation models. As a result of our simulation we can restrict the supernova direction to be within a circle of radius 7°. Especially in several neutrino oscillation models, this accuracy is enhanced to 6°.

1. Introduction

A future core-collapse supernova explosion in our Galaxy is expected to be detected by several neutrino detectors around the world. When it occured, neutrinos produced in the core can escape from the supernova immediately, because of their very weak interaction with matter. On the other hand, photons do not get out until the shock wave travels from the core through the stellar envelope and breaks out of the stellar photosphere. So we can catch the neutrino signal several hours earlier than the light. So, if we could determine the direction of the supernova explosion by its neutrinos, a lot of astronomical observations of its early state would be possible.

We simulate a supernova explosion in the Galactic plane ($D = 10$ kpc) and discuss the statistical error of its direction. In addition, we also consider the more realistic case, or "neutrino oscillation", which is supported by solar and atmospheric neutrino data.

2. Methods

We assume that the supernova explosion occured at $D = 10$ kpc. To get the information about the direction of the supernova, we have to set the coordinate at SK. Here, we set z-axis to the upward-direction. And we use spherical coordinate

(θ, ϕ) to determine the direction

We also assume here that the direction of the supernova is $(\theta, \phi) = (135°, 270°)$. In this case, the peak position should be $(\theta, \phi) = (45°, 90°)$ according to the elastic scattering of electrons and neutrinos. (Of course when we analyze the events, we must not use this peak information.)

First, we consider the case of "no oscillation".

To analyze the event distribution systematically, we divide $\cos\theta$ direction into 20 bins and sum up in each bin. Then, we rotate the coordinate for the events to be seen as in the case of $\theta = 0°$ peaked by using the least-square method. This result is obtained at coordinate rotation angle $(\theta, \phi) = (47.7°, 80.0°)$.

This result is near the "true" value $(45°, 90°)$, but this only one simulation is not sufficient to estimate the errors. So, we simulated 1,000 times under the same condition and obtained the distribution of the best-fitted coordinate rotation angle. We fitted these results to Gaussian. These one sigma errors are $(\delta\theta, \delta\phi) = (6.8°, 10.0°)$. The one sigma error in two-dimensions is $\sigma = (\delta\theta \sin\theta_0 \delta\phi)^{1/2} = 7.0°$. Then in this method we can determine the supernova direction with $\sigma = 7.0°$.

3. Results

We adopted the same simulations to the other four "neutriono oscillation" cases. We summerize these results in Table 1.

Table 1. Best-fitted angles of coordinate rotation and these errors (°)

model	LMA-L	LMA-S	SMA-L	SMA-S	no osc
θ_0	45.8	46.1	45.5	45.8	46.3
ϕ_0	89.7	89.9	90.4	90.0	90.0
$\delta\theta$	5.7	8.0	6.0	7.0	6.8
$\delta\phi$	7.9	11.0	7.8	9.1	10.0
σ	5.7	8.0	5.9	6.7	7.0

For details, see Ref. [1].

References

[1] S. Ando and K. Sato, hep-ph/0110187.

Astrometric Microlensing due to an Extrasolar Planet

Hideki ASADA
Faculty of Science and Technology, Hirosaki University, Hirosaki 036-8561, Japan

Abstract

In expectation of dramatic improvements in the precision of the future astrometric measurements, we develop a method of analytically investigating the astrometric microlensing due to an extrasolar planet. We have shown that the distortion of an ellipse of the light centroid shifts is dependent on the mass ratio and separation between the star and the planet. The maximum angular size of the distortion is about a micro arcsec. for the Jupiter mass planet. A time scale of the distortion is of the order of months, depending on the projected separation and the transverse velocity.

1. Introduction

Extrasolar planets searches are successfully going on [6]. Astrometry is considered as a supplementary method to find out planets by measuring the transverse motion of a star. Indeed, we can expect dramatic improvements in the precision of the future astrometric measurements by FAME, SIM, DIVA and GAIA. Their precision will achieve a few micro arcseconds.

Also by these future astrometry missions, the astrometric microlensing will become the third method for searching extrasolar planets: The typical scale in the gravitational lensing is the Einstein ring radius, which is of the order of AU for a lensing star within our galaxy. Hence, the microlensing is quite effective even for earth-type planets [5, 2]. In the planetary lens case, we can determine its true mass through observation of spikes in the light curve, which are produced by the magnification effect of the gravitational lensing. On the other hand, the astrometric microlensing is a consequence of a combination of the magnification and the position shift of the image. The numerical results [7, 3] have shown that the photo centroid shifts provide us a clue for extrasolar planets, though their study is limited only for some values of parameters. Therefore, it is a main purpose of this work to derive the analytic formulae and clarify the parameter dependence [1].

2. Conclusion

With respect to the unlensed position, the photo center of the images due to the gravitational lensing moves on an ellipse, when the stellar motion in our galaxy is approximated over several decades by a straight line [8, 4].

In the planetary lens case, we find a correction to the photo centroid as [1]

$$\delta\boldsymbol{\theta}_C = \frac{\mu}{\beta^6}\Big(-2\boldsymbol{\epsilon}(\boldsymbol{\beta}\cdot\boldsymbol{\epsilon})\beta^2 - \boldsymbol{\beta}\epsilon^2\beta^2 + 4\boldsymbol{\beta}(\boldsymbol{\beta}\cdot\boldsymbol{\epsilon})^2 + O(\beta^3)\Big) + O(\mu^2), \tag{1}$$

where μ and $\boldsymbol{\epsilon}$ are the mass ratio and the angular separation vector between the star and the planet, and $\boldsymbol{\beta}$ is the impact parameter of the light ray.

For stellar cases in our galaxy, the maximum angular size of the distortion is estimated as

$$\frac{\mu\theta_E}{\beta} \sim \text{micro arcsec.}\Big(\frac{\mu}{10^{-3}}\Big)\Big(\frac{1}{\beta}\Big)\Big(\frac{\theta_E}{\text{mas}}\Big), \tag{2}$$

where θ_E is the Einstein ring radius angle. The maximal distortion occurs at

$$t_C \sim 10^6\text{s}\Big(\frac{\epsilon_{\parallel}}{\text{AU}}\Big)\Big(\frac{100\text{km/s}}{v_{\perp}}\Big), \tag{3}$$

namely about a few months before/after, depending on a location of the planet, the source passes the point closest to the lensing star. Here, $\epsilon_{\parallel}$ is the projection of $\boldsymbol{\epsilon}$ onto the source trajectory, and $v_{\perp}$ is the transverse relative velocity between the lens and the source.

References

[1] H. Asada, submitted to ApJ (2001).
[2] A. Gould and A. Loeb, ApJ 396 (1992) 104.
[3] C. Han and C. Lee, MNRAS (2001) *in press* (astro-ph/0108157).
[4] Y. Jeong, C. Han and S. Park, ApJ 511 (1999) 569.
[5] S. Mao and B. Paczynski, ApJ 374 (1991) 37L.
[6] G. W. Marcy and R. P. Butler ARA&A, 36 (1998) 57.
[7] N. Safizadeh, N. Dalal and K. Griest, ApJ 522 (1999) 512.
[8] M. A. Walker, ApJ 453 (1995) 37.

From Theoretical to Observational Via Selection Effects

Paul Bristow
ST-ECF, ESO, Karl-Schwarzschild Str. 2, D-85748, Garching, Germany

Abstract

A suite of software tools for modeling the catalogues and statistics that may be expected to be observed for a given model cosmology and galaxy distribution is presented. The aim is to make accurate predictions for the model that take into account selection effects, crowding, data processing and detector limitations via complete simulation of the detection process. Hence we may hope to determine with more certainty if a given catalogue really does rule out a given model.

1. Introduction

The most obvious measurements that we can make of the universe at large are simply obtained by counting objects on the sky. However, as we have become able to make ever deeper and increasingly sophisticated surveys, we have also become aware that the selection of objects for inclusion in a given catalogue needs very careful attention. Moreover, the selection criteria can effect statistical results in subtle ways as well as obvious ones eg. Disney &Phillipps (1983).

It is now standard practice to correct for selection effects when deriving galaxy distribution statistics. However, such is the complexity of the effects when extracting low signal to noise objects from crowded fields in data manipulated with advanced resolution enhancement techniques such as drizzling (Fruchter & Hook 2001), in modern ultra deep surveys, that it would seem prudent to investigate these effects empirically. Many authors (eg. Chokshi et al 1994) have developed software which creates an "artificial sky" which contains a set of objects with known intrinsic properties. The selection effects incurred by running a detection algorithm can then be quantified by comparing the results with the input.

2. The Numerical Universe Software

I present here briefly one such implementation of this idea (see Bristow 1996).

1. A volume (defined by Ω, Λ etc.) of space is populated, Monte Carlo fashion, with galaxies following an input LF and an evolutionary scheme.

2. All galaxies are assigned spectra (SEDs), brightness profiles and a range of rest frame central brightnesses appropriate to their types and redshifts.

3. The K/extinction corrections are computed from the SED, the filter response function and extinction as a function of redshift and wavelength.

4. The corresponding apparent surface brightness profiles in the required band pass and the specified evolutionary model are calculated.

5. Each galaxy is assigned a position on the sky (no clustering), an inclination, position angle and a simple two-dimensional structure.

6. The galaxy images are mapped onto a 2D array (definable CCD)

7. A stellar LF, Sky noise, readout noise and gain are incorporated appropriate to a survey and filter/detector combination of interest.

8. The frame is created several more times with all galaxy positions offset (i.e. dithered) The simulated frames are then "drizzled" (STSDAS) together.

9. SExtractor (Bertin & Arnouts, 1996) is used to extract and measure faint images (with parameters set to match a real catalogue of interest).

10. The properties of the extracted objects are compared with the known intrinsic properties (known, that is, from steps 1 to 3 above). In this way it is not only possible to derive the $n(m)$ and $n(z)$, but also to gain an insight into the way the observing process has affected the underlying distribution.

The same artificial sky can be "observed" with various filters, in this way colours, spectroscopic redshifts etc. can be added to the catalogues.

3. Current work and Summary.

The detectability of an E/S0 progenitor population in deep field data is currently being investigated in this way, however, the early results can't be discussed properly in the space available.In summary, this approach correctly deals with selection and measurement uncertainties that are difficult to account for analytically.

References

[1] E. Bertin and S. Arnouts A&AS 117 (1996) 393.
[2] P. Bristow, *Ph.D. Thesis.* (1996). University of Wales, College of Cardiff
[3] M. Disney and S. Phillipps MNRAS 205 (1983) 1253.
[4] A. Fruchter and R. Hook PASP Submitted (2001).

Light Curve Models for the Hypernova SN 1999as

Jinsong Deng, Kazuhito Hatano, Hideyuki Umeda, and Ken'ich Nomoto
Research Center for the Early Universe (RESCEU), School of Science, The University of Tokyo, 7-3-1 Hongo, Bunkyo-ku, Tokyo 113-0033, Japan

Abstract

We present light curve models for SN 1999as, the brightest supernova ever discovered. About $4M_\odot$ ^{56}Ni is needed to explain its brilliance and hence the explosion is hyper-energetic. We make parameter studies to constrain the ejecta mass, M_{ej}, and explosion energy, E_K. A realistic C+O progenitor is exploded to match our best fit parameters, i.e. $M_{ej} \sim 20M_\odot$ and $E_K \sim 3 \times 10^{52}$ergs.

1. Introduction

SN 1999as was discovered on 1999 February 18 UT by Supernova Cosmology Project. Its peak absolute visual magnitude of ~ -21.5 makes it the most luminous supernova ever recorded, 9 times brighter than SN 1998bw (see Figure 1). Spectroscopically, it can be classified as a type Ic supernova (see Figure 2).

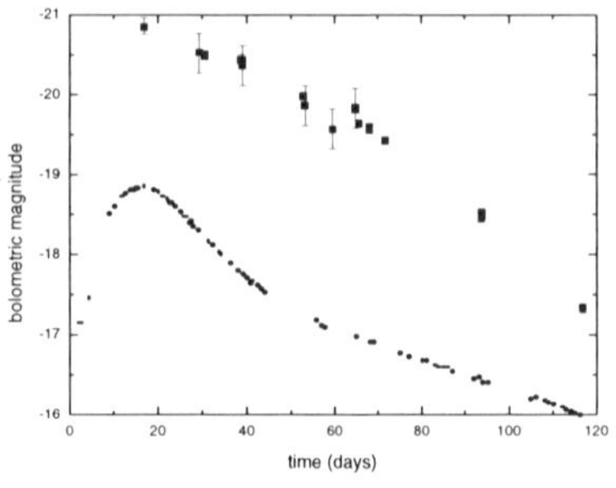

Fig. 1. Bolometric light curve of SN 1999as (squares) compared with that of SN 1998bw (circles).

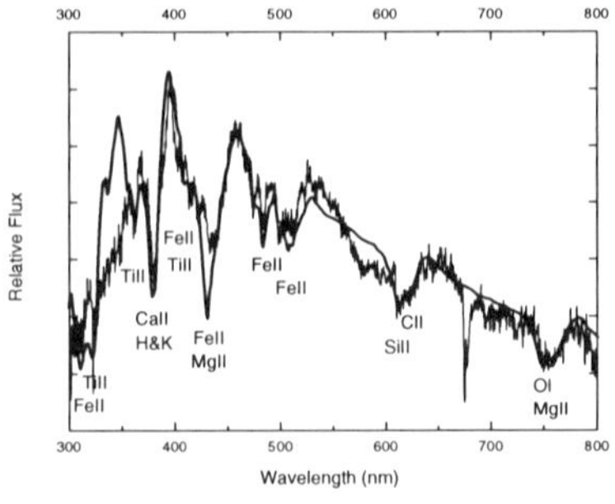

Fig. 2. March 16 spectrum of SN 1999as (thin line) compared with a SYNOW synthetic one (thick line).

2. A Hypernova Scenario

SN 1999as is a hypernova, i.e. the hyper-energetic explosion of a massive star with explosion energy of $E > 10^{52}$ergs. The amount of ^{56}Ni required to power its light curve is about 4 $M_\odot$. In a spherical explosion induced by core-collapse, ^{56}Ni is produced roughly inside a radius, $R_{\rm Ni} \approx 3700{\rm km} \cdot (E/10^{51}{\rm ergs})^{1/3}$, of the progenitor [1]. Thus as shown in Figure 3, to have $\sim 4M_\odot$ ^{56}Ni ejected, explosion energy as high as $\sim 3 \times 10^{53}$ergs is expected.

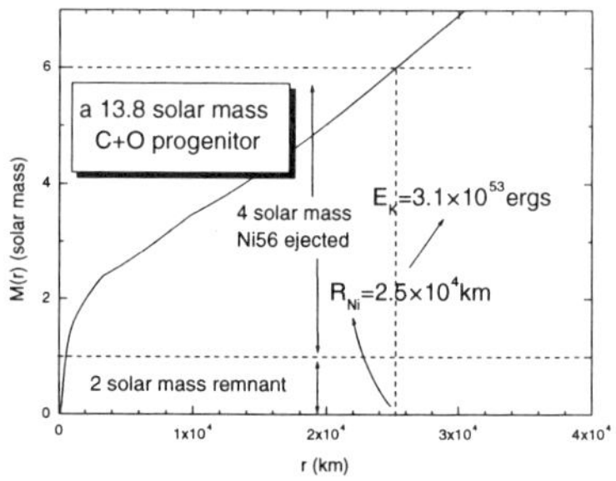

Fig. 3. $E_{\rm K}$ required to eject 4 $M_\odot$ ^{56}Ni for a 13.8 $M_\odot$ C+O progenitor.

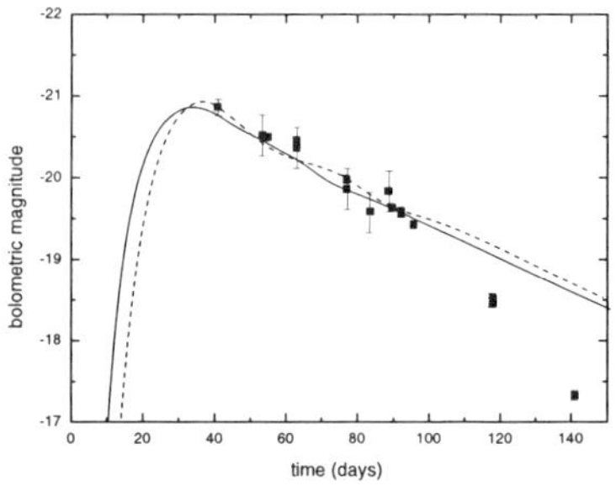

Fig. 4. Model light curves of SN 1999as compared with observations

3. Light Curve Models

We try to constrain $M_{\rm ej}$ and $E_{\rm K}$ by reproducing the bolometric light curve with our radiative transfer code [2]. We parameterize the post-explosion hypernova model CO138E30 [3] to mimic explosion results of different $E_{\rm K}$ and $M_{\rm ej}$. Out best fit parameters are $M_{\rm ej} \sim 20$ $M_\odot$ and $E_{\rm K} \sim 3 \times 10^{52}$ ergs (see Figure 4; dotted line).

We construct a 24 $M_\odot$ C+O progenitor from an evolved 32 $M_\odot$ He star [4] by removing the outer layer and then explode it with $E_{\rm K} = 3 \times 10^{52}$ ergs, assuming a mass cut of 4 $M_\odot$. The model light curve, with $M_{\rm Ni} = 4$ $M_\odot$, is shown in Figure 4 as a solid line.

Our best fit parameters seem insufficient to produce 4 $M_\odot$ ^{56}Ni. One remedy is to make the core of a progenitor much denser than usual, e.g. via the collapsar mechanism [5]. We also artificially explode an evolved helium star of 47 $M_\odot$ with $E_{\rm K} = 5 \times 10^{52}$ ergs and find that the synthesized ^{56}Ni is of 16 $M_\odot$!

References

[1] F.-K. Thielemann, K. Nomoto, and M. Hashimoto, ApJ 460 (1996) 408.
[2] K. Iwamoto et al., ApJ 534 (2000) 660.
[3] T. Nakamura et al., ApJ 550 (2001) 991.
[4] M. Hashimoto, Prog. Theor. Phys. 94 (1995) 663.
[5] A. I. MacFadyen, and S. E. Woosley, ApJ 524 (1999) 262

Effects of Formation Epoch Distribution on X-ray Luminosity and Temperature Functions of Galaxy Clusters

Motohiro ENOKI (1,2), Fumio TAKAHARA (1) and Yutaka FUJITA (2)
1 Department of Earth and Space Science, Graduate School of Science, Osaka University, Toyonaka, Osaka, 560-0043, Japan
2 National Astronomical Observatory, Osawa 2-21-1, Mitaka, Tokyo, 181-8588, Japan

Abstract

We investigate statistical properties of galaxy clusters in the context of hierarchical clustering scenario. Using the formalism which discriminates between major merger and accretion, the epoch of a cluster formation is identified with that of the last major merger. The properties of X-ray emission from clusters are determined by the total mass and density at their formation time. Under these assumptions, we calculate X-ray luminosity and temperature functions of galaxy clusters.

1. Introduction

The properties of galaxy clusters are considered to depend on cosmology and to reflect the structure formation history. Fujita and Takahara [2] have found that clusters form a plane (the fundamental plane), suggesting that the physical quantities of clusters are determined by the halo mass and density at the time of their formation. Motivated by the finding of the fundamental plane, we present a formulation of statistics of clusters in the context of hierarchical structure formation theory taking account of formation epoch distribution. So far, Press-Schechter (PS) mass function has been widely used to calculate temperature and luminosity functions. A conventional way to compute the temperature and luminosity functions is simply to assume that the formation time is the same as the observed time. In this study, we adopt the formulation of the formation epoch distribution proposed by Salvador-Solé et al. [3] (SSM). Their formulation differentiates merger from accretion and define the formation epoch as the epoch of the last major merger. Following their formalism, we assume that the temperature and luminosity of galaxy clusters are determined in terms of the halo mass and density at their formation and calculate $L-T$ distribution, the temperature and luminosity functions. The detailed model description and discussion can be found in [1].

2. Results and discussion

First, we examine the case where the gas temperature is equal to the virial temperature. We calculate the temperature and luminosity functions based on the SSM formalism and compare them with those based on the PS formalism. About the temperature functions and their redshift evolution, there is little difference in prediction of two formalisms. When we take account of the distribution of the formation time, physical quantities of clusters are affected by two factors; (1) if the object formed earlier, its virial density becomes larger and (2) if the object formed earlier, its mass at the formation becomes smaller. Gas temperature depends on the virial density and mass at the formation time as $T \propto \rho_f^{1/3} M_f^{2/3}$, and these two factors tend to cancel out. On the other hand, about the luminosity function and their redshift evolution, SSM formalism predicts larger number density than PS formalism and shows little evolution in contrast with the PS formalism. Since $L \propto \rho_f^{7/6} M_f^{4/3}$, the luminosity depends more strongly on the virial density than the temperature. Therefore, earlier formed dense clusters contribute to the increase of the luminosity function. Next, we investigate the distribution on the $L - T$ relation. The luminosity behaves as $L \propto \rho_f^{1/2} T^2$. In the PS formalism it is assumed that the observed clusters have the same virial density Therefore, in the PS formalism, clusters form a one-parameter family, which is shown by the straight line. On the other hand, in the SSM formalism, the virial densities take a wide range of values. Thus, clusters form a two-parameter family. The predicted $L - T$ relation has the scatter which reflects the dispersion of the halo formation time. Since ρ_f tends to be distributed more widely for smaller clusters, the scatter of $L - T$ relation is larger at lower T and L. Therefore, the slope of $L - T$ relation becomes shallower than $L \propto T^2$ and this is in conflict with the observed correlation $L \propto T^3$. To resolve this discrepancy, we consider preheating models of intracluster gas, in which intracluser gas had already been heated before the cluster formed. The predicted $L - T$ relation can match the observed relation rather well.

References

[1] Enoki, M., Takahara, F. and Fujita, Y. 2001, ApJ, 556, 77
[2] Fujita, Y., and Takahara, F. 1999, ApJ, 519, L51
[3] Salvador-Solé, E., Solanes, J. M., and Manrique, A. 1998, ApJ, 499, 542 (SSM)

A Non-Separable Solution of the Vacuum 5D Space-Time-Matter Universe

Takao FUKUI
Department of Language and Culture, Dokkyo University, 1-1 Gakuencho, Soka, Saitama 340-0042, Japan, tfukui@dokkyo.ac.jp
Sanjeev S. SEAHRA and Paul S. WESSON
Department of Physics, University of Waterloo, Ontario N2L 3G1, Canada, ssseahra@sciborg.uwaterloo.ca and wesson@astro.uwaterloo.ca

Abstract

Singularities and Geodesics are studied in an exact class of solutions of the 5D vacuum Einstein field equations.

1. Non-separable Solution

The vacuum EFEs for space-time plus an extra dimension $l(= mG/c^2)$ are given in terms of the Ricci tensor by $R_{AB} = 0$ $(A, B = 0-3, 5)$. These contain the field equations of general relativity, $G_{\alpha\beta} = 8\pi T_{\alpha\beta}$ $(\alpha, \beta = 0-3)$[1].

A class of solutions has a 5D line element given by

$$dS^2 = \sigma(l)\left\{dt^2 - a(t,l)^2\left[\frac{dr^2}{(1-kr^2)} + r^2 d\Omega^2\right]\right\} + \epsilon b(t,l)^2 dl^2. \tag{1}$$

Here $\epsilon(=\pm 1)$ describes the nature of the extra dimension.

A simple class of solutions of $G_{AB} = 0$ is obtained if we write $\sigma \equiv \sigma_0 (\equiv \Sigma^2)$, $\hat{a} \equiv \partial a/\partial l = f(l)b$. The scale factors $a(t,l)$ and $b(t,l)$ for ordinary space-time and the extra dimension are

$$a^2 = -\frac{(2Ft - g)^2 + \kappa}{4F}, b = -\frac{4\epsilon\sigma_0 f\hat{f}Ft^2 - 2\hat{g}F^2 t + g\hat{g}F - \epsilon\sigma_0 f\hat{f}(g^2+\kappa)}{4afF^2}. \tag{2}$$

Here $F \equiv \epsilon\sigma_0 f^2 + k$, $f = f(l)$, $g = g(l)$ and κ is a constant. On the hypersurface $(\equiv \Sigma_l)$ of $l = const.$, the scale factor a shows that the constants in the scale factor of 4D cosmology are given by the fifith coordinate.

For perfect fluid, an equation of state with the solution is

$$p = \frac{\rho}{3}\left[1 - \frac{8\epsilon\sigma_0 a^2 f^2}{4\epsilon\sigma_0 a^2 f^2 + \kappa}\right]. \tag{3}$$

	$F<0$	$F>0$
$\kappa>0$ and $R^A{}_{BCD}\neq 0$ in general	no big bang	$a(t,l)$ is complex
$\kappa=0$, $R^A{}_{BCD}=0$	one big bang	$a(t,l)$ is complex
$\kappa<0$ and $R^A{}_{BCD}\neq 0$ in general	two big bangs	big bang and big crunch

Table 1. Characteristics of the 5D manifold and the 4D cosmologies embedded on the Σ_l hypersurfaces

On a given Σ_l hypersurface, $p \to -\rho/3$ as $t \to \infty$, $a(t,l) \to \infty$. This is the equation of state of the Milne model. Hence, it might be understood that the 4D cosmologies on the Σ_l are asymptotically empty as $t \to \infty$. In the neighborhood of the big bang on the other hand, we recover $p = \rho/3$. So, near the big bang(s), the induced matter behaves like radiation.

2. Singularities and Geodesics

A direct attack on singularities could entail the calculation of the 5D Kretschmann curvature invariant $K \equiv R^{ABCD}R_{ABCD}$ for (1). We summarize in Table 1 all the cases studied. The expansion Θ of a congruence of 5D geodeisics with one of the tangent vecors to null geodeiscs is given by

$$\Theta = \frac{3}{a}\left(\frac{1}{\sqrt{2\Sigma}}\frac{\partial a}{\partial t} + \frac{1}{\sqrt{2b}}\frac{\partial a}{\partial l}\right) = \frac{3}{a}\frac{da}{d\lambda}, \tag{4}$$

where $\lambda \equiv \sqrt{2\Sigma}t$. An observer in a galaxy traveling along a 5D null-comoving geodesic will see the other galaxies receding away in a Hubble-like expansion with scale factor $a(\lambda)$.

3. Comments

Geometrical effects of 5D cosmologies on the 4D big bang are notable, for example in regard to particle masses[1], the thermalization of photons[2] and 4D big bang akin to a 5D shock wave[3]. For more details, see Ref.[4].

References

[1] P. S. Wesson, *Space-Time-Matter* (1999) World Scientific, Singapore. J. M. Overduin and P. S. Wesson, Phys. Reports **283** (1997) 303.
[2] F. Hoyle, Astrophys. J. **196** (1975) 661.
[3] P. S. Wesson, H. Liu and S. S. Seahra, Astron. Astrophys. **358** (2000) 425.
[4] T. Fukui, S. S. Seahra and P. S. Wesson, J. Math. Phys. **42** (2001) 5195.

Stability Criterion for Self-Similar Solutions in General Relativity

Tomohiro HARADA
Department of Physics, Waseda University, Shinjuku, Tokyo 189-0022, Japan

Abstract

A stability criterion is derived for self-similar solutions with perfect fluids which obey the equation of state $P = k\rho$ in general relativity. By the obtained criterion, we analyze the stability of known self-similar solutions. It has important implications to convergence phenomena and critical phenomena in gravitational collapse.

1. Introduction

Invariance under scale transformation is one of the important features of gravitational physics. It implies the existence of *self-similar solutions*, for which all dimensionless quantities are functions of $z \equiv r/t$. Among them, the most widely researched system in general relativity is a spherically symmetric self-similar spacetime with a perfect fluid. The existence of self-similar solutions requires the equation of state to be the form $P = k\rho$ if it is barotropic. Self-similar solutions have called attention as a critical solution [1, 2] and as an attractor. [3] For Newtonian self-similar solutions, which are obtained as the limit $k \to 0$, Ori and Piran [4] pointed out the existence of *kink instability* in a wide class of self-similar solutions. Unfortunately, their analysis was restricted to Newtonian gravity. The analysis is generalized to fully general relativistic case. See Harada [5] for details.

2. Kink Instability

A self-similar solution may have a sonic point. The sonic point is a stationary point of an autonomous system of ordinary differential equations. When the set of ordinary differential equations is linearized around the sonic point, it has two nontrivial eigenvalues with the corresponding eigenvectors. Self-similar solutions are allowed to cross the sonic point only along these two eigenvectors. Sonic points are classified into nodes, degenerate nodes and saddles in terms of these two eigenvalues. We call the directions which correspond to the larger and smaller eigenvalues "+" and "−" directions, respectively.

We consider perturbations which satisfy the following initial conditions in the background of a self-similar solution: (1) The initial perturbations vanish inside (outside) the sonic point for $t < 0$ ($t > 0$). (2) The initial density profile is continuous everywhere, in particular at the sonic point. (3) The initial density gradient is discontinuous at the sonic point, although it has definite one-sided values there.

Then, we can derive a full-order evolution equation for the injected discontinuity. The evolution equation is completely local at the sonic point. Indeed, this equation is easily integrated. The result is summarized in Table 1., where + and − directions for nodal sonic points are called primary and secondary directions, respectively.

Table 1. Stability for the kink mode and the class of sonic points

t	Primary node	Secondary node	Degenerate node	Saddle: +	Saddle: −
> 0	Stable	Unstable	Unstable	Stable	Unstable
< 0	Unstable	Stable	Unstable	Unstable	Stable

3. Applications

The sonic points for known numerical solutions have been classified by several authors. The results of the applications of the criterion to the known solutions are the following: the expanding flat Friedmann solution for $1/3 \leq k < 1$ and the collapsing one for $0 < k \leq 1/3$ are unstable; the static self-similar solution is unstable; the Larson-Penston (attractor) solution is stable for this mode for $0 < k \leq 0.036$, while it is unstable for $0.036 \leq k$; the Evans-Coleman (critical) solution is stable for this mode for $0 < k \leq 0.89$, while it is unstable for $0.89 \leq k$. The last application suggests that the Evans-Coleman solution for $0.89 \leq k$ is *not critical* because it has at least two unstable modes.

References

[1] M. W. Choptuik, Phys. Rev. Lett. 70 (1993) 9.
[2] C. R. Evans and J. S. Coleman, Phys. Rev. Lett. 72 (1994) 1782.
[3] T. Harada and H. Maeda, Phys. Rev. D63 (2001) 084022.
[4] A. Ori and T. Piran, MNRAS 234 (1988) 821.
[5] T. Harada, Class. Quantum Grav. 18 (2001) 4549.

Non-Linear Stochastic Biasing Effect in Genus Statistics for Dark Halo Distributions

Chiaki HIKAGE, Atsushi TARUYA, and Yasushi SUTO
Department of Physics, School of Science, University of Tokyo, Tokyo 113-0033, Japan

Abstract

We explore the statistical characterization of the biasing of dark matter halos with the simulated halo catalogue by the Hubble volume simulation. In particular, we pay attention to the effects of nonlinearity and stochasticity of the halo biasing on the genus statistics. We find that a perturbation theory developed by [1] describes the non-Gaussian feature of the genus statistics of dark halos in the simulation data fairly accurately.

1. Introduction

To understand the universe from the distribution of galaxies and clusters it is necessary for us to understand the behavior of biasing, i.e., the statistical description of the spatial distribution between luminous objects and dark matter. Dark matter halos are the most likely sites for galaxies and clusters. In particular, dark halos and clusters are usually supposed to have a fairly good one-to-one correspondence in a statistical sense. In this spirit, [2] derived an analytical formula for dark halos in consideration into the nonlinear effect of biasing. Here we perform direct analysis of cosmological numerical simulation of Hubble scale made by the Virgo consortium, which enables us to obtain liable results for the statistics of rich clusters of galaxies whose mean separation is $\sim 50h^{-1}$Mpc (Hikage et al. in preparation). We focus on the biasing effects on genus statistics which characterizes the topological nature in the density field. Note that this is the first systematic study of the genus statistics of halos directly evaluated from numerical simulations.

2. Details of the Simulation and Methodology of Genus Calculation

Specifically we use the ΛCDM (Lambda Cold Dark Matter) data at $z = 0$ (ΛHVS, hereafter), which employ $N = 10^9$ particles with each mass of $2.25 \times 10^{12}h^{-1}M_\odot$ in a boxsize of $3h^{-1}$Gpc. The number of dark halos identified with the friends-of-friends finder amounts to 1.5×10^6 in total with the mass range from $6.7 \times 10^{13} h^{-1} M_\odot$ to $6.0 \times 10^{15} h^{-1} M_\odot$. The more details of the simulation data is written in [3] etc.. After assigning densities of halos and dark matter particles on 128^3 grids using the cloud-in-cell method, we smooth data and compute genus in each isodensity surface with the CONTOUR 3D code [4] by integrating the deficit angles over the surface.

3. Results and Conclusion

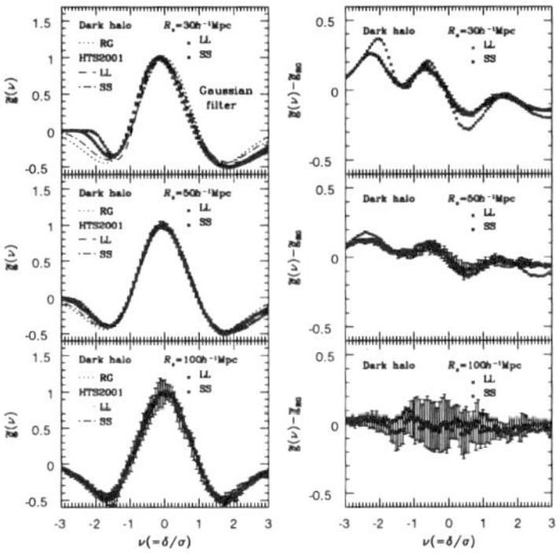

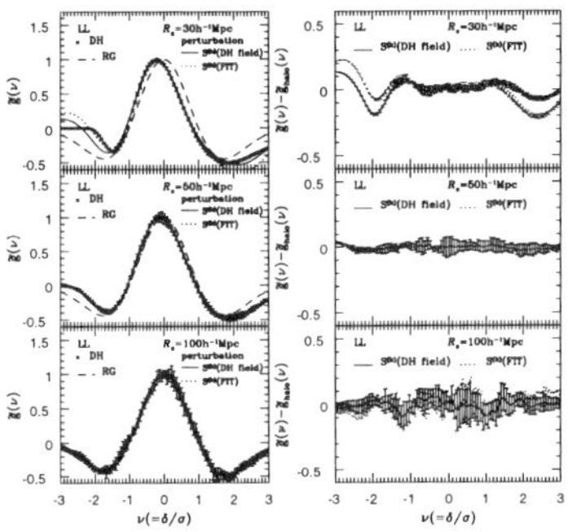

Left figure shows genus densities(left) from ΛHVS for halos and its difference from genus predicted in random-Gaussian statistics(right) with $R_{\rm G} = 30, 50 and 100 h^{-1}$Mpc from top to bottom. The analytical prediction by [2] is written with dashed lines for the most massive halo subgroups and with dot-dashed lines for the least massive halo subgroups. The model prediction is qualitatively well agreed with the simulation results. In right figure, we fit the results for the simulated dark halos against the perturbation model by [1] treating the skewness parameters $S_{\rm h}^{(k)} (k = 1, 2, 3)$ as free parameters. The agreement is quite good, which suggests that the perturbative formula is also a useful approximation for dark halos in a weakly nonlinear regime and reproduces the best-fit values for those skewness parameters from independent theoretical model and/or simulation data.

References

[1] T. Matsubara, ApJL434 (1994) 43.
[2] C. Hikage, A. Taruya, and Y. Suto, ApJ556 (2001) 641.
[3] A. Jenkins, et al. MNRAS321 (2001) 372.
[4] R.J. Gott, et al. ApJ340 (1989) 625.

Influence of Inflation on Parametric Resonance During Preheating

Shiro Hirai
Osaka Electro-Communication Junior College Neyagawa, Osaka 572-8530, Japan

Abstract

The influence of inflation on parametric resonance during preheating is investigated. It is found that the preheating scalar field is squeezed on both small and large scales during inflation. However, the positive-frequency solution is usually adopted in the initial condition for the scalar field during preheating. Here it is shown that while the difference in the comoving occupation number of particles between two initial conditions is not large, the ratio of comoving occupation numbers changes from 0.2–5.0, dependent on k. Such a situation is examined by calculating the squeeze parameter r for the period from inflation to preheating.

1. Introduction and Model

One of the most interesting problems in cosmology is that of parametric resonance during preheating, and it has been extensively investigated in all aspects. In particular, Kofman, Linde and Starobinsky [1] discovered that the comoving occupation number of particles n_k exhibits parametric resonance behavior when the positive-frequency solution is used as the initial condition for the preheating field χ. This introduces the problem of whether the initial state of the field χ is really the positive-frequency solution at the starting of preheating. As inflation exists before preheating, it is necessary to take into account the influence of inflation on the field χ. We use a model similar to that of Kofman et al. in order to clarify the differences between initial conditions.
Here, we consider the Lagrangian

$$L = \frac{1}{2}(\partial_\mu \phi)^2 + \frac{1}{2}(\partial_\mu \chi)^2 - \frac{1}{2}m^2\phi^2 - \frac{1}{2}g^2\phi^2\chi^2, \tag{1}$$

where ϕ is the inflation scalar field and χ is a preheating scalar field, and the chaotic inflation model is considered ($m = 10^{-6}m_{\rm p}$, and $g = 5 \times 10^{-4}$). For simplicity, we approximate the scale factors of inflation and preheating to $a^I = b_0 e^{H(t-t_1)}$ and $a^R = b_0(3H(t-t_1)/2+1)^{2/3}$, respectively. The inflation starts at

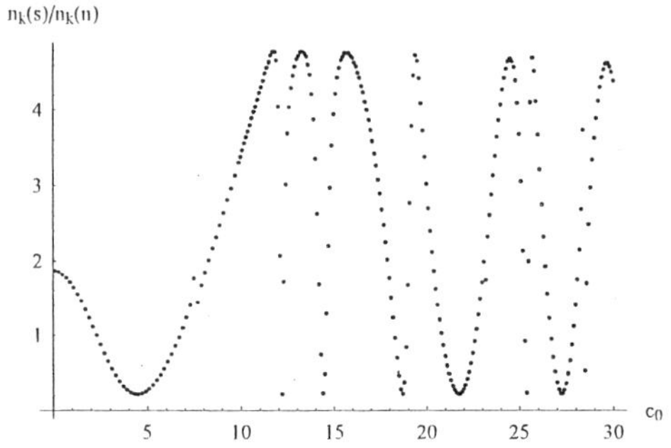

Fig. 1. Ratio $n_k(s)/n_k(n)$

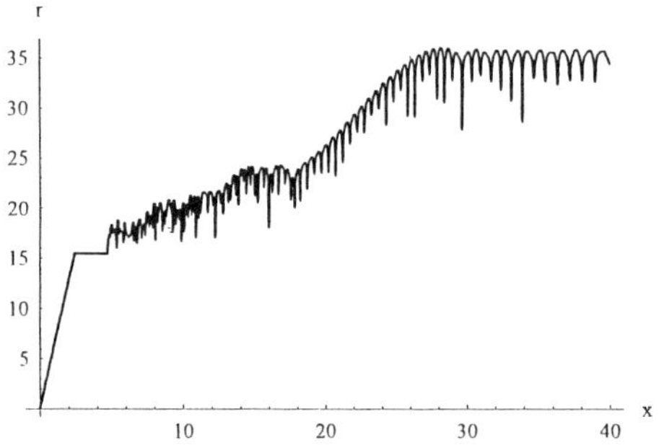

Fig. 2. Evolution of squeeze parameter r from inflation to preheating

$t = 0$ and ends at $t = t_1$ ($= 60/H$) and preheating starts at $t = t_1$. Assuming $\phi \approx m_{\rm p}$, the equation of motion for χ during inflation can be solved. The two solutions can then be obtained with some approximations, i.e.,

$$f_k^1(t) = \frac{1}{\sqrt{2gm_p}} \exp(-igm_p t), \tag{2}$$

$$f_k^2(t) = \frac{1}{\sqrt{2k}} (a^I)^{3/2} \frac{H}{k} (i + \frac{k}{b_0 H} e^{-H(t-t_1)}) \exp[i \frac{k}{b_0 H} e^{-H(t-t_1)}], \tag{3}$$

where $f_k(t) = a^{3/2}(t)\chi_k(t)$. For the majority of the inflation period, the solution of χ can be taken as $f_k^2(t)$, which at the end of inflation becomes $f_k^1(t)$. Then, the solution can be written as $f_k^3(t) = \alpha_k f_k^1(t) + \beta_k f_k^{1*}(t)$, which we call the squeezed state. Therefore, $f_k^3(t)$ should be used as the initial solution for preheating.

2. Summary

We have investigated the influence of inflation on parametric resonance during preheating. From an analysis of the behavior of the preheating scalar field

χ during inflation, we have found that the scalar field becomes squeezed on both large and small scales. Using this initial condition, we calculated the comoving occupation number of particles n_k. We found that while the difference in n_k between the two initial conditions, i.e., the positive-frequency solution ($n_k(n)$) and the squeezed solution ($n_k(s)$), is not particularly large, a ratio of $n_k(s)/n_k(n) \approx$ 0.2–5.0 is derived at $x = 20$ in the range $10^{-5} \leq c_0 \leq 20$, where $x = mt/2\pi$ and $k = c_0 b_0 m$ (Fig. 1). This implies that more accurate values can only be obtained if we use the squeezed initial condition. Here, we investigate such a situation using the squeeze formulation [2]. Calculating the squeeze parameter r from inflation to preheating, it is possible to understand the behavior of parametric resonance from inflation to preheating. The evolution of the squeeze parameter r from inflation to preheating is given in Fig. 2 as function of x $(= mt/2\pi)$ in the case of $c_0 = 10^{-10}$. It is noted that the range $0 \leq x \leq 4.666$ represents the inflation epoch and $4.666 \leq x$ represents the preheating epoch.

References

[1] L. Kofman, A. Linde and A. A. Starobinsky, Phys. Rev. **D56** (1997), 3258.
[2] S. Hirai, Prog. Theor. Phys.**103** (2000), 1161

Cosmic Acceleration With a Positive Cosmological Term

Arbab I. Arbab
Comboni College for Computer Science, P.O. Box 114, Khartoum, Sudan

Abstract

We have considered a cosmological model with a cosmological term of the form ($\Lambda = \beta\frac{\ddot{R}}{R}$, $\beta =$ cont.). For age parameter consistent with observational data the Universe must be accelerating in the presence of a positive cosmological term. The minimum age of the Universe is H_0^{-1}, where H_0 is the Hubble constant. The cosmological term decreases as t^{-2}. Allowing the gravitational constant to change with time leads to an ever increasing gravitational constant at the present epoch. In the presence of a viscous fluid this decay law for Λ is equivalent to the one with $\Lambda = 3\alpha H^2$ ($\alpha =$ const.) provided $\alpha = \frac{\beta}{3(\beta-2)}$. The inflationary solution obtained from this model is that of the de-Sitter type.

1. Introduction

One of the puzzling problems in standard cosmology is the cosmological term problem. Observational data indicate that $\Lambda \sim 10^{-55}\text{cm}^{-2}$ while particle physics prediction for Λ is greater than this value by a factor of order 10^{120}. This discrepancy is known as the cosmological constant problem. A point of view which allows Λ to vary in time is adopted by several workers. The point is that during the evolution of the universe the energy density of the vacuum decays into particles thus leading to the decrease of the cosmological term. As a result one has the creation of particles although the typical rate of the creation is very small. The entropy problem which exists in the Standard Model can be solved by the above mechanism. One of the motivations for introducing Λ term is to reconcile the age parameter and density parameter of the universe with current observational data. Recent observations of Type 1a supernovae which indicate an accelerating universe draw once more the attention to the possible existence, at the present epoch, of a small positive cosmological term (Λ). The purpose of this work is to study the phenomenological decay law for Λ that is proportional to the deceleration parameter. This is parameterized as $\Lambda = \beta\frac{\ddot{R}}{R}$, where β is a constant. This law finds little attention among cosmologist. The cosmological consequences of this decay law are very attractive. This law provides a reasonable solution

to the cosmological puzzles presently known. We have found that a resolution to these problems is possible with a positive cosmological term ($\Lambda > 0$). This requires the deceleration parameter to be negative ($q < 0$). Usually people invoke some kind of a scalar field that has an equation of state of the form $p < 0$, p the pressure of the scalar field. A more recent review for the case of a positive cosmological term is found in [6].
A variable gravitational constant G can also be incorporated into a simple framework in which Λ varies as well, while retaining the usual energy conservation law [1,10,11]. The above decay law leads to a power-law variation for G. Inflationary solutions are also possible with this mechanism thus solving the Standard Model problems. We have recently shown that a certain variation of G may be consistent with palaeontological as well as geophysical data [15].

2. The Model

For the Friedmann–Robertson-Walker cosmologies the Einstein's field equations with the variable cosmological term and a source term given by a stress-energy tensor of a perfect fluid read

$$3\frac{\dot{R}^2}{R^2} + \frac{3k}{R^2} = 8\pi G\rho + \Lambda \ , \tag{1}$$

$$2\frac{\ddot{R}}{R} + \frac{\dot{R}^2}{R^2} + \frac{k}{R^2} = -8\pi G p + \Lambda \tag{2}$$

where ρ is the fluid energy density and p its pressure. The equation of the state is taken in the form

$$p = (\gamma - 1)\rho \tag{3}$$

where γ is a constant. From eqs.(1) and (2) one finds

$$\frac{d(\rho R^3)}{dt} + p\frac{dR^3}{dt} = -\frac{R^3}{8\pi G}\frac{d\Lambda}{dt} \tag{4}$$

We propose a phenomenological decay law for Λ of the form [4,5]

$$\Lambda = \beta\frac{\ddot{R}}{R} \tag{5}$$

where β is a constant. Overdin and Cooperstock have pointed out that there is no fundamental difference between the first and second derivatives of the scale factor that would preclude the latter from acting as an independent variable if the former is acceptable [4]. In fact this form is equivalent to modifying the General

Relativity [17] and having a built-in cosmological constant, i.e., $\Lambda \propto \mathcal{R}$, where $\mathcal{R}$ is the Ricci scalar. Moreover, from eqs.(1) and (4), one can write

$$\ddot{R} = \frac{8\pi G}{3}(1-\frac{3\gamma}{2})\rho R + \frac{\Lambda}{3}R \; . \tag{6}$$

Thus writing Λ in the above for is equivalent to considering the universe to be filled with a fluid characterized by $\gamma = \frac{2}{3}$. For the matter-dominated universe $\gamma = 1$ and therefore eqs.(1), (2),(3), and (5) yield (for $k = 0$)

$$(\beta - 2)\ddot{R}R = \dot{R}^2 \; , \tag{7}$$

which can be integrated to give

$$R(t) = [\frac{A(\beta-3)}{(\beta-2)}t]^{(\beta-2)/(\beta-3)} \; , \; \beta \neq 3 \; , \tag{8}$$

where $A =$ consatnt. It follows from eq.(5) that

$$\Lambda(t) = \frac{\beta(\beta-2)}{(\beta-3)^2}\frac{1}{t^2} \; , \; \beta \neq 3 \; . \tag{9}$$

Using eqs.(1), (5) and (8) the energy density can be written as,

$$\rho(t) = \frac{(\beta-2)}{4\pi G(\beta-3)}\frac{1}{t^2} \; , \; \beta \neq 3 \; . \tag{10}$$

The vacuum energy density (ρ_v) is given by

$$\rho_v(t) = \frac{\Lambda}{8\pi G} = \frac{\beta(\beta-2)}{8\pi G(\beta-3)^2}\frac{1}{t^2} \; , \quad \beta \neq 3 \; . \tag{11}$$

The deceleration parameter (q) is defined as

$$q = -\frac{\ddot{R}R}{\dot{R}^2} = \frac{1}{2-\beta} \; , \; \beta \neq 2 \; . \tag{12}$$

We see from eqs.(9) and (12) that for a positive Λ the deceleration parameter is negative (for $\beta > 2$). For $\beta < 2$ the cosmological term is negative, $\Lambda < 0$. It has been recently found that the universe is probably accelerating at the present epoch. The density parameter of the universe (Ω_m) is given by

$$\Omega_m = \frac{\rho}{\rho_c} = \frac{2\,(\beta-3)}{3\,(\beta-2)} \quad , \tag{13}$$

where $\rho_c = \frac{3H^2}{8\pi G}$ is the critical energy density of the universe. We notice that the Standard Model formula $\Omega_m = 2q$ is now replaced by $\Omega_m = \frac{2}{3}q + \frac{2}{3}$. However,

both models give $q = \frac{1}{2}$ for a critical density. This relation has been found by several authors [1,3].
The density parameter due to vacuum contribution is defined as $\Omega^\Lambda = \frac{\Lambda}{3H^2}$. Employing eqs.(8) and (9) this gives

$$\Omega^\Lambda = \frac{\beta}{3(\beta - 2)}, \quad \beta \neq 2 . \tag{14}$$

We shall define $\Omega_{\rm total}$ as

$$\Omega_{\rm total} = \Omega_m + \Omega^\Lambda \tag{15}$$

Hence eqs.(1), (13) and (14) give $\Omega_{\rm total} = 1$. This setting is favored by the inflationary scenario.

The present value of the age of the universe, deceleration parameter and the cosmological term are obtained from eqs.(8), (9) and (13)

$$t_0 = \frac{(\beta - 2)}{(\beta - 3)} H_0^{-1}, \quad \Omega_{m0} = \frac{2\,(\beta - 3)}{3\,(\beta - 2)}, \quad \Lambda_0 = \frac{\beta}{(\beta - 2)} H_0^2, \quad \beta \neq 2, \quad \beta \neq 3 . \tag{16}$$

(the subscript '0' denotes the present value of the quantity and H is Hubble constant). Inasmuch as $q < 0$, $\Omega_m < 1$ hence the low density of the universe is no longer a problem. Moreover, in order to solve the age problem we require $\beta > 3$. Thus the constrain $\beta > 3$ may resolve both problems. Observational evidence, however, does not rule out the negative deceleration parameter and the stringent limits on the present value of q_0 are $-1.25 \leq q_0 \leq 2$ [8].
The case $\beta = 2$ represents an empty static universe with $\Lambda = 0$. For $\beta = 0$, the usual expression for FRW models are recovered. For $\beta = 4$ one obtains $t_0 = 2H_0^{-1}, \Omega_{m0} = \frac{1}{3}, \Omega_0^\Lambda = \frac{2}{3}$; $\beta = 12$, $t_0 = \frac{10}{9} H_0^{-1}, \Omega_{m0} = \frac{3}{5}, \Omega_0^\Lambda = \frac{2}{5}$. It is interesting to note that when $\beta \to \infty$, all parameters are finite, namely, $R_0 \to t_0$, $t_0 \to H_0^{-1}, \Lambda \to H_0^2$, $\Omega_{m0} \to \frac{2}{3}, \Omega_0^\Lambda = \frac{1}{3}$. We have recently [15] investigated the implications of a variable G on the Earth-Sun system, contrary to what have been believed, we have found that the palaeontological data are consistent with a variable G provided that the age of the universe is $t_0 \sim 11 \times 10^9$ years and that $G \propto t^{1.3}$. However, a recent value for the age of the universe from gravitational lensing suggests $t_0 \sim 11 \times 10^9$ years. Recent estimates from observations of galaxy clustering and their dynamics indicate that the mean mass density is about one-third of the critical value [9]. Thus if $\beta \neq 0$ then the present age of the universe can not be less than H_0^{-1}. This constrain represents our strongest prediction for the age of the Universe.

3. A Model with Variable G

We now consider a model in which both G and Λ vary with time. Imposing the usual energy conservation law one obtains [1,10,11]

$$\dot{\rho} + 3\gamma H\rho = 0 \ , \tag{17}$$

and

$$\dot{\Lambda} + 8\pi\dot{G}\rho = 0 \ . \tag{18}$$

Using eqs.(8) and (9), eqs.(19) and (20) yield

$$\rho(t) = Dt^{-3(\beta-2)/(\beta-3)} \ , \quad D = \text{const.} \tag{19}$$

and

$$G(t) = [\frac{(\beta-2)}{4\pi A(\beta-3)}]t^{\beta/(\beta-3)} \quad \beta \neq 3 \ . \tag{20}$$

For $\beta = 0$, G=const. and $\rho = Dt^{-2}$ and $R = [\frac{3}{2}At]^{2/3}$, which is the usual FRW result. Clearly for $\beta > 3$ the gravitational constant increases with time while for $\beta < 3$ it decreases with time. Again the constraint $\beta > 3$ considered before implies that gravitational constant increases with time. An increasing gravitational constant is considered by several workers [1,10,11,14].
In a recent work we have shown that the variation of the gravitational constant is consistent with palaeontological data [15]. The gravitational constant might have had a very different value from the present one. This depends strongly on the value of β assumed at a given epoch. The development of the large-scale anisotropy is given by the ratio of the shear σ to the volume expansion ($\theta = 3\frac{\dot{R}}{R}$) which evolves as [16]

$$\frac{\sigma}{\theta} \propto t^{(3-2\beta)/(\beta-3)} \ , \tag{21}$$

and since $\beta > 3$ this anisotropy decreases as the universe expands and this explains the present observed isotropy of the Universe.
For example, if $\beta = \frac{3}{2}$ then $G \propto t^{-1}, R \propto t^{1/3}, \rho \propto t^{-1}$. This behavior of G was considered by Dirac in his Large Number Hypothesis (LNH) model [12]. In an earlier work we shown that some non-viscous models are equivalent to bulk viscous ones. This behavior is also manifested in our present model provided one takes $\beta = \frac{3(2n-1)}{(3n-2)}$, where the bulk viscosity (η) is defined as $\eta = \text{const.}\rho^{\text{n}}$, where $0 \leq n \leq 1$ [1]. The two models, though different in the form of Λ, are equivalent if one puts $\alpha = \frac{\beta}{3(\beta-2)}$. Thus the decay law $\Lambda = \beta\frac{\ddot{R}}{R}$ and $\Lambda = 3\alpha H^2$, where α = const., are identical in the presence of a cosmic fluid.

4. An Inflationary Solution

This solution is obtained from eqs.(1), (2), (3) and (5) with $\beta = 3$. One then gets $R\ddot{R} = \dot{R}^2$, which may integrated to give

$$R = \text{const.}\exp(\text{Ct}) \tag{22}$$

where $H = C =$ const.. Applying eq.(24) to eqs.(5), (20), employing eq.(1), we get

$$\Lambda = 3H^2 \quad , \quad \rho = 0 \quad , \quad G = \text{const.} \,. \tag{23}$$

This is the familiar de-Sitter inflationary solution (in the matter dominated epoch). A similar inflationary solution is obtained with $\beta = 3$ in the radiation dominated epoch. The inflationary solution resolves some of the outstanding issues of standard cosmology.

5. Conclusion

In this paper we have considered the cosmological implications of a decay law for Λ that is proportional to the deceleration parameter. The model is found to be very interesting and apparently a lot of problems can be solved. To solve the age parameter and the density parameter one requires the cosmological term to be positive or equivalently the deceleration parameter to be negative. This implies an accelerating universe. However, the strongest support for an accelerating universe comes from intermediate redshift results for Type 1a supernovae. The model predicts that the minimum age of the Universe is H_0^{-1}. The behavior that $\Lambda \propto t^{-2}$ is found by several authors. The gravitational constant is found to increase with time at the present epoch. Our model predicts an inflationary phase in the matter dominated epoch as well as in the radiation dominated epoch. The cosmological tests for this model can be obtained from those ones already investigated by us [7]. The choice among these models awaits the emergence of the new data.

Acknowledgements

My ideas on this subject have benefited from discussions with a number of friends and colleagues. I am grateful to all of them.
I wish to thank Omdurman Ahlia University for financial support of this work.

References

[1] Arbab, A. I., 1997. *Gen. Rel. Gravit.* **29**, 61.
[2] Beesham, A., 1993. *Phys. Rev.* **D48**, 3539.

[3] Matyjasek, J., 1995. *Phys. Rev.***D51**, 4154.

[4] Overdin, J. M., and Cooperstock, F.I., 1998. *Phys. Rev.***D58**, 043506.

[5] Al-Rawaf, A. S., 1998. *Mod. Phys. Lett.***A 13**, 429.

[6] Sahni, V., and Starobinsky, A., 1999. *Los Alamos preprint* astro-ph/9904398.

[7] Arbab, A. I., 1998. *Astrophys. Space Science Vol.***259**, 371.

[8] Klapdor, H. W., and Grotz, K., 1986. *Astrophys. J.***301**, 139.

[9] Peebles, P. J. E., 1986. *Nature* **321**, 27.

[10] Abdel-Rahman, A. -M. M., 1990. *Gen. Rel. Gravit.***22**, 655.

[11] Beesham, A., 1986. *Nouvo Ciment.***B96**, 17.

[12] Dirac, P. A. M., 1937. *Nature* **139**, 323.

[13] Kalligas, D., Wesson, P., and Everitt, C. W., 1992. *Gen. Rel. Gravit.***24**, 351.

[14] Abussattar, A., and Vishwakarma, R. G., 1997. *Class. Quantum Gravit.***14**, 945.

[15] Arbab, A. I., 1998. *Los Alamos preprint* physics/9811024.

[16] Barrow, J. D., 1978. *Mon. Not. astr. Soc. 184*, 677.

[17] Al-Rawaf, A.S., and Taha, M.O., *Gen. Rel. Gravit.***28**,935, *Phys. Lett.***B366**, 69.

Stability of AdS $\times$ S $\times$ S

Daisuke IDA, Tetsuya SHIROMIZU, Takashi TORII
Research Center for the Early Universe (RESCEU), The University of Tokyo, 7-3-1 Hongo, Bunkyo-ku, Tokyo 113-0033, Japan
Hirotaka OCHIAI
Department of Physics The University of Tokyo, 7-3-1 Hongo, Bunkyo-ku, Tokyo 113-0033, Japan

Abstract

We have shown the stability of $AdS_p \times S^n \times S^{(q-n)}$ for $q \geq 9$ under non-linear dilatonic deformations by estimating the total gravitational energy.

The Randall-Sundrum-type brane-world scenario has been received much attention. This model assumes five-dimensional AdS space at the begining. One might expect more fundamental theories serve as the origin of the AdS space. For instance, the supergravity has an $AdS \times S$ vacuum, which is given by the Freund-Rubin solution. It is important question whether the AdS vacuum emerges in a natural way. We here consider the stability problem of the Freund-Rubin vacuum.

The bosonic part of supergravities is typically described by the following action,

$$S_{p+q} = \int d^{p+q}x\sqrt{|\hat{g}|}\left(\frac{1}{2}\hat{R} - \frac{1}{4q!}F_q^2\right), \tag{1}$$

where $F_q = dA_{q-1}$ represents the q-form field. A simple solution is given by the Freund-Rubin $AdS_p \times S^q$ solution

$$ds^2 = g^{(AdS)}{}_{\mu\nu}dx^\mu dx^\nu + g^{(S)}{}_{ij}dx^i dx^j, \tag{2}$$

$$F_q = f *_q 1, \tag{3}$$

where f is a constant. The metric $g^{(S)}$ is the metric of S^q, however it can be replaced by that of $S^n \times S^{q-n}$. This realize the AdS_p space in a natural way, which gives the basis of the Randall-Sundrum-type brane-world. Moreover, the S^q might be relevant for the geometric realization of the gauge symmetry of the standard particles. We are interested in the stability of this solution. The perturbative analysis shows that $AdS_p \times S^n \times S^{q-n}$ is stable if $q \geq 9$. This is valid only when the background geometry is exactly $AdS_p \times S^n \times S^{q-n}$. However, from the viewpoint of the brane-world cosmology, the stability of $AdS_p - Schwarzschild \times S^n \times S^{q-n}$ is also important. We therefore take into account more generic situation including $AdS_p - Schwarzschild \times S^n \times S^{q-n}$ background and consider the stability under non-linear fluctuations.

We assume the following form of the metric as a generalization of the Freund-Rubin metric,

$$g = g_{\mu\nu}(x)dx^\mu dx^\nu + e^{2\phi_1(x)}d\Omega_n^2 + e^{2\phi_2(x)}d\Omega_{q-n}^2, \tag{4}$$

where $d\Omega_n^2$ and $d\Omega_{q-n}^2$ are the metric of S^n and S^{q-n}, respectively. Then, the following form of the q-form field

$$F_q = f *_n 1 \wedge *_{q-n} 1 \tag{5}$$

solves the field equations. This form of the metric is rather restrictive, however it can express the non-linear regime of the unstable mode previously considered.

We have shown that the total gravitational energy is always positive for arbitrary fluctuation of $\Phi \propto \phi_1, \phi_2$ if $q \geq 9$ is satisfied, which is consistent with the linear analysis.

Thus, we expect that the $\mathrm{AdS}_p \times \mathrm{S}^n \times \mathrm{S}^{q-n}$ for $q \geq 9$ is stable even for large fluctuations. The advantages of the use of the positive energy theorem are (i) one can confirm stability of the background geometry under large fluctuations, (ii) one can argue the semi-classical or quantum mechanical stability without topology change of the space-time, and (iii) the background geometry need not be exactly $\mathrm{AdS}_p \times \mathrm{S}^n \times \mathrm{S}^{q-n}$; the positive energy theorem can be applied to the case of the non-trivial spatial topology such as black hole space-times.

Is dark energy the only solution to the apparent acceleration of the present universe?

Hideo IGUCHI
Department of Physics, Tokyo Institute of Technology Oh-Okayama, Meguro-ku, Tokyo 152-8550, Japan
Takashi NAKAMURA
Yukawa Institute for Theoretical Physics, Kyoto University, Kyoto 606-8502, Japan
Ken-ichi NAKAO
Department of Physics, Osaka City University, Osaka 558-8585, Japan

Abstract

Even for the observed luminosity distance $D_L(z)$ which suggests the existence of the dark energy, we show that the inhomogeneous dust universe solution without the dark energy is possible in general. Future observation of $D_L(z)$ for $1 < z < 1.7$ may confirm or refute this possibility.

1. Introduction

Recent measurements of the luminosity distance $D_L(z)$ using Type Ia supernovae [1, 2, 3] suggest that accurate $D_L(z)$ may be obtained in near future. Especially SNAP [4] will give us the luminosity distance of ~2000 Type Ia supernovae with an accuracy of a few % up to $z \sim 1.7$ every year. On the other hand from the observation of the first Doppler peak of the anisotropy of CMB, it is now suggested that the universe is flat [5, 6], which may be proved in future by MAP and Planck. Under the assumption of the homogeneity and the isotropy of our universe, these observations suggest that the dark energy is dominant at present. However at present we do not have a firm and reliable theoretical basis to discuss such a small amount of energy scale compared with the Planck one. Recently we have considered to construct a possible inhomogeneous dust universe derived from the observed $D_L(z)$ without any dark energy [7].

2. Results and conclusion

The main question of [7] is whether or not the dark energy is the only solution to the apparent acceleration of the universe. Our answer for this question is "no dark energy". To derive this conclusion we have performed the "reconstruction" of the universe from the $D_L(z)$ which is considered to be plausible from the SNIa observations. The assumptions in this analysis are that the $D_L(z)$ is correct for all z and that the dark energy does not exist. The result is that we can reconstruct the universe up to $z \sim 1$ without difficulty and the crucial difference appears at $1 < z < 1.6$. Therefore we conclude that if there is not dark energy, D_L(z) should be different from Eq. (1) in [7] and these differences would be observed by future or ongoing projects.

Recently, the SNIa at the redshift of ~ 1.7 was found [8, 9] with rather large error bars. However only a single SNIa at the redshift of ~ 1.7 is not enough to construct the accurate $D_L(z)$. The first Doppler peak as well as the higher ones will give us another constraints to the inhomogeneous universe models. One may suspect that the existing observations in $0 < z < 1$ such as : (i) evolution of cluster abundance, (ii) lensing rate, (iii) ages of stellar populations already rule out the inhomogeneous models. The model dependence including various undetermined parameters and the observational uncertainty are much larger than the dependence on the cosmological parameters. Therefore we think that these observations can not easily rule out the inhomogeneous model.

In conclusion, the dark energy is not the only solution to the apparent acceleration of the present universe but inhomogeneous dust models can also explain the observations.

This work was partially supported by the Grant in Aid for Scientific Research (No. 11217, Nos.11640274, 09NP0801) from the Japanese Ministry of Education, Science, Sports, and Culture.

References

[1] B. P. Schmidt *et al.*, Astrophys. J. 507, 46 (1998).
[2] A. G. Riess *et al.* [Supernova Search Team Collaboration], Astron. J. 116, 1009 (1998).
[3] S. Perlmutter *et al.* [Supernova Cosmology Project Collaboration], Astrophys. J. 517, 565 (1999).
[4] Supernova/Acceleration Probe, http://snap.lbl.gov
[5] de Bernardis, P. *et al.*, Nature (London) , 404, 955 (2000).
[6] Lange, A.E. *et al.*, Phys. Rev. D 63, 042001 (2001).
[7] H. Iguchi, T. Nakamura and K. Nakao, arXiv:astro-ph/0112419.
[8] R. L. Gilliland, P. E. Nugent and M. M. Phillips, Astrophys. J. 521, 30 (1999).
[9] A. G. Riess *et al.*, astro-ph/0104455.

Non-Gaussianity as a Probe of Topology of the Space

Kaiki Taro INOUE
National Astronomical Observatory, Division of Theoretical Astrophysics, 2-21-1 Osawa, Mitaka, Tokyo 181-8588, Japan, tinoue@th.nao.ac.jp

Abstract

Inhomogeneous and anisotropic Gaussian fluctuations for a particular observer can be regarded as non-Gaussian fluctuations for a homogeneous and isotropic ensemble of the observer. If non-Gaussianity with vanishing skewness but non-vanishing kurtosis are found only on large angular scales in the cosmic microwave background (CMB) then it will be a strong sign that the spatial geometry has an non-trivial topology.

The methods of observationally proving the topology of the spatial geometry of the universe are: the "direct" methods advocated by many authors in which one looks for some periodic patterns either in the CMB fluctuations, or the distribution of high-z clusters; the "indirect" methods in which one looks for an unusual correlation patterns. The latter method is firstly recognized by [1] but no statistical tool that does not depend on a particular topology has been developed. Here I propose the Non-Gaussianity as a probe of topology of the spatial hypersurface of our universe.

Let us consider closed cosmological models that has locally homogeneous and isotropic spatial 3-hypersurface H. Then H should be globally anisotropic except for the projective space P^3 [2]. If the initial fluctuations obey the Gaussian distribution as predicted by the "standard" inflationary scenario, then the CMB temperature fluctuations are described by an anisotropic Gaussian random field for a particular choice of location and orientation of the observer. Marginalizing over the orientation and the position, then the ensemble of fluctuations can be regarded as a homogeneous and isotropic non-Gaussian random field [3]. Suppose a temperature fluctuation in the sky written in terms of spherical harmonics $\Delta T = \sum a_{lm} Y_{lm}$. Then, for spatially closed models, a_{lm} is written as a discrete sum of products of three quantities: initial perturbation $\Phi_\nu(0)$, the expansion coefficient $\xi_{\nu lm}$ of mode functions (eigenfunctions of the Laplace-Beltrami operator) and the temperature transfer function $F_{\nu l}$ as $a_{lm} = \sum_\nu \Phi_\nu(0) \xi_{\nu lm} F_{\nu l}$. Here ν denotes

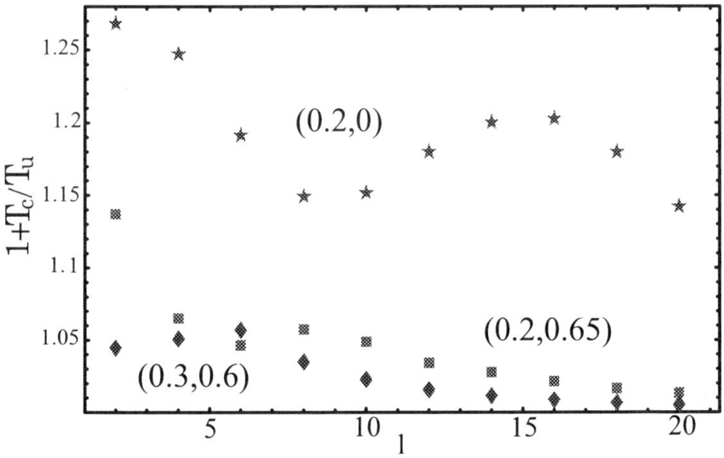

Fig. 1. The ratio of ensemble averaged total trispectrum $\langle T(l,l,l,l,2)\rangle$ to unconnected trispectrum $\langle T(l,l,l,l,2)\rangle_u$ in the smallest known compact hyperbolic model with various cosmological parameters $(\Omega_m, \Omega_\Lambda)$. This calculation is carried out by using the analytic formula [4]. Deviation from 1 indicates the degree of non-Gaussianity.

the wave number. In closed hyperbolic models, for random choices of location and orientation of the observer, it is conjectured that a set of $\xi_{\nu lm}$ can be well approximated by Gaussian random numbers with vanishing mean and variance $\propto \nu^{-2}$. In closed flat models, we have $\xi_{\nu lm} \propto \sum_{g_i} Y_{lm}(g_i(\hat{\mathbf{k}}))$, where g_i denotes the generators of the discrete isometry group of a toroidal 3-space T^3.

Then one can show that the ensemble average of n-point correlation of a_{lm} vanishes for *odd* n's inasmuch $\Phi_\nu(0)$ obeys a symmetric probability distribution function. Thus, in closed locally homogeneous and isotropic models we expect to observe vanishing connected bispectrum $\langle B_{l_1 l_2 l_3}\rangle_c$ but non-vanishing connected trispectrum $\langle T_{l_1 l_2 l_3 l_4 L}\rangle_c$ provided that $\Phi_\nu(0)$'s are Gaussian.

References

[1] J. Levin, E. Scannapieco, G. de Gasperis, and J. Silk, Phys. Rev. D58 (1998) 123006.
[2] J. Wolf, *Space of Constant Curvature* (1967) McGraw-Hill, New York.
[3] G. Ferreira and J. Magueijo, Phys. Rev. D56 (1997) 4578.
[4] E. Komatsu, PhD thesis, Tohoku University.

Peculiar Velocities of the Galaxies beyond the Local Void

Ikuru IWATA and Kouji OHTA
Department of Astronomy, Faculty of Science, Kyoto University, Kitashirakawa-Oiwakecho, Sakyo-ku, Kyoto 606-8502, Japan
Kouichiro NAKANISHI
Nobeyama Radio Observatory, National Observatory of Japan, Minamimaki, Minamisaku, Nagano, 384-1305, Japan
Adel T. ROMAN
National Research Institute for Astronomy and Geophysics, Helwan, Cairo, Egypt
Pierre CHAMARAUX
Observatoire de Paris - Section de Meudon, 5, Place Jules Janssen 92195 Meudon Cedex, France

Abstract

The Local void is the nearest void from us and is thought to be playing an important role in the kinematics of the local universe, especially as one of the suspected source of the motion of the Local group. The proximity of the Local void provides us the exclusive opportunity to investigate the kinematics around a void, using Tully-Fisher relation. We are executing a project aiming at examining the systematic motion of the galaxies beyond the Local void, using the near-infrared Tully-Fisher relation.

1. Introduction

It is known that the Local group (LG) has a motion against the microwave background with a velocity of $\sim$ 630 km/s. During the past decade, several extensive projects aimimg at understanding the distribution of the galaxies and the velocity fields in the local universe have been executed. Although it is claimed that results of these projects explain the motion of the LG for a certain degree [1], it should be noted that these studies are based on the sample which is *not* distributed on the whole sky and it is deficient of galaxies located in the sky area obscured by the Milky Way. Especially, we know little about the origin of the LG's motion perpendicular to the Supergalactic plane, which is considered to be more than 300 km/s, and the extent of this motion; specific to the LG, or common in the Local Supercluster? [2]. Understanding this motion is crucial for the study of the velocity fields in the local universe.

Faber and Burstein (1988)[3] suggested that the Local void is a main contributor of the LG's motion perpendicular to the Supergalactic plane. Since the Local void is the nearest void from us and the LG is on the edge of the Local void, it is natural to consider the anti-gravity of the Local void effects the motion of the LG. This effect can be considered to be a part of the evolution of the large-scale structure in the local universe; higher mass density regions are contracting and voids are expanding. If this is the case, the galaxies located at the far-side boundary of the Local void must have excess of receding motions in addition to the normal Hubble flow, caused by the mass concentration behind them. Measuring the motion of the galaxies located at the far-side boundaries of the void is a simple and effective way to evaluate its gravitational effect on the LG. Since measuring the distances and peculiar velocities of the galaxies have large uncertainties and the errors are getting larger propotional to their distances, all other more distant voids cannot provide us a chance to investigate their detailed kinemtatics.

2. Current Status of the Project

As a sample for the Tully-Fisher relation, we have selected 57 spiral galaxies with $cz \leq 5000$ km/s around the North Supergalactic pole. We have tested and confirmed the feasibility of the detection of the excess of receding motions with this number of sample using results of a numerical simulation on the formation of the large-scale structure. Near-infrared (H band) photometry have been made using University of Hawaii 2.2m telescope, and HI line widths have been measured at Nancay radio observatory, France.

The data processing is still underway, and currently we have obtained redshift-independent distances for ~20 galaxies among our sample. We can see the systematic excess of receding motions for galaxies closer than 50 h^{-1} Mpc. Increase of the number of sample will improve the statistical significance and confirm the reality of expansion of the Local void.
(Comments are welcome. Please contact *iwata@kusastro.kyoto-u.ac.jp*.)

References

[1] E. Branchini, L. Teodoro, C.S. Frenk et al., MNRAS 308 (1999) 1.

[2] D. Burstein, in *Cosmic Flows 1999: Towards an Understanding of Large-Scale Structure* (2000) ASP Conf. Ser, vol. 201, 178.

[3] S.M. Faber and D. Burstein, in *Large-Scale Motions in the Universe* (1988) Princeton University Press, NJ, 115.

Strong Correlation between the High-latitude Steady Unidentified Gamma-ray Sources and Possibly Merging Clusters of Galaxies

Wataru KAWASAKI
Department of Astronomy, School of Science, The University of Tokyo, 7-3-1 Hongo, Bunkyo-ku, Tokyo 113-0033, Japan
Tomonori TOTANI
Princeton University Observatory, Peyton Hall, Princeton, NJ 08544-1001, USA

Abstract

We report an indirect but the first evidence that merging clusters of galaxies are a promising candidate of physical counterpart of the steady unidentified EGRET gamma-ray sources in the high galactic-latitude sky. We assessed the projected number density excess of clusters and merging cluster candidates around 7 such EGRET sources. A strong excess of 3.7 σ is found for merging cluster candidates while, in contrast, not for single clusters.

1. Introduction

Nearly three decades after the beginning of systematic exploration for astronomical gamma-ray sources, still there is a great mystery that over half ($\sim$ 170 out of 271) of the gamma-ray sources listed in the third EGRET catalog [1], the largest catalog to date, are unidentified in other wavelengths. Though about 1/3 of the unidentified EGRET sources are thought to have extragalactic origin, no powerful candidate other than hidden AGNs has been proposed as the physical counterpart of the (likely) extragalactic sources consisting of 19 sources at $|b| > 45°$ ($\sim$ 65 in the whole sky). Totani & Kitayama (2000) [2] proposed that shock-heated region in forming clusters of galaxies could be an origin of the extragalactic steady EGRET sources. They showed such nearby "gamma-ray clusters" can explain well both the number and the flux distribution of the unidentified steady EGRET sources at $|b| > 45°$. If their hypothesis is the case, gamma-ray clusters should practically be observed as apparent merging clusters in other wavelengths.

Here we report the early results from our project to systematically examine the gamma-ray cluster hypothesis using optical galaxy data (see [3] for detail).

2. Data and Analysis

Among the 19 unidentified EGRET sources at $|b| > 45°$, we focus here on the 7 sources classified as 'steady' (D. Macomb 2000, private communication) since the remaining 12 variable ones might have different origin such as flaring AGNs. To obtain a sample of clusters and of merging cluster candidates collected in a uniform and objective manner, we make a new, quasi-3D cluster sample using the APM galaxy catalog and a matched-filter cluster-finding software by [4], for the seven $4° \times 4°$ areas centered at each EGRET source and a 100 deg^2 area near the South Galactic Pole (183 deg^2 in total). After having a sample of 154 clusters, complete out to $z = 0.15$, we selected a sample of 18 merging cluster candidates. Then we assessed the projected number density excess of clusters and merging cluster candidates within 1° from the 7 EGRET sources.

3. Results

A strong number density excess of 3.7σ is seen for the merging cluster candidates. Moreover, five among the seven EGRET sources, namely, all except for 3EG J1235+0233 and 3EG J1337+5029 have one or two merging cluster candidates within 1° from them. While, in contrast, only weak excess (1.1σ) is seen when using all clusters. Since the sample of all clusters contain also the member clusters of merging cluster candidates, actually there is almost no correlation between the 7 EGRET sources and the sample of all clusters.

4. Discussion and Conclusion

Though indirect, our results support the "gamma-ray cluster" hypothesis proposed by [2] and imply that bulk of the steady unidentified EGRET sources at high-latitude might be associated with merging cluster candidates, not with single clusters, showing the first observational evidence for the existence of a new type of extragalactic gamma-ray sources. More direct verification for the existence of violent shock regions in these merging cluster candidates, such as hard X-ray observations, should be made in the near future.

References

[1] R. C. Hartman et al., ApJS 123 (1999) 79.
[2] T. Totani and T. Kitayama, ApJ 545 (2000) 572.
[3] W. Kawasaki and T. Totani, ApJ (2001) submitted. (astro-ph/0108309)
[4] W. Kawasaki et al., A&AS 130 (1998) 567.

Log-normal Property of Cosmological Density Fluctuations

Issha KAYO
Department of Physics, School of Science, The University of Tokyo, 7-3-1 Hongo, Bunkyo-ku, Tokyo 113-0033, Japan

Abstract

We quantitatively study the probability distribution function (PDF) of cosmological nonlinear density fluctuations from N-body simulations with Gaussian initial condition[1][2]. In particular, we examine the validity and limitations of one-point and two-point log-normal PDF models against those directly estimated from the simulations. We find that the one-point log-normal PDF describes very accurately the cosmological density distribution even in the nonlinear regime (the rms variance $\sigma_{\rm nl} < 4$ and the over-density $\delta < 100$). Furthermore the two-point log-normal PDFs are also in good agreement with the simulation data from linear to fairly nonlinear regime, while slightly deviate from them for $\delta < -0.5$. Thus the log-normal PDF can be used as a useful empirical model for the cosmological density fluctuations. While this conclusion is fairly insensitive to the shape of the underlying power spectrum of density fluctuations $P(k)$, models with substantial power on large scales, i.e., $n \equiv d\ln P(k)/d\ln k < -1$, are better described by the log-normal PDF. Spherical collapse model is also discussed as a physical candidate of an alternative method for deriving good expression of PDF, because the lognormal model is mere an emprical ones. The result is winning of lognormal PDF. On the other hand, we note that the one-to-one mapping of the initial and the evolved density fields consistent with the log-normal model does not approximate the broad distribution of their mutual correlation even on average. Thus the origin of the phenomenological log-normal PDF approximation still remains to be understood.

Lognormal PDF

One-point lognormal PDF is discribed as

$$P^{(1)}_{\rm LN}(\delta) = \frac{1}{\sqrt{2\pi\sigma_1^2}} \exp\left[-\frac{\{\ln(1+\delta)+\sigma_1^2/2\}^2}{2\sigma_1^2}\right]\frac{1}{1+\delta}, \quad \sigma_1^2 \equiv \ln(1+\sigma_{\rm nl}^2). \quad (1)$$

where $\sigma_{\rm nl}$ is rms of field δ. Two-point "lognormal" PDF can be derived on the analogy of one-point [3]:

$$P^{(2)}_{\rm LN}(\delta_1, \delta_2; r) = \frac{1}{2\pi\sqrt{S^2 - E^2}} \exp\left[-\frac{S\{L_1^2 + L_2^2\} - 2EL_1L_2}{2\{S^2 - E^2\}}\right] \frac{1}{(1+\delta_1)(1+\delta_2)}, \tag{2}$$

where

$$E \equiv \ln(1+\xi_{\rm nl}), \qquad L_1 \equiv \ln\left\{(1+\delta_1)\sqrt{1+\sigma_{\rm nl}^2}\right\},$$
$$S \equiv \ln(1+\sigma_{\rm nl}^2), \qquad L_2 \equiv \ln\left\{(1+\delta_2)\sqrt{1+\sigma_{\rm nl}^2}\right\},$$

and $\xi_{\rm nl}$ is the two-point correlation function at separation r.

References

[1] Y. P. Jing and Y. Suto, ApJ 494 (1998) L5
[2] Y. P. Jing, ApJ 503 (1998) L9
[3] I. Kayo, A. Taruya, and Y. Suto, ApJ 561 (2001) 22

Formation and Chemodynamical Evolution of Elliptical Galaxies: Monolithic Collapse and Major Merger

Chiaki KOBAYASHI
Department of Astronomy, School of Science, University of Tokyo, 7-3-1 Hongo, Bunkyo-ku, Tokyo 113-0033, Japan

Abstract
I simulate the chemodynamical evolution of elliptical galaxies using the GRAPE-SPH code. By reproducing the detailed observations, I discuss the origin of elliptical galaxies: Elliptical galaxies form through the successive merging of galaxies with various masses, which looks like a major merger on one extreme end, and a monolithic collapse of slow-rotating gas cloud in the other extreme end.

1. Introduction

Two competing scenarios of the formation of elliptical galaxies have so far been proposed: [A] Elliptical galaxies should form monolithically by gravitational collapse of gas cloud with considerable energy dissipation (e.g., Larson 1974; Arimoto & Yoshii 1987), or [B] alternatively ellipticals should form via mergers of relatively small galaxies (e.g., Kauffmann & White 1993; Cole et al. 1994).

I construct the GRAPE-SPH chemodynamical model that includes various physical processes associated with the formation of stellar systems; radiative cooling, star formation, feedback of Type II and Ia supernovae, and of stellar winds, chemical enrichment, and UV background radiation (see Kobayashi 2002, PhD thesis, for the detail). By comparing the theoretical predictions with the detailed observations, I discuss the formation and evolution of elliptical galaxies.

2. Results

I simulate 72 slow-rotating spherical fields (spin parameter $\lambda \sim 0.02$), and obtain 124 galaxies (82 ellipticals and 42 dwarf galaxies) from the CDM initial fluctuation. Most stars of ellipticals form with the initial star burst at $z \gtrsim 2$, while dwarfs undergo relatively continuous star formation. All simulated galaxies have the de Vaucouleurs' surface brightness profiles. In my scenario, galaxies form through the successive merging of sub-galaxies. The merging history is various and the difference lies in the seed in the initial conditions. According to the merging histories, I classify ellipticals into 2 classes; [A] *Monolithic-like* (i.e., non-major merger): Galaxies form at $z \sim 3$ through the assembly of many subgalaxies with the stellar masses of $M \sim 10^{8-9}$. [B] *Major merger*: Galaxies undergo the major merger at $z \lesssim 3$ with the mass ratios of merging galaxies being $f \gtrsim 0.2$.

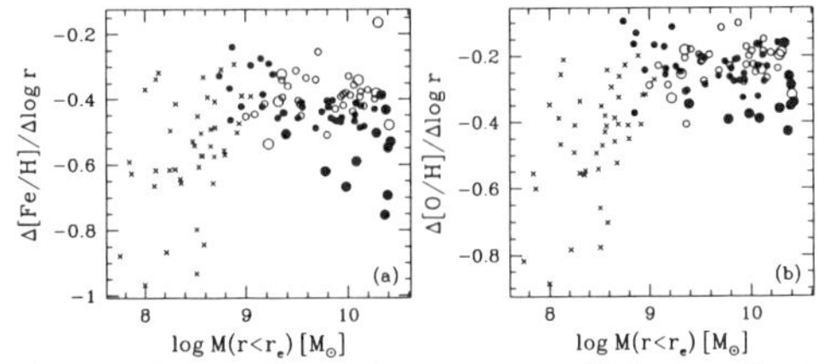

Fig. 1. No correlation between metallicity gradients vs. masses. Symbols show the merging histories of elliptical galaxies: non-major merger (filled circles) and major merger (open circles). The crosses are for dwarf galaxies.

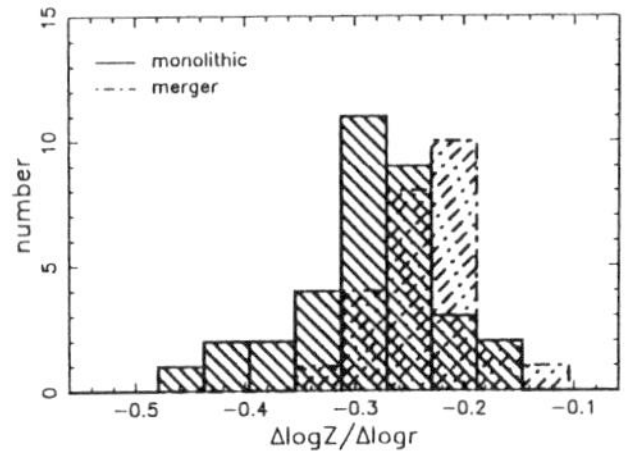

Fig. 2. The histograms of the metallicity gradients for non-major merger (solid line) and major merger galaxies (dash-dotted line).

I reproduce the slopes of the relations among global properties, although the luminosities and surface brightnesses of the simulated galaxies are 10 times smaller than the observation: i) the Faber-Jackson relation, ii) the effective radius-surface brightness relation, iii) the fundamental plane, and iv) the mass/luminosity-metallicity relations.

I success in reproducing the observations of the radial metallicity gradients: i) The average of the oxygen gradients is ~ -0.3 and the dispersion is ~ 0.2, which are both consistent with the observations with the Mg_2 index (Kobayashi & Arimoto 1999). ii) For iron, the dispersion is ~ 0.2, and the average is ~ -0.4, which is steeper than oxygen. This is because the late star formation increases the iron abundance at the center. iii) No correlation between gradients and masses is produced (Fig.1). The metallicity gradients do not depend on the galaxy mass, and the variety of the gradients stems from the difference of the merging histories. At $M > 5 \times 10^9 M_\odot$, the galaxies that form monolithically (filled circles) have steeper gradients, and the galaxies that undergo major mergers (open circles) have shallower gradients. The distributions for [A] non-major merger (solid line) and [B] major merger galaxies (dash-dotted line) are quite different (Fig.2).

3. Conclusions and Discussion

I simulate the chemodynamical evolution of a hundred elliptical galaxies using the GRAPE-SPH code. I reproduce the slopes of the correlations among global properties such as the fundamental plane and the mass-metallicity relation. Simultaneously, I reproduce the variety of the internal structures, i.e., no correlation between metallicity gradients and galaxy masses.

The global properties of elliptical galaxies are mainly determined from the galaxy masses, while the metallicity gradients are much affected by the merging events. Therefore, the merging histories can in principle be inferred from the observed metallicity gradients of the present-day galaxies. The observed variation in the metallicity gradients cannot be explained by either the *monolithic collapse* only or the *major merger* only. It is well reproduced in the present model where both formation processes arise based on the CDM scenario.

Equation of State of Electron in Strong Magnetic Fields and Neutrino Driven Wind in Supernovae

Kazunori KOHRI
Yukawa Institute for Theoretical Physics (YITP), Kyoto University, Kyoto, 606-8502, Japan
Shoichi YAMADA
Institute of Laser Engineering (ILE), Osaka University, Osaka 565-0871, Japan
Shigehiro NAGATAKI
Department of Physics, School of Science, The University of Tokyo, 7-3-1 Hongo, Bunkyo-ku, Tokyo 113-0033, Japan

Abstract

In this study, we investigate the equation of state of electron in strong magnetic fields which are greater than the critical value $B_c \simeq 4.4 \times 10^{13}$ Gauss. We find that such a strong magnetic field induces the anisotropic pressure of electron. We apply the result to the neutrino-driven wind in core-collapse supernovae and find that it can produce large entropy per baryon, $S \sim 400 k_B$. This mechanism might successfully account for the production of the heavy nuclei with mass numbers A = 80 – 250 through the r-process nucleosynthesis.

1. Anisotropic pressure in strong magnetic field

Recent years strongly magnetized neutron stars ($B > 10^{15}$ Gauss) called "magnetar" have been reported. Their magnitude is much stronger than the critical value $B_c \equiv m_e^2/e \simeq 4.4 \times 10^{13}$ Gauss if we assume that their spin-down originates from the energy loss of the electromagnetic dipole radiation. Such a strong magnetic field is intriguing in the astrophysical context because unlike the weak magnetic fields, some extraordinary phenomena would be expected from large energy splitting of Landau levels, large refractive indices of photons, photon splitting effect and so on, (see Refs. [1, 2] and references therein). In this paper, we study the equation of state of electron in strong magnetic fields and show that the electron pressure can become anisotropic. In addition, we show that the anisotropic pressure can generate entropy and that it can have an implication for the r-process nucleosynthesis in the neutrino driven wind in core-collapse supernovae. Such a large entropy production is required for the successful r-process nucleosynthesis in the wind to account for the observational solar abundances.

In Fig. 1 we plot the extra entropy production $\Delta S/k_B$ as a function of the magnitude of the magnetic field. As a result, we obtain a large entropy $S \sim 400k_B$ for $B \simeq (4-6) \times 10^{16}$ Gauss [3]. This mechanism of the entropy production might successfully give an account of the observational solar abundances of the heavy nuclei through the r-process nucleosynthesis in the magnetized neutrino wind.

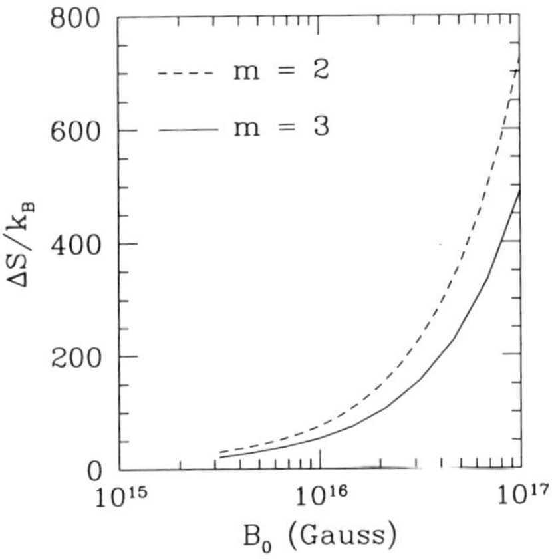

Fig. 1. Plot of the extra entropy production $\Delta S/k_B$ as a function of the magnitude of the magnetic field at the surface of the neutron star. The dashed line represents the case of $m = 2$ which corresponds to the configuration which means the flux conservation of the magnetic field. The solid line represents the case of $m = 3$ which corresponds to the configuration of the dipole.

References

[1] S. L. Adler, Annals Phys. **67**, 599 (1971).
[2] K. Kohri, and S. Yamada, to appear in Phys. Rev. D, astro-ph/0102225.
[3] K. Kohri, S. Yamada, and S. Nagataki, hep-ph/0106271.

Iron line in the Afterglow of GRB 000214 and Environment around the GRB.

Kei Kotake and Shigehiro Nagataki
Department of Physics, School of Science, The University of Tokyo, 7-3-1 Hongo, Bunkyo-ku, Tokyo 113-0033, Japan

Abstract

We present a model for explaining the Fe Kα line in the afterglow of GRB 000214. In this study, we pose the importance to seek the physically natural environment around GRB 000214. For reproducing the observation, we need a ring-like circumstellar matter (CSM) around the progenitor, like that of SN 1987A produced by the mass-loss of the progenitor and the fireball spread over all directions. We conclude that the line in GRB 000214 requires the CSM whose number density of : $n_{\rm CSM} \sim 10^{10} - 10^{11}$ cm^{-3}. This agrees with previous observations of $n_{\rm CSM}$ ([1]). The mass loss model seems more natural than the supranova model ([2]) in the sense that the mass loss model naturally explains the origin of the CSM. So far, hypernova or collapsar model ([5]) (hereafter we call them jet models) have been tried to explain the birth of GRB and the stellar death of a massive star at the same time. The mass loss model requires large opening angle of the jet, on the other hand, the jet models require small opening angle. If the correlation between the iron line and the opening angle will be found by the observation (by HETE, SWIFT), it will constrain the GRB models, such as the mass loss model and the jet models.

1. Introduction

The possible association of GRB 980425 with the type Ic supernova 1998bw ([3]) has been reported. The observation imply that the GRBs are related with the death of the massive stars. Thus, we consider that the death of the massive star and the origin of the GRB are associated.

Recently, four GRBs displaying iron lines in the X-ray afterglow have been discovered. The most recent one is in GRB 000214 ([4]). What are the implications of the observation ? How can we explain the lines in the framework of the death of massive stars ?

The supranova model tackles to explain the problem. In the model, the system is assumed to explode twice. The ring-like remnant should be produced so as to realize a GRB and an afterglow, although it seems a severe requirement. It is because the usual supernova remnants have shell structures. In addition, it seems difficult to explain the association between SN 1998bw and GRB 980425 within one day. The region around the GRB after one day of the supernova will be very gaseous, then the fireball created by the second explosion will be stopped by the interaction between the fireball and the gaseous environment.

Jet models have been intensively investigated to explain the origin of the GRB and the massive star death at the same time. Then which is correct between the mass loss model and jet models ? We consider the problem should be settled down by observations. Here, we tabulate the difference between them.

	opening angle (Small))	opening angle (Large)	Correlation
Jet models	○	○	NO
Mass loss model	×	○	YES

The "○" in the table means the line can be detected, "×" in the table is on the contrary. "Correlation" means the correlation of the line and the opening angle. For the jet models, the line should be produced within the jet, irrespective of the opening angle of the jet, the line should be detected. In the sense, the correlation of the line and the opening angle of the jet does not exist. On the other hand, for the mass loss model, only when the opening angle of the jet is large, the lines can be produced. In the sense, the correlation of the line and the jet opening angle does exist. Recently, the opening angle of the jet has been observed in some GRBs. Unfortunately, no observation exist suggesting the correlation. Thus, in the near future, by HETE or SWIFT, we hope that the correlation will be detected to determine whether the mass loss model or the jet model is correct.

References

[1] K. Nomoto et al. 2000, *Supernovae and Gamma Ray bursts ed, M.Livio et al. (Cambridge University Press)*

[2] Vietri, M., Perola, C., Piro, L., & Stella L. 1999, MNRAS , 308, L29

[3] Galama T.J., Vreeswik, P. M., van Paradijs, J., Konvelioton, C., Augusteijn, T., Bohnhardt, H., Brewer, J.P., & Doublier, V. 1998, Nature 395, 670

[4] Antonelli, L. A., Piro,L., Vietri, M., Costa, E., Soffita, P., Feroci, M., Amati, L., & Frontera, F. 2000, APJ, 545, L39

[5] Woosley, S. E. 1993, Astrophys. J, 405, 273

Nonlinear Gravitational Interactions in the Stabilized Randall-Sundrum Two Branes Model

Hideaki KUDOH and Takahiro TANAKA
Yukawa Institute for Theoretical Physics, Kyoto University, Kyoto 606-8502, Japan

Abstract

The second order perturbation is investigated in the Randall-Sundrum two branes model with the radius stabilization mechanism. As the stabilization model, we assume the Goldberger-Wise stabilization model. We discuss the gravity on each brane induced by the matter on its own side, taking the limit of the large coupling. It is shown that the 4D Einstein gravity is approximately recovered in the second order perturbations.

1. Introduction

The Randall and Sundrum two branes model is constructed in a 5D AdS spacetime. Apart from the fine tuning of the brane tension that is necessary to solve the cosmological constant problem, one of the significant points in discussing the consistency of this model is whether the 4D Einstein gravity is recovered on the brane. The stabilization mechanism plays an important roll in the recovery of the 4D linear Einstein gravity.

A large number of studies have been made on the gravity in the brane world model. Although the model does not have a drawback in the linear perturbation, it is not trivial whether the second order gravitational perturbation works as well [1, 2]. Hence we study the second order gravitational perturbation of the RS two branes model with the stabilization mechanism due to a bulk scalar field.

2. Model and Discussion

Our analysis gives the result for the temporal component on the negative tension brane [3];

$$\triangle \bar{A}_{0-}^{(2)}(r, y_-) = 8\pi G(\rho_-^{(2)} + 3P_-^{(2)}) - 4\Phi_- \triangle \Phi_- + O\left(\frac{\beta_- \Phi_-^2}{r_\star^2}\right). \tag{1}$$

where $\rho_-^{(2)}$ and $P_-^{(2)}$ are second order quantities, and they are respectively the energy density and the pressure of the nonrelativistic matter on the negative tension brane. The Newton's potential induced by the first order of the energy density is written by Φ_-. The first two terms are identical with the 4D Einstein gravity, and the last term is the correction for it. The suppression factor is evaluated as

$$\beta_- := \frac{\ell^2}{a_-^4 r_\star^2} = \left(\frac{0.1mm}{r_\star}\right)^2 \left(\frac{10^{-16}}{a_-}\right)^4 \left(\frac{\ell}{\ell_{Pl}}\right)^2, \quad (y = y_-), \tag{2}$$

where ℓ is the AdS curvature, which is assumed to be Planck scale. Hence the 4D Einstein gravity is approximately recovered on the negative tension brane. The detail discussion including the analysis of the perturbation on the positive tension brane is given in Ref. [3].

We summarize our results. When we consider the case in which the matter fields are on the negative tension brane, the correction to the 4D Einstein gravity appears at the relative order of $O((a_+/a_-)^4(\ell/r_\star)^2)$, where ℓ is the AdS curvature scale, $r_\star$ is the typical length scale of the perturbation and (a_+/a_-) is the ratio between the warp factors on the positive and the negative tension branes. When this ratio (a_+/a_-) is $O(10^{16})$, the hierarchy between Planck and electro-weak scales can be explained. With this choice of the hierarchy, the correction to the metric in the linear perturbation becomes comparable to the usual Newtonian potential when $r_\star < 0.1$ mm. However, this correction does not give contribution to the force outside the matter distribution. Hence, it seems to be harmless to reproduce the predictions of the 4D Einstein gravity. We have not confirmed if this feature remains in the second order perturbation, but the correction is suppressed by the above factor compared to the usual post-Newtonian correction. Hence, the effect due to this correction is almost impossible to detect.

When we consider the case in which the matter fields are on the positive tension brane, the correction to the 4D Einstein gravity in the linear perturbation appears at the relative order of $O((\ell/r_\star)^2)$. While the correction in the second order perturbation is $O((a_+/a_-)^2(\ell/r_\star)^2)$ compared to the usual post-Newtonian terms. Hence, it seems that the deviation from the 4D Einstein gravity appears at a larger scale in the second order perturbation. However, this is very likely to be an artifact due to the limitation of our approximation scheme.

References

[1] I. Giannakis and H. C. Ren, Phys. Rev. D **63**, 024001 (2001)
[2] H. Kudoh and T. Tanaka, Phys. Rev. D **64**, 084022 (2001)
[3] H. Kudoh and T. Tanaka, hep-th/0112013.

Quasi-normal Modes of D-brane Black Holes

Yasunari KURITA
Graduate School of Human and Environmental Studies, Kyoto University, Sakyo-ku, Kyoto 606-8501, Japan
Masa-aki SAKAGAMI
Faculty of Integrated Human Studies, Kyoto University, Sakyo-ku, Kyoto 606-8501, Japan

Abstract

We investigate the scattering problem of a D3-brane black hole in type IIB String theory with wave interpretation, and review that of black 3-brane in Supergravity. Their absorption cross sections are in exact agreement. Therefore, we expect Quasi-normal modes of D3-brane black holes also agree in two theories.

1. Introduction

In String theory, some D-brane systems are thought of as black holes, because they become black hole solutions in Supergravity limit. A few years ago, some physical quantities, such as entropy and absorption cross section, were compared in String theory and Supergravity. In most cases, they are in exact agreement, which imply AdS/CFT correspondence. Absorption cross section of D3-brane black holes were calculated in terms of quantum field theory [1] [3], in which absorption process of a dilaton is its transition into two gauge fields. In black 3-brane cases, absorption probability is given by the solution of scattering problem on black 3-brane background. Two absorption cross sections agree in case of low energy incidental particles. This imply that field theory on D3-brane can describe bulk space-time in supergravity. Then we expect quasi-normal modes(QNM) of black 3-brane can be calculated by the theory on D3-brane. We, however, don't know how to calculate QNM by quantum field theory on D3-brane. In order to do it, we work out a new method to calculate scattering problem of D-brane black holes.

2. The wave scattering of D3-brane

In this section, we introduce our new calculation method briefly. It must give the same absorption cross section with quantum field theory. To show it, we consider scattering problem of a D3-brane black hole. The low energy effective action is Dirac-Born-Infeld action with flat induced metric and bulk action with free dilaton in ten dimensions. We interpret incidental particles as waves, and consider D3-brane scattering as wave scattering. The time evolution of this waves is given by their wave equations, which show a wave interact with gauge fields on D3-brane. We can solve the wave equation by perturbation about interaction with gauge fields. At lowest order, dilaton waves don't interact with D3-brane, and pass through it. And we take negative energy gauge field solutions, which means Dirac's sea or vacuum on D3-brane. At first order, dilaton waves interact with negative energy gauge fields and excite them into positive energy ones on D3-brane. At second order, a excited gauge field and negative one annihilate(they are interpreted as a particle-hole pair) and emit a dilaton wave into bulk. Then, amplitude of emitted waves decreases that of reflection waves in lowest order. This is the absorption process, and we can calculate absorption cross section, which agree with the results of the quantum field thory and supergravity.

3. The condition of QNM and Discussion

Our new calculation methods of D-brane scattering problems give dilaton wave functions. Then we can treat the D3-brane scattering problem as wave scattering. Now, we can also obtain the condition of QNM in D3-brane picture as usual wave scattering analysis. In supergravity, a scattering problem of massless scalar fields on black 3-brane background can be solved in terms of Mathieu's functions, and the asymptotic form of them gives absorption probability [2]. It also gives the condition of QNM that they are decaying modes of no incidental waves. Our interest is whether QNM of D3-brane agree with those of black 3-brane. We can't exactly solve the condition equation of QNM yet and don't compare two QNMs concretly. Moreover, the fact that the effective parameter region of D-brane black holes is different with that of black branes makes it difficult to compare them. We leave this problem as future work here.

References

[1] Igor R.Klebanov, Nucl. Phys. B496 (1997) 231.

[2] Steven S.Gubser, and Akikazu Hashimoto, hep-th/9805140.

[3] Steven S.Gubser et.al., Nucl. Phys. B526 (1998) 393.

The Pairwise Peculiar Velocity Distribution Function from the Halo Density Profile

Takeshi KUWABARA
Department of Physics, School of Science, The University of Tokyo, 7-3-1 Hongo, Bunkyo-ku, Tokyo 113-0033, Japan

Abstract

We derive the pairwise peculiar velocity distribution function of dark matter particles applying the dark matter halo approach. Unlike the previous work, we do not assume a Gaussian velocity distribution function of dark matter in a single halo, but compute it self-consistently with the assumed density profile for dark matter halo. The resulting distribution function is well approximated by an exponential distribution which is consistent with the previous observational, numerical and theoretical results. We also compute the pairwise peculiar velocity dispersion for different density profiles, and provide a practical fitting formula. We apply an empirical biasing scheme into our model and present prediction for pairwise peculiar velocity dispersion of *galaxies*, and reproduce the previous results of simulations using our semi-analytical method.

1. Introduction

Since analyzed the anisotropy in the galaxy distribution from Center for Astrophysics (CfA) redshift catalog [1], it has been recognized that the peculiar velocity field of galaxies induces a significant systematic effect in statistics of observed galaxy distributions in redshift space. In particular, virialized random motion of galaxies produces an elongated pattern of galaxy distribution along the line of sight, called *finger-of-God*. This effect significantly suppresses the amplitude of the two-point correlation function of galaxies in redshift space, especially on scales below $\sim 3h^{-1}$Mpc. Some observations[1] and simulations[2] already showed that the pairwise velocity distribution function are describes by the exponential form, instead of a Maxwellian. Theoretical models[3][4] phenomenologically introduced a nonlinear model of the pairwise peculiar velocity distribution function using the Press–Schechter formalism. To be more specific, they assume that any dark matter particle belongs to one of virialized clumps (dark halos) with the 1-point velocity distribution function being a Maxwellian form. Based on Jeans theorem, we self-consistently derived the pairwise velocity distribution function,

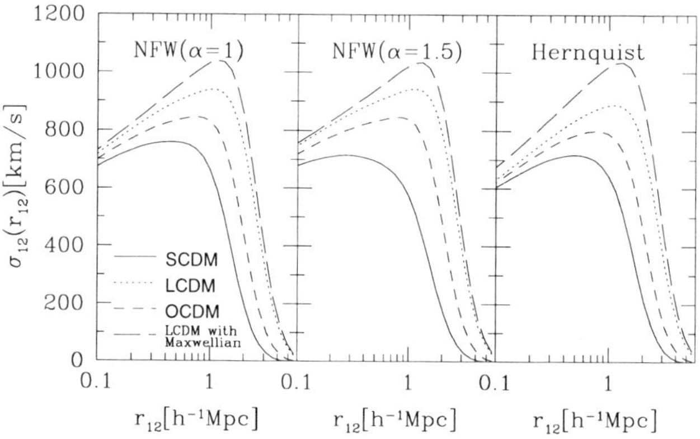

Fig. 1. The pairwise velocity dispersion.

adopting same assumptions of previous theoretical work excluding that 1-point velocity distribution function is a Maxwellian form.

2. Analysis and result

The pairwise velocity distribution function is computed by multiplying follow three elements and integrating it of mass: the pairwise velocity distribution in a halo, the halo mass function, and the density profile of halos.

The result of the pairwise velocity distribution function becomes nearly exponential form consistent with observations and simulations. Figure 1. shows the pairwise velocity dispersion, integrated pairwise velocity distribution, for some cosmological models and the density profile of halos. In addition, we give the fitting formula of the pairwise velocity dispersion for the purpose of analysis of redshift distortion[5].

References

[1] M. Davis, and P. J. E. Peebles, ApJ 267 (1983) 465.
[2] G. P. Efstathiou, C. S. Frenk, S. D. M. White, and M. Davis, MNRAS 235 (1988) 715.
[3] R. K. Sheth, MNRAS 279 (1996) 1310
[4] R. K. Sheth, L. Hui, A. Diaferio, and R. Scoccimarro, MNRAS 325 (2001) 1288
[5] T. Kuwabara, A. Taruya, and Y. Suto, PASJ in submitted

Electromagnetic Field Structure of a Force-Free Black Hole Magnetosphere

Chul H. Lee
Theoretical Physics Institute, Department of Physics, University of Alberta, Canada T6G 2J1, and
Department of Physics, Hanyang University, Seoul 133-719, Korea

Abstract

The angular momentum of the black hole and the magnetic field lines threading the event horizon are the two essential ingredients for the Blandford-Znajek process to extract the rotational energy from a black hole. The effectiveness of the process could be jeopardized by the fact that the rotation of the black hole tends to expel magnetic field lines from the horizon. However, with the net electric charge induced on the black hole, the magnetic flux on the horizon can be retained. Assuming the force free condition in the plasma around the black hole, we discuss how to calculate the amount of electric charge induced on the horizon.

1. Introduction

R. Wald [1] considered the case of a black hole immersed in an originally uniform magnetic field whose direction is aligned with the rotation axis of the black hole, and showed that the magnetic flux through the upper half of the horizon decreases as the angular momentum of the black hole increases. It becomes zero when the black hole is maximally rotating. This could jeopardize the Blandford-Znajek process [2] [3] as a viable mechanism of extracting energy from a black hole. The rescue is provided by the fact that, if the rotating black hole immersed in a magnetic field is surrounded by plasma, it accretes net electric charge until the amount of the accreted charge reaches the equilibrium value. With a net elctric charge induced on the black hole, the magnetic flux on the event horizon does not vanish completely even if the hole is maximally rotating. Section 2 extends this analysis to the case where the asymptotic direction of the magnetic field is not aligned with the rotation axis of the black hole. Section 3 treats a more realistic case where the magnetic field is supported by an accretion disk and shows how to estimate the amount of electric charge accreted into the black hole.

2. A black hole immersed in an originnally uniform magnetic field whose direction is not aligned with the rotation axis of the hole

We consider the general case where the asymptotically uniform stationary magnetic field happens to be "oblique", i.e., aligned at some angle to the hole's axis of rotation. Indeed the case of uncharged Kerr hole was studied long ago by Bicak and Janis [4]. We denote their solution by F^{BJ}. The new solution in the presence of the accreted charge Q can be obtained by $F = F^{BJ} - \frac{Q}{2M}d\xi$ where $\xi = \frac{\partial}{\partial t}$ is the time translational Killing vector field. The magnetic flux on a generally oriented hemisphere of the event horizon can be calculated from the field configuraton thus obtained and the results, presented in Ref. [5], show that, with the net charge accreted on the hole, the magnetic flux threading the hole does not vanish even in the maximally rotating case.

3. A black hole surrounded by a force-free magnetosphere

We now consider a more realistic case where the magnetic field is supported by an accretion disk. The magnetic field is assumed to be strong enough so that the inertia of the plasma around the black hole can be ignored to make the force-free condition a good approximation. Then the structure of the electromagnetic field in the magnetosphere can be characterized by the three scalar functions, the stream function Ψ, the current potential I and the angular velocity of the magnetic field lines Ω_F, satisfying the "stream" equation [6]. Starting from the simple solution of the euation, discussed in Ref's [2] and [7], that can represent the magnetic field supported by an accretion disk, and assuming the optimal case where $\Omega_F \approx \frac{\Omega_H}{2}$, the net electric charge accreted into the black hole is calculated in Ref. [8]. For the case of a rapidly rotating black hole of $M = 3M_\odot$ and $a = 0.5M$ with a strong magnetic field of $B = 10^{15}G$ at the horizon, for example, the charge Q is estimated to be about $10^{15}C$.

References

[1] R. Wald, Phys. Rev. D10 (1974) 1680.
[2] R. Blandford and R. Znajek, MNRAS 179 (1977) 433.
[3] H. K. Lee, R. Wijers, and G. Brown, Phys. Rep. 325 (2000) 83.
[4] J. Bicak and V. Janis, MNRAS 212 (1985) 899.
[5] H. Kim, C. H. Lee and H. K. Lee, Phys. Rev. 63 (2001) 064037.
[6] D. Macdonald, and K. Thorne, MNRAS 198 (1982) 345.
[7] P. Ghosh and M. Abramowicz, MNRAS 292 (1997) 887.
[8] H. K. Lee, C. H. Lee and M. van Putten, MNRAS 324 (2001) 781

Aspherical Hypernova Explosions and the Early Galactic Chemical Evolution

Keiichi Maeda[1], Ken'ichi Nomoto,[1,2] and Paolo A. Mazzali[2,3]
[1]*Department of Astronomy, School of Science, The University of Tokyo, 7-3-1 Hongo, Bunkyo-ku, Tokyo 113-0033, Japan*
[2]*Research Center for the Early Universe (RESCEU), School of Science, The University of Tokyo, 7-3-1 Hongo, Bunkyo-ku, Tokyo 113-0033, Japan*
[3]*Osservatorio Astronomico di Trieste, via G. B. Tiepolo 11, I-34131 Trieste, Italy*

Abstract

We present aspherical explosion models for hypernovae (hyper-energetic supernovae), which can reproduce the observed late time spectra of SN1998bw (a prototype of a hypernova) better than spherical models. Possible contribution of such explosions on the early Galactic chemical evolution is discussed.

1. Introduction

Recently we have realized the existence of very energetic core-collapse supernovae, i.e., 'hepernovae' (see Nomoto et al. 2002 [1] for a review). SN1998bw was the first one identified as a hypernova with kinetic explosion energy as $E \gtrsim 10^{52}$ erg (e.g., Nakamura et al. 2001 ; [2]). The (spherical) hypernova model was very successful to explain the early phase observations of SN1998bw, while in the late light curve and spectra Nakamura et al. (2001; [2]) and Mazzali et al. (2001; [3]) have identified some signatures of an asymmetric explosion.

2. Aspherical Model for SN1998bw

We have examined the effect of aspherical (jet-like) explosions on nucleosynthesis in hypernovae. The progenitor model is the 16 $M_\odot$ He core of the 40 $M_\odot$ star and the explosion energy is $E = 1 \times 10^{52}$ ergs (Maeda et al. 2002; [4]).

Figure 1 (left) shows the 2D distribution of ^{56}Ni and ^{16}O in the homologous expansion phase. In the z-direction, where the ejecta carry more kinetic energy, the shock is stronger and post-shock temperatures are higher. Therefore, larger amounts of α-rich freeze-out elements, such as ^{4}He, ^{44}Ti, and ^{56}Ni are produced in the z-direction than in the r-direction. The calculated line profiles of the Fe-dominated blend near 5200Å, and of O I] 6300, 6363Å(Figure 1: right) well

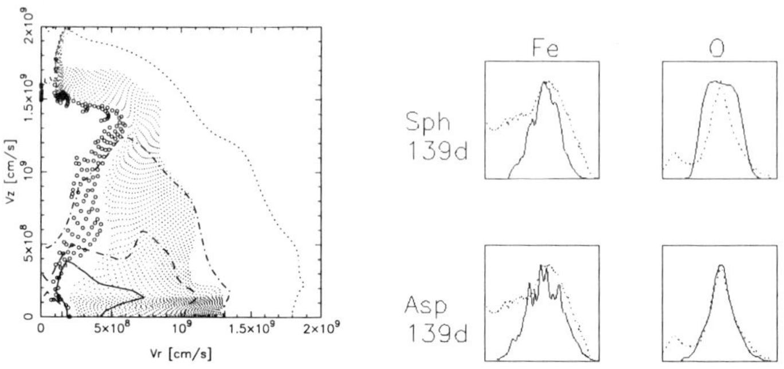

Fig. 1. Left: The distribution of ^{56}Ni (open circles) and ^{16}O (dots). The open circles and the dots denote test particles in which the mass fraction of ^{56}Ni and ^{16}O, respectively, exceeds 0.1. The lines are density contours ([4]). Right: The profiles of the Fe-blend (left panels) and of O I] 6300, 6363 Å (right panels) viewed at 15° from the jet direction ([4]) with the late time spectra of SN1998bw. The top panels show the profiles of the spherically symmetric model.

reproduce the observed late time spectral feature, which could not be explained by spherical models [3].

3. Possible Contribution on the Early Galactic Chemical Evolution

In the spherical case, Zn is produced only in the deepest layer, while in the aspherical model, the complete silicon burning region is elongated to the z (jet) direction, so that [Zn/Fe] is enhanced irrespective of the mass cut. On the other hand, ^{55}Mn, which is produced by incomplete silicon burning, surrounds ^{56}Fe and located preferentially in the r-direction.

In this way, larger asphericity in the explosion leads to larger [Zn/Fe] and [Co/Fe], but to smaller [Mn/Fe] and [Cr/Fe]. Then, if the degree of the asphericity tends to be larger for lower [Fe/H], the trends of [Zn, Co, Mn, Cr/Fe] follow the ones observed in metal-poor stars (e.g., Nomoto et al. 2002: [1]).

References

[1] Nomoto, K., Maeda, K., Umeda, H. 2002, in this volume

[2] Nakamura, T., Mazzali, P. A., Nomoto, K., & Iwamoto, K. 2001a, ApJ, 550, 991

[3] Mazzali, P.A., Nomoto, K., Patat, F., & Maeda, K. 2001, ApJ, 559, 1047

[4] Maeda, K., Nakamura, T., Nomoto, K., Mazzali, P.A., & Hachisu, I. 2002, ApJ, 565, 405

Quintessence in a brane world

Shuntaro MIZUNO
Department of Physics, Waseda University, 3-4-1 Okubo, Shinjuku-ku, Tokyo 169-8555, Japan
and Kei-ichi MAEDA
Department of Physics, Waseda University, 3-4-1 Okubo, Shinjuku-ku, Tokyo 169-8555, Japan
and Advanced Research Institute for Science and Engineering, Waseda University, Shinjuku, Tokyo 169-8555, Japan

Abstract

We reanalyze a new quintessence scenario in a brane world model, assuming that a quintessence scalar field is confined in our 3-dimensional brane world. We first study three typical quintessence models : (1) an inverse-power-law potential, (2) an exponential potential, and (3) kinetic-term quintessence model. After that we investigate the naturalness for each model and evaluate the advantage in a brane world.

1. Introduction

Quintessence is one of the candidates for the observed dark energy [1]. There seems to be, however, still a kind of fine-tuning problem as far as the conventional cosmology. We have tried to improve the situation by considering the brane cosmology[2]. Here we adopt the Randall-Sundrum type II brane scenario[3], because the model is simple and concrete. We have prepared [4] for those who want to know more.

2. Basic Equations

We shall use the following equations.

$$H^2 = \frac{1}{3m_4^2}\rho + \frac{1}{36m_5^6}\rho^2, \tag{1}$$

$$\ddot{\phi} + 3H\dot{\phi} + \frac{dV}{d\phi} = 0, \tag{2}$$

where P and ρ are the total pressure and total energy density of matter fields, respectively and V is a potential of the scalar field. The origin of a scalar field confined in the brane might be a condensation of some fermions confined in the brane. The 5-dimensional Planck mass m_5 could be much smaller than 4-dimensional Planck mass m_4.

3. Results and Conclusion

With an inverse power law potential model ($V(\phi) = \mu^{\alpha+4}\phi^{-\alpha}$),which may appear as an effective potential for a kind of fermion condensation in a supersymmetric QCD model, we show that in the quadratic dominant stage, the density parameter of a scalar field Ω_ϕ decreases as $a^{-4(\alpha-2)/(\alpha+2)}$ for $2 < \alpha < 6$, which is followed by the conventional quintessence scenario. This feature provides us wider initial conditions for a successful quintessence. In fact, even if the universe is initially in a scalar-field dominant, it eventually evolves into a radiation dominant era in the ρ^2-dominant stage. Assuming an equipartition condition, we discuss constraints on parameters, resulting that $\alpha \geq 4$ is required. This constraint also restricts the value of the 5-dimensional Planck mass, e.g. $4 \times 10^{-14} m_4 \leq m_5 \leq 3 \times 10^{-13} m_4$ for $\alpha = 5$.

For an exponential potential model $V = \mu^4 \exp(-\lambda\phi/m_4)$, which is often found in unified theories of fundamental interactions of particles such as supergravity theory, this provides an extremely flat potential in ρ^2-dominant stage, for λ to be successful quintessence. Therefore we cannot find a natural and successful quintessence scenario as it is.

While, for a kinetic-term quintessence, whose "pressure" and "energy density" is given by $p_\Phi = \mu^{4-\alpha}\Phi^{-\alpha}(-X + X^2)$, $\rho_\Phi = \mu^{4-\alpha}\Phi^{-\alpha}(-X + 3X^2)$, with $X = (1/2)\dot{\Phi}^2$, we find a tracking solution even in ρ^2-dominant stage, rather than the Ω_ϕ-decreasing solution for an inverse-power-law potential. Then we do find a little advantage in a brane world. Only the density parameter increases more slowly in the ρ^2-dominant stage, which provides a wider initial condition for a successful quintessence. (Note that in this model. Φ does not have mass dimension but inverse-mass dimension.)

References

[1] R. R. Caldwell, R. Dave and P. J. Steinhardt, Phys. Rev. Lett. **80**, 1582 (1998).

[2] T. Shiromizu, K. Maeda, and M. Sasaki, Phys. Rev. D **62**, 024012 (2000).

[3] L. Randall and R. Sundrum, Phys. Rev. Lett. **83**, 4690 (1999).

[4] S. Mizuno, and K. Maeda, Phys. Rev. D **64**, 123521 (2001).

Fluctuation-Dissipation Relation of the Vacuum Noise in Curved Spacetime

Tsukasa MURATA, Keiji TSUNODA and Kazuhiro YAMAMOTO
Department of Physics, Hiroshima University, Higashi-Hiroshima 739-8526, Japan

Abstract

Motivated by a recent work by Terashima(Phys. Rev. D**60** 084001), we revisit the fluctuation-dissipation (FD) relation between the dissipative coefficient of a detector and the vacuum noise of quantum fields in curved spacetime. By applying his investigation to the N-dimensional de Sitter spacetime, we show that his prescription properly reproduces the vacuum noise in de Sitter spacetime. This practice is useful to understand that the essential condition for the FD relation comes from a periodicity in time of the Euclidean section of curved spacetime.

1. Introduction and Method

It is well-known that the vacuum fluctuations of the Minkowski spacetime can be a thermal ambiance when viewed by uniformly accelerated observers. This effect, which is called the Unruh effect, has been investigated by many authors. Recently Terashima has proposed a new method to evaluate the Rindler noise indirectly by using the fluctuation-dissipation (FD) relation (or theorem) formulated by Callen and Welton [1]. He has explicitly shown that the FD relation holds between the Rindler noise and the dissipative coefficient of the De Witt detector which is obtained from classical equations of motion of the detector and the (scalar and Dirac) fields in N-dimensional Rindler spacetime.

Though several authors have investigated the vacuum noise of quantum fields in curved spacetime from a view point of the FD relation, however, Terashima's approach is slightly different from those previous approaches. Following the Terashima's prescription, the dissipative coefficient of the detector's inner motion is explicitly calculated from the classical equations of motion of the detector and the fields. Then the Rindler noise is simply obtained by inserting the dissipative coefficient into the FD relation, which relates the dissipative coefficient to the noise of vacuum fluctuations. He has shown that this method is successful in the system of the De Witt detector linearly coupled to the scalar field in N-dimensional Rindler spacetime, and he has generalized the method to the Dirac field by using the fermionic FD relation.

In 2- and 4-dimensional spacetimes, it is well-known that the spectrum of the Rindler noise is completely equivalent to the thermal noise spectrum with the temperature $T = \alpha/(2\pi)$, where α is the acceleration of the detector. Therefore one might think that the existence of the FD relation is trivial. In odd spacetime dimensions, however, it is known that the statistical inversion occurs. This suggests that the FD relation originates from other essential conditions rather than the similarity between the vacuum noise spectrum and the thermal one.

The Rindler spacetime is quite different to the de Sitter spacetime in spacetime character, however, the vacuum noises of these spacetimes are very similar. We have explicitly shown the existence of the FD relation using the Green function in de Sitter spacetime [2]. The Wightman functions in the de Sitter spacetime and the Rindler spacetime have the same periodicity in the direction of the imaginary time.

$$G^{+}(\Delta\tau - i\beta) = G^{-}(\Delta\tau) \qquad \text{(for scalar field)}$$

where $G^{+}(\Delta\tau) = \langle 0|\phi(x)\phi(x')|0\rangle$, $G^{-}(\Delta\tau) = \langle 0|\phi(x')\phi(x)|0\rangle$ and $\Delta\tau = \tau - \tau'$. Hence the periodicity in time of the Euclidean section is understood as an essential condition for the FD relation. The periodicity is traced back to the symmetry of the Euclidean spacetime. Thus, if a Euclidean spacetime possesses a periodicity, the FD relation might exist for the scalar and the Dirac fields [2].

2. Summary and Discussions

We have investigated the FD relation between the dissipative coefficients of a detector and the vacuum noise of the scalar and the Dirac fields which are linearly coupled to each other in the N-dimensional de Sitter spacetime explicitly. From a consideration of general curved spacetime, essential conditions for the FD relation have been summarized. The periodicity in the Wightman functions (the KMS condition) is essential for the existence of the FD relation. Hence the FD relation exists irrelevant to the statistical inversion on the contrary to naive expectation associated with the statistical inversion phenomenon. Though our result might depend on the model of the couplings between the detector and fields, this investigation is interesting because information of quantum fluctuations can be obtained by observing the classical motion of the detector in principle.

References

[1] H. B. Callen and T. A. Welton, Phys. Rev. **83**, 34 (1951)

[2] T. Murata, K. Tsunoda and K. Yamamoto, Int. J. Mod. Phys **16**, 2841 (2001)

Hoop Conjecture and Brane World Gravity

Kouji NAKAMURA
Division of Theoretical Astrophysics, National Astronomical Observatory, Mitaka, Tokyo 181-8588, Japan.
Ken-ichi NAKAO
Department of Physics, Graduate School of Science, Osaka City University, Osaka 558-8585, Japan.
Takeshi MISHIMA
Laboratories of Physics, College of Science and Technology, Nihon University, Narashinodai, Funabashi, Chiba 274-0063, Japan.

Abstract

One of the initial data of gravity for a cylindrical matter distribution confined on the brane is studied in the framework of the single brane Randall-Sundrum scenario. It is shown that the hoop conjecture does not hold in the Randall-Sundrum scenario; Even if a mass M does not get compacted into a region whose circumference in every direction is $\mathcal{C} \leq 4\pi GM$, black holes with horizons can form in the Randall-Sundrum scenario.

1. Summary

Within 4-dimensional general relativity, Thorne proposed the conjecture for the necessary and sufficient condition on black hole formation[1]; *Black holes with horizons form when and only when a mass M gets compacted into a region whose circumference in every direction is $\mathcal{C} \leq 4\pi GM$*[1]. This is well known as hoop conjecture. No counter example for this conjecture has been presented in 4-dimensional general relativity. However, strictly speaking, we do not know whether general relativity can describe the situation of strong gravity in our real universe even for classical situations. We have no experimental evidence for it.

As an alternative theory of gravity, recently, Randall and Sundrum (RS) proposed a scenario of the compactification of a higher dimension without compact manifold[2]. They considered 5-dimensional spacetime with negative cosmological constant $\Lambda < 0$ including a single 3-brane with the positive tension λ. In their scenario, it is assumed that all the physical fields, except for gravity, are confined on the brane and gravity is governed by 5-dimensional Einstein gravity. Using the cancellation of the long range forces due to the negative cosmological constant and the brane tension, i.e., $\sqrt{-6/\Lambda} = 1/\lambda =: l$, they showed that 4-dimensional Newtonian and general relativistic gravity on the brane is reproduced to more than adequate precision. The deviation in the gravitational force from Newtonian gravity appears in the scale less than l and this scenario is not ruled out by the recent experimental tests if the length scale of l is less than millimeter scale, since the corrections to the Newtonian gravitational potential has not proven in the sub-millimeter scale (see e.g.[3]).

We can easily expect that the hoop conjecture does not hold in the RS-scenario. In the 5-dimensional Einstein gravity, there are so called black string solutions those are cylindrically symmetric black holes. In the higher dimensional spacetime, gravity may become so strong that non-compact black holes can form. It is natural to expect that this 5-dimensional aspect will also appear in the short-range (the scale less than l) force of gravity in the RS-scenario, since the short-range gravity is not so sensitive to the cosmological constant. Then the hoop conjecture may not hold in the RS-scenario.

We have confirmed this speculation by showing a counter example, which is a time-symmetric initial data of gravity including a single brane with a cylindrically symmetric matter distribution on the brane. In this initial data, the matter is confined in the infinite cylinder with the radius $< 0.01l$ and the cylindrically symmetric (noncompact) marginal surfaces exist both in the bulk and brane, nevertheless the matter distribution is noncompact. Further, we can easily show that the inequality $\mathcal{C} > 4\pi GM$ is hold in this example. Then we conclude that this initial data shows the hoop conjecture does not hold in RS-scenario.

Our conclusion implies that more black holes may forms in RS-scenario than in 4-dimensional general relativity. This fact will become more important in the very early universe if the RS-scenario is true in our universe. Details of our discussion will be seen in Ref.[4].

References

[1] K. S. Thorne, in *Magic Without Magic; John Archibald Wheeler*, edited by J. Klauder (Frieman, San Francisco, 1972), p231.

[2] L. Randall and R. Sundrum, Phys. Rev. Lett. 83 (1999) 4690.

[3] G. C. Long, H. Chang, and J. Price, Nucl. Phys. B529 (1999) 23. C. D. Hoyle, U. Schmidt, B. R. Heckel, E. G. Adelberger, J. H. Gundlach, D. J. Kapner, and H. E. Swanson, Phys. Rev. Lett. 86 (2001) 1418.

[4] K. Nakao, K. Nakamura, and T. Mishima, *in preparation*.

Galaxy Number Counts in the Subaru Deep Field: A Constraint on a Hierarchical Galaxy Formation Model

Masahiro NAGASHIMA
National Astronomical Observatory, Mitaka, Tokyo 181-8588, Japan
Yuzuru YOSHII
Institute of Astronomy, The University of Tokyo, Mitaka, Tokyo 181-8588, Japan
Research Center for the Early Universe, The University of Tokyo, Tokyo 113-0033, Japan
Tomonori TOTANI and Naoteru GOUDA
National Astronomical Observatory, Mitaka, Tokyo 181-8588, Japan

Abstract

We reanalyze the faint galaxy number counts by using a semi-analytic galaxy formation model based on the hierarchical clustering scenario. We find that our model show a good agreement with observed galaxy counts both in near-infrared and optical pass-bands simultaneously.

1. Introduction

It is well known that the number of faint galaxies in a given area of sky strongly constrains the cosmological parameters because it depends on the geometry of the Universe. In recent analyses of galaxy counts by using the traditional monolithic cloud collapse models, it has been found that the standard cold dark matter (CDM) universe, or the Einstein-de Sitter (EdS) universe, is not reconciled with the observed high counts of faint galaxies and that a low-density universe is preferred (Totani & Yoshii 2000; Totani et al. 2001).

Recently, it is widely accepted that the hierarchical clustering scenario becomes a standard one. Semi-analytic modeling of galaxy formation based on the hierarchical clustering scenario also has been developed. Therefore it is important for constructing a correct galaxy formation model to compare the model with observed properties of galaxies. As for the optical galaxy counts, we have shown that our model is consistent with the HDF results (Nagashima et al. 2001). Here we investigate faint galaxy number counts both in optical and near-infrared pass-bands. Because galaxy counts in different pass-bands reflect different stellar populations and constrain star formation histories, a correct galaxy formation model must be consistent with those.

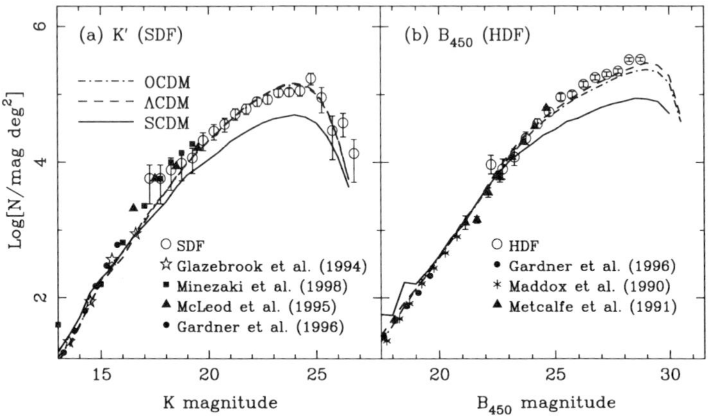

Fig. 1. Galaxy number counts in K'-band (SDF; left), in B-band (HDF; right).

2. Results and Conclusions

Our model includes the following physical processes: merging histories of dark halos, gas cooling, star formation, supernova feedback, and mergers of galaxies. Moreover some processes related with observational galaxy properties are taken into account: internal dust extinction, absorption of emitted light by intervening HI clouds, and selection effects by cosmological dimming of surface brightness. In Figure 1 we show the galaxy counts in the K'-band in the Subaru Deep Field (SDF) and in the B-band in the Hubble Deep Field (HDF). The solid, dot-dashed and dashed lines indicate the standard CDM, an open CDM ($\Omega_0 = 0.3$) and a ΛCDM ($\Omega_0 = 0.3$ and $\Lambda_0 = 0.7$). As clearly shown, the low density models agree well with the observed SDF and HDF counts and the SCDM model does not agree with those.

We showed that our semi-analytic model reproduce the multi-band observed galaxy counts simultaneously in low-density universes. This may means that our hierarchical galaxy formation model satisfies a important necessary condition, that is, faint galaxy number counts, to be consistent with observations.

References

[1] T. Totani and Y. Yoshii, Ap. J. 557 (2000) 505.
[2] T. Totani et al., Ap. J. 557 (2001) 505.
[3] M. Nagashima, T. Totani, N. Gouda and Y. Yoshii, Ap. J. 557 (2001) 505.

Dynamical Evolution of Monopoles and Anti-Monopoles

Michiyasu NAGASAWA *Department of Information Science, Faculty of Science, Kanagawa University, Kanagawa 259-1293, Japan*

Abstract

The formation and evolution of monopoles and anti-monopoles produced at TeV energy scale are considered. Particularly the efficiency of the pair annihilation and the resulting equilibrium number density is estimated. It is shown that the baryon number generation using the string-monopole connected system has a difficulty since the number density of such defects is too low.

In the model which we employ, at the first phase transition monopoles are produced and then the second phase transition occurs and strings appear between monopoles and anti-monopoles[1]. Subsequently at the final phase transition below the electroweak scale, strings become unstable and disappear so that the baryon asymmetry will be produced. Using the appropriate parameter values, the monopole mass, m_M, can be evaluated as

$$m_M \sim 2400 \text{ GeV} , \tag{1}$$

which is comparable to the electroweak energy scale rather than the GUT scale.

The produced baryon number density must be deduced by the string density. It would be calculated using the monopole number density at the string appearance, n_M, which can be written under reasonable approximations as

$$\frac{n_M}{s} \sim 10^{-16} , \tag{2}$$

where s is the cosmological entropy density. The basic feature of the monopole interaction in this approximation has been confirmed numerically by means of a particle simulation as you can see in Figure 1.

Thus it may be difficult to produce sufficient baryon number in this scenario. However, the fact that the monopole number density is very small means that the string production process would be approximately identical to the conventional topological string case. Therefore the usual topological string loop baryogenesis scenario can be also applied to this model. The ordinary topological string

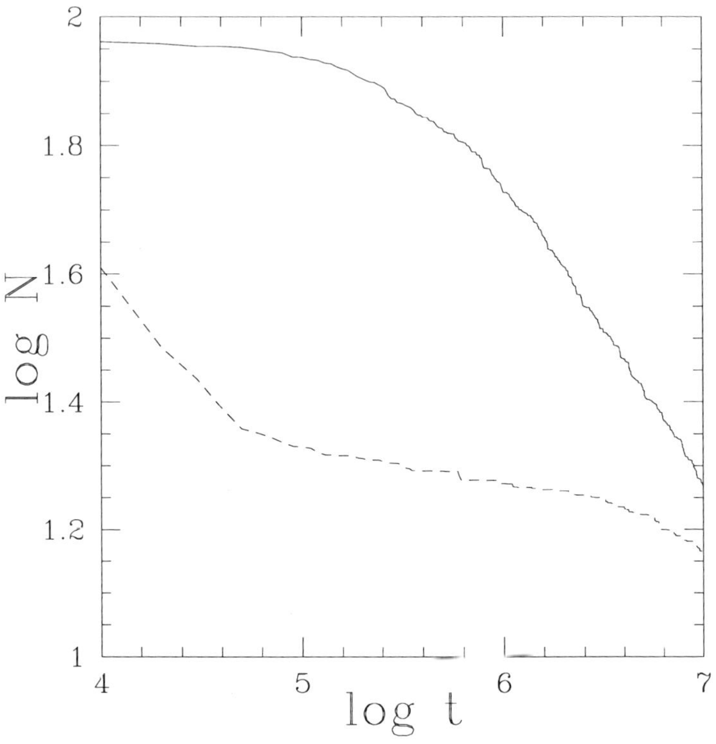

Fig. 1. Time evolution of total monopole and anti-monopole number is depicted. The solid line corresponds to the case $T = 10^{-3} m_M$ and the dashed one is for the case $T = 10^{-6} m_M$. The initial monopole number is equal to 100.

produced at TeV scale[2] or the sphaleron bound state on the string[3] would be more useful in order to generate the baryon asymmetry.

Note that in contrast to the GUT monopole case($m_M \sim 10^{16}$GeV), the cosmic energy density should not be dominated by monopoles in this model.

References

[1] T. W. Kephart and T. Vachaspati, Phys. Lett. **B388** (1996) 481.
[2] R. H. Brandenberger and A.-C. Davis, Phys. Lett. **B332** (1994) 305.
[3] V. Soni, Phys. Lett. **B394** (1997) 275.
C. S. Aulakh, M. Nagasawa, and V. Soni, Phys. Lett. **B471** (1999) 13.

Global Foliations of Gowdy Symmetric Spacetimes with Stringy Matter

Makoto Narita
Astrophysics Group, Physics Department and Advanced Research Institute for Science and Engineering, Waseda University, Okubo 3-4-1, Shinjuku, Tokyo, 169-8555, Japan

The strong cosmic censorship conjecture proposed by Penrose is an unsolved and the most important problem in classical general relativity. The idea is that naked singularities should not evolve from regular initial data. More precisely, this conjecture can be stated as saying that for generic initial data sets, the corresponding maximal globally hyperbolic extension spacetime is inextendible [AL99, CD, KN, ME]. To answer this problem we must solve global existence problems for Einstein-matter equations and inextendibility of domain of dependence of initial surfaces.

In order to attack the above issue, we will consider T^3 Gowdy symmetric spacetimes with stringy matter fields [NTM]. These are Maxwell-dilaton-axion fields which arise naturally from low-energy effective superstring theory [STW]. Let us call (γ, π) Gowdy symmetric initial data on T^3 if (γ, π) are invariant under an effective T^2 action and satisfy the constraint equations. For simplicity, we will assume that (γ, π) are C^∞ on T^3 and one component of the Maxwell field is trivial. In this circumstance we can show the following *global existence theorem.*

Theorem. *Let (γ, π) be a smooth Gowdy symmetric initial data set on T^3. For some nonnegative constant c there exists a globally hyperbolic spacetimes $(\mathcal{M}, g, \phi, \kappa, \chi)$ such that*

1. $\mathcal{M} = (c, \infty) \times T^3$.
2. (g, ϕ, κ, χ) *satisfies the Einstein-Maxwell-Dilaton-Axion (EMDA) equations.*
3. $\mathcal{M}$ *is covered by areal coordinate (t, θ, x, y) with $t \in (0, \infty)$.*
4. $(\mathcal{M}, g, \phi, \kappa, \chi)$ *is isometrically diffeomorphic to the maximal globally hyperbolic development of the initial data (γ, π),*

where ϕ, κ, χ are dilaton, axion and Maxwell fields, respectively.

The local existence and uniqueness of smooth solutions of this system follows from standard results for hyperbolic system [CBG, ?]. Then, it is sufficient to verify that for any globally hyperbolic subset of the (t, θ) cylinder on which they exist as a solution to this system, the functions $(Z, X, \omega, \chi, \phi, \kappa)$ and their first and second derivatives are uniformly bounded. To do this, we will use the standard technique that is *Moncrief's light cone estimate* [AH, BCIM, M81, M97].

To end the strong cosmic censorship conjecture we must prove inextendibility of domain of dependence of initial surfaces. To do this, we need to know the nature of the singularities. Using *the Fuchsian algorithm* developed by Kichenassamy and Rendall, we have a conclusion that, in general, Gowdy symmetric spacetimes in the EMDA system have Asymptotically Velocity-Terms Dominated (AVTD) singularities [AR, KR, NTM]. Since spacetime curvature is blow up near the AVTD singularities in the spacetime, inextendibility is shown. Then the strong cosmic censorship conjecture is valid if the spacetime has an AVTD singularity.

References

[AL99] L. Andersson, *The global existence problem in general relativity*, gr-qc/9911032.

[AR] L. Andersson and A. Rendall, Commun. Math. Phys. **218**, (2001) 479-511.

[AH] H. Andreasson, Commun. Math. Phys. **206**, (1999) 337-365.

[BCIM] B. K. Berger, J. Isenberg, P. T. Chruściel and V. Moncrief, Annals Phys. **260**, (1997) 117-148.

[CD] D. Christodoulou, Class. Quantum Grav. **16**, (1999) A23-A35.

[CBG] Y. Choquet-Bruhat and R. Geroch, Commum. Math. Phys. **14**, (1969) 329-335.

[FR] H. Friedrich, A. D. Rendall, Lect. Notes Phys. **540**, (2000) 127-224.

[KR] S. Kichenassamy and A. Rendall, Class. Quantum Grav. **15**, (1998) 1339-1355.

[KN] S. Klainerman and F. Nicolo, Class. Quantum Grav. **16**, (1999) R73-R157.

[M81] V. Moncrief, Annals Phys. **132**, (1981) 87-107.

[M97] V. Moncrief, *Spacetime singularities and cosmic censorship*, in "*Proceedings of the 14th International Conference on General Relativity and Gravitation*", edited by M. Francaviglia, G. Longhi, L. Lusanna and E. Sorace (World Scientific, 1997) pp. 259-276.

[ME] V. Moncrief and D. M. Eardley, Gen. Rel. Grav. **13**, (1981) 887-892.

[NTM] M. Narita, T. Torii and K. Maeda, Class. Quantum Grav. **17**, (2000) 4597-4613.

[STW] A. Shapere, S. Trivedi and F. Wilczek, Mod. Phys. Lett. **A6**, (1991) 2677.

Measuring the time evolution of the redshift-distortion parameter

Hiroaki NISHIOKA
Department of Physics, Hiroshima University, Higashi-Hiroshima 739-8526, Japan

Abstract

The method to measure the time evolution of the redshift-space distortion parameter $\beta(z)$ is presented. To measure the time evolution of $\beta(z)$, we take the following strategy. First, we express $\beta(z)$ as a power series in z. Next, we derive the formulas to determine their coefficients with the observable statistical quantities from the redshift-space galaxy distribution on a light cone. Our formulas are not constrained to be applied in the distant-observer limit so that we can apply them to the large scale clustering where the nonlinear evolution of the density fluctuations are negligible. We also estimate the systematic error of this method by the numerical computation assuming some models for the time evolution of bias. We find that by using the formulas derived here up to the second order in z we can reconstruct the time evolution of $\beta(z)$ as a function of z within 2% systematic error when the galaxy samples within the depth $z \leq 0.5$ is used.

1. Introduction

It is well known that the anisotropic character of the galaxy clustering in redshift space can be used to measure the redshift-distortion parameter $\beta(z)$. In the present paper, we explore the method to measure the time evolution of $\beta(z)$ with the galaxy redshift map which has the nonzero redshift range as the extension of the method of [1] to the one for the galaxy distribution on a light cone. Our approach is the following. First, we express $\beta(z)$ as a power series in z. Next, we derive the formulas to determine their coefficients with the observable statistical quantities from the redshift-space galaxy distribution on a light cone. In this paper we assume we know the values of the cosmological parameters exactly. Thus we does not mention the anisotropy caused by the geometric distortion effect which means that our method presented in this paper can measure the time evolution of $\beta(z)$ in an assumed cosmological model. Throughout the present paper, we use the unit in which c is unity.

2. Formulation

Throughout the present paper, we assume the spatially flat Friedman-Lemaitre Universe. In liner theory, the analytic expression of the two-point correlation function in redshift space of the two objects located at the real-space positions $\mathbf{r}_1$ and $\mathbf{r}_2$, the angle between which is not constrained to be the small angle, is given by [2]. We reformulate this expression in terms of the shape functions $B_{i,j}(\triangle)$ defined by the shape of the triangle which consists of the observer and two objects, and their coefficients $\xi_{i,j}(R, z_1, z_2)$ as $\xi^s(\mathbf{r}_1, \mathbf{r}_2) = b(z_1)b(z_2)D(z_1)D(z_2)\sum_{i,j} B_{i,j}(\triangle)\xi_{i,j}(R, z_1, z_2)$, where the shape functions are defined by $B_{0,1}(\triangle) = 1$, $B_{1,1}(\triangle) = R\mu_1/r_1$, $B_{1,2}(\triangle) = R\mu_2/r_2$, $B_{2,1}(\triangle) = P_2(\mu_1)/2$, $B_{2,2}(\triangle) = P_2(\mu_2)/2$, $B_{3,1}(\triangle) = R\mu_1(1-\mu_2^2)/r_1$, $B_{3,2}(\triangle) = R\mu_2(1-\mu_1^2)/r_2$, and $B_{4,1}(\triangle) = (35\mu_1^2\mu_2^2 - 15\mu_1^2 - 15\mu_2^2 + 3)/8$, and their coefficients $\Xi_{i,j}(R, z_1, z_2)$ are the functions of the mass two-point correlation function at present $\xi(R)$ and $\beta(z)$. Here $\mu_1 = \cos\gamma_1$, $\mu_2 = \cos\gamma_2$ and $\mu = \cos\theta$, where γ_1, γ_2, and θ are the same definition as that in [2]. Next, we express $b(z)D(z)$ and $\beta(z)$ as a power series in z as $b(z)D(z) = \sum_l A_l z^l$ and $\beta(z) = \sum_m \beta_m z^m$, respectively. By the similar discussion of [1], we find some ratios of the observable statistical quantities,

$$\Xi^{(i)}(R) = \sum_j M_{ij}^{-1}(R) \iint_V d^3\mathbf{r}_1 d^3\mathbf{r}_2 W(\mathbf{r}_1, \mathbf{r}_2; R)\xi^s(\mathbf{r}_1, \mathbf{r}_2)\bar{B}^{(j)}(\triangle, z_1, z_2), \quad (1)$$

where $M_{ij}(R) = \iint_V d^3\mathbf{r}_1 d^3\mathbf{r}_2 W(\mathbf{r}_1, \mathbf{r}_2; R)\bar{B}^{(i)}(\triangle, z_1, z_2)\bar{B}^{(j)}(\triangle, z_1, z_2)$, $W(\mathbf{r}_1, \mathbf{r}_2; R)$ is the arbitrary weighting function, V is the region of the survey, $\bar{B}^{(i)}(\triangle, z_1, z_2)$ can be expressed in terms of $B_k(\triangle)(k = 0, ..., 4)$ defined in [1], z_1, and z_2, yield the measure of β_m independent of the shape of the two-point correlation function.

3. Discussion

We examined how accurate $\beta(z)$ can be reconstructed by this method using the formulas up to the second order in z. We adopted the parametrized model for bias to which we give the enough range for the parameters to have the enough variation of the rapidity of the time evolution and amplitude of bias. And we assumed the maximum redshift of the galaxy sample to be 0.5. In this case, we found the systematic error to reconstruct $\beta(z)$ is within 2%.

References

[1] A. J. S. Hamilton and M. Culhane, MNRAS 278 (1996) 73.
[2] T. Matsubara, ApJ 535 (2000) 1.

Quantum Creation of the Randall-Sundrum Bubble

Hirotaka OCHIAI
Department of Physics, School of Science, The University of Tokyo, 7-3-1 Hongo, Bunkyo-ku, Tokyo 113-0033, Japan

Abstract

We discuss the semiclassical instability of the Randall-Sundrum brane-world model. We present the bubble solution with the Randall-Sundrum background which expresses the decay of the brane-world. The decay probability is estimated following the Euclidean path integral approach to quantum gravity.

1. Introduction

In the nonperturbative approach to string theory, the brane-world scenario has received much attention. The simplest model has been proposed by Randall-Sundrum (RS). [1] Although RS model has great success, there seems to be a crucial problem of the stability. It is well known in the standard Kaluza-Klein theory that the Kaluza-Klein vacuum is unstable against the decay channel to the so-called Kaluza-Klein bubble [2]. We will present an explicit example describing a sort of the Kaluza-Klein bubble (RS bubble) in the section 2 and evaluate the decay probability in the section 3 [3, 4].

2. Randall-Sundrum Bubble

The RS model with two 3-branes is given by the AdS_5 metric of the form

$$g = dy^2 + a^2 \eta_{\mu\nu} dx^\mu dx^\nu, \tag{1}$$

where $a = e^{-y/l}$, $\eta_{\mu\nu}$ is the 4-dimensional Minkowski metric and $0 \le y \le y_0$. The positive and negative tension branes are located at $y = 0$ and $y = y_0$, respectively. We here consider a generalization with a same asymptotics (RS bubble):

$$g = \left[\frac{1-(\rho_*/ar)^2}{1+(\rho_*/ar)^2}\right] dy^2 + a^2 \left[1 + \left(\frac{\rho_*}{ar}\right)^2\right]^2 (-r^2 d\tau^2 + dr^2 + r^2 \cosh^2 \tau d\Omega_2^2), \tag{2}$$

where $\rho_* \ge 0$ is a constant. The induced metric h on the branes is given by

$$h = \left[1 + \left(\frac{\rho_*}{r}\right)^2\right]^2 (-r^2 d\tau^2 + dr^2 + r^2 \cosh^2 \tau d\Omega_2^2), \tag{3}$$

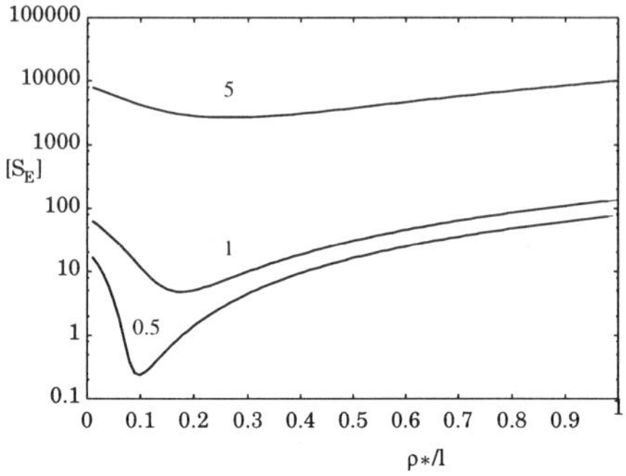

Fig. 1. $[S_E]$ for $l = 1$ and $y_0 = 0.5$, 1, 5 in the unit the 5-dimensional gravitational scale $\kappa_5^2 = 1$.

where $0 \le r \le \infty$. The induced geometry of the brane looks like the Einstein-Rosen bridge, which connects the positive and negative tension brane. The bubble rapidly expands and there are a trapped region also forms around it. The RS model has a decay channel to the RS bubble. To be a good model of the universe, the RS model should be stabilized against this kind of instability.

3. The Decay Rate from RS Model to RS Bubble

We evaluate the decay probability from the RS model to the RS bubble:

$$P \sim e^{-(S_E^{bubble} - S_E^{vac})} := e^{-[S_E]}, \tag{4}$$

where S_E^{vac} and S_E^{bubble} are the Euclidean action of the RS model and the RS bubble, respectively. In general, we need numerical computation to evaluate it (See Fig. 1.). If the distance y_0 between the branes is small, the RS bubble will be relatively nucleated. For the hierarchy problem ($y_0/l \sim 10$), the transition rate is significantly suppressed.

References

[1] L. Randall and R. Sundrum, Phys. Rev. Lett. 83 (1999) 3370.
[2] E. Witten, Nucl. Phys. B195 (1982) 481.
[3] D. Ida, T. Shiromizu, and H. Ochiai, Phys. Rev. D65 (2002) 023504.
[4] H. Ochiai, D. Ida and T. Shiromizu, to appear.

Time Delay Statistics of Strong Gravitational Lensing and the Density Profile of Dark Halos

Masamune OGURI, Atsushi TARUYA, Yasushi SUTO
Research Center for the Early Universe (RESCEU) and Department of Physics, School of Science, The University of Tokyo, 7-3-1 Hongo, Bunkyo-ku, Tokyo 113-0033, Japan
Edwin L. TURNER
Princeton University Observatory, Peyton Hall, Princeton, NJ 08544, USA

Abstract

We derive an explicit expression for the conditional probability distribution function of time delays $P(\Delta t \,|\, \theta)$, given an image separation between multiple images θ. We also show that the time delay distribution is very sensitive to the density profile of the lens object and thus can be used as a good probe of the density profile.

1. Introduction

So far, time delays between two images have been used primarily to estimate the Hubble constant H_0 using a detailed lens model of each system [1]. But the time delay is also sensitive to the mass distribution of lens objects. This led to the recognition that the mass distribution in estimating Hubble constant is important [2]. Inversely, we consider the statistics of the time delay effect and examine its possibility to use as a discriminator of the density profile of the lens object. Moreover, the statistical treatment of time delays will also be useful in designing the observing program of future large scale synoptic variability surveys and for evaluating possible selection biases operating against large splitting lens systems.

2. Result

Figure 1 shows the predicted median time delay $\Delta t_{\rm med}$ as a function of separations θ using our expression for the conditional probability distribution function of time delays $P(\Delta t \,|\, \theta)$ [3]. We consider lensing halos described by the singular isothermal sphere (SIS) approximation, $\rho \propto r^{-2}$, and by its generalization as proposed by Navarro, Frenk, & White [4] which has a density profile $\rho \propto$

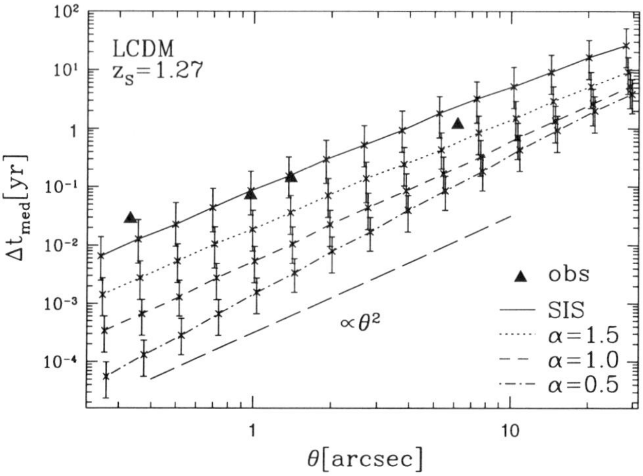

Fig. 1. Median time delays, $\Delta t_{\rm med}$, as a function of separations θ, for various density profiles; SIS (*solid*), generalized NFW with $\alpha = 1.5$ (*dotted*), $\alpha = 1.0$ (*short dashed*), and $\alpha = 0.5$ (*short dash-dotted*) [3]. The source galaxies are located at $z_{\rm S} = 1.27$ and the cosmological model is fixed to $(\Omega_0, \lambda_0, h) = (0.3, 0.7, 0.7)$. The error-bars in this plot are defined by the $\pm 25\%$ level. A line with $\Delta t \propto \theta^2$ is also shown for reference. Some observational data are indicated by triangles.

$r^{-\alpha}(r + r_{\rm s})^{\alpha-3}$. Our result clearly shows that the time delay is very sensitive to the density profile of the lens object; steeper inner slopes tend to produce larger time delays. The difference of time delays between different density profiles we examined becomes more than one order of magnitude. On the other hand, the effects of cosmological models, source redshifts or magnification bias are rather small [3]. Therefore, we expect to constrain the density profile from the time delay data. Comparison with the meager existing observational data suggests that the density profile has a rather steep cusp, $\alpha \sim 2$, although it is already known that the current observed systems are dominated by baryonic component. To see the density profile of *dark halos*, time delay data for separations with $\theta \sim 10''$ will be essential.

References

[1] N. A. Grogin, and R. Narayan, ApJ 464 (1996) 92
[2] C. R. Keeton et al., ApJ 542 (2000) 74
[3] M. Oguri, A. Taruya, Y. Suto, and E. L. Turner, ApJ, in press
[4] J. F. Navarro, C. S. Frenk, and S. D. M. White, ApJ 490 (1997) 493

Chemical Enrichment in Damped Lyman α Systems on Hierarchical Galaxy Formation

Katsuya Okoshi, Masahiro Nagashima, Naoteru Gouda, and Satoshi Yoshioka[1]
National Astronomical Observatory, Mitaka, Tokyo 181-8588, Japan;
Department of Physics, Tokyo University of Mercantile Marine, Tokyo, Japan[1]

Abstract

We investigate that the chemical enrichment in Damped Lyman α (DLA) systems on the hierarchical structure formation scenario by using a semi-analytic model of galaxy formation which can show the fundamental properties of galaxies.

We find good agreement with a reasonable range of parameters for reproducing both the metallicity evolution of DLA systems and the observational properties of local galaxy population. Moreover, we also investigate the effect of multi-components in neutral gas on the evolution of DLA systems to estimate the amount of hydrogen molecule in these systems.

1. Introduction

The absorption-line systems observed in the spectra of background quasars have proved to be a strong probe to the physical conditions at high redhifts. Especially, DLA systems can offer some possibilities to measure abundances of various elements and good clues to infer the metallicity evolution. Here we focus on this advantage for the observation and apply our semi-analytic methods in order to explore the metallicity evolution in DLA systems at high redshifts.

2. Model

We use the semi-analytic models for galaxy formation at high redshifts based upon the work in NTGY. Here, we apply two assumptions for star-formation process in DLA. The first one is that the star-formation timescale obeys single power law: $\tau_* \propto V_c^{\alpha_*}$ (e.g.CLBF and NTGY). The second one is that DLA systems have two component gas: cold(hydrogen molecules) and warm(neutral hydrogen) gas. Therefore, we assume that absorption lines of DLA systems mainly arise from warm(HI) gas in protogalactic disks which also contain small amount of hydrogen molecule(cold gas).

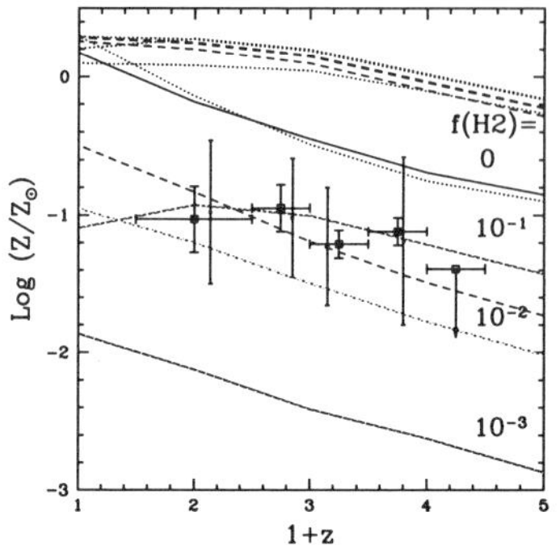

Fig. 1. Metallicity evolution of warm(solid) and cold(dotted) gas for various H_2 fraction in DLA systems. The numbers are H_2 fraction in warm gas component. Also shown for the model (dashed):$\tau_* \propto V_c^{\alpha_*}$ $\alpha_* = -2$.

3. Results

Fig.1 shows the metallicity evolution of DLA systems. We can find that $\alpha_* \sim -2$ is the best when τ_* is dependant on V_c. Moreover, for multi-component model, we also find that the metal abundance constrains the H_2 fraction $f(H_2)(= M(H_2)/(M(H_2) + M(HI))) > 10^{-2}$ in DLA, while the stellar abundance can be much closer to the solar ones at present.

4. Conclusion and Discussion

We can find that the metallicity evolution requires the relation between the star-formation timescale and circular velocity of the halo, or new assumption that DLA systems mainly consist of warm gas which contains small amount of hydrogen molecules. By introducing warm phase in DLA systems, we can constrain the abundance of hydrogen molecules without violating the metal abundance of the other components such as stars at present. Radio observations(e.g.GMRT) have recently suggested multi-component gas in DLA systems, and would suggest that this problem deserves futher investigation.

References

[1] Cole, S., Lacey, C. G., Baugh, C. M., & Frenk, C. S. MNRAS, 319, 168 (2000) (CLBF)

[2] Nagashima, M., Totani, T., Gouda,N., & Yoshii, Y. ApJ, 557, 505 (2001) (NTGY)

Two Classes of Cluster Size and the Connection to the $L-T$ Relation

Naomi OTA
Department of Physics, Tokyo Metropolitan University, 1-1 Minami-Osawa, Hachiouji, Tokyo 192-0387, Japan
Kazuhisa MITSUDA
Institute of Space and Astronautical Science (ISAS), 3-1-1 Yoshinodai, Sagamihara, Kanagawa 229-8510, Japan

Abstract

We discuss the core radius distribution and its connection to the X-ray luminosity-temperature ($L-T$) relation of clusters of galaxies based on the systematic X-ray analysis on 79 clusters of galaxies by [4, 5]. Assuming the β-profile for the intracluster gas distribution, we show that the model in which the core radius takes one of the two typical values can consistently explain the observed $L-T$ relation.

1. Introduction

The luminosity-temperature ($L-T$) relation of X-ray clusters is one of the fundamental correlations among cluster parameters. Observationally, the correlation is well approximated with a power-law function; $L_X \propto T^\alpha$ with $\alpha \sim 3$ (e.g. [1]). However, the self-similar model ([2]) predict $\alpha = 2$. The discrepancy between the model prediction and the observations indicates that some of the fundamental assumptions may be invalid for real clusters.

2. Observational results and a model $L-T$ relation

From the systematic analysis of 79 clusters by [4, 5], we discovered that the core radius distribution shows high concentrations at $60\,h_{50}^{-1}$kpc and $220\,h_{50}^{-1}$kpc (Fig.1.). We also find that the core radius does not show strong correlation with the virial radius (Fig.1.), and that the normalization of the $L-T$ relation depends on the core radius. Thus the luminosity can be approximated with $L_X \propto T^{2.9} r_c^{-1}$. These suggest that the core radius is determined by some physical processes in the cluster central region, almost independent of the virial mass. Assuming the β-profile for the intracluster gas distribution, we show that the model in which

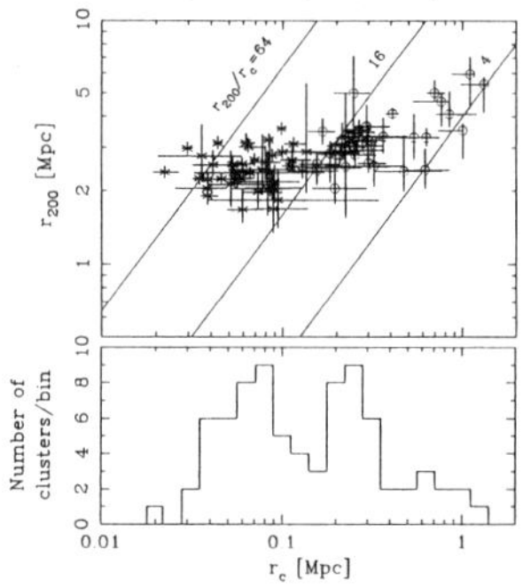

Fig. 1. Relation of the core radius, r_c and the virial radius, r_{200} and histogram of r_c for 79 clusters. $H_0 = 50\,\mathrm{km/s/Mpc}$, $\Omega_0 = 0.3$ and $\Lambda = 0.7$ are used throughout the paper.

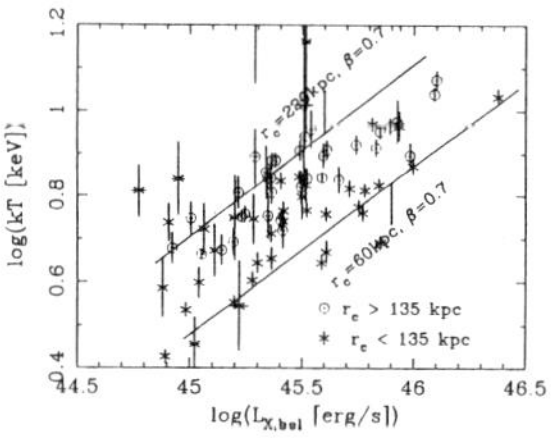

Fig. 2. $L_X - T$ relations expected for clusters of a constant core radius, $r_c = 60\,h_{50}^{-1}$ or $220\,h_{50}^{-1}$kpc with $\beta = 0.7$ shown with the observed data points. These two model curves quantitatively explain the $L_X - T$ relations of cluster subgroups for $r_c < 135\,h_{50}^{-1}$ and $r_c > 135\,h_{50}^{-1}$kpc, respectively.

the core radius takes one of the two typical values can consistently explain the observed $L-T$ relation (Fig.2.). In this model, the $L-T$ relation is expected to show very small dependency on the cluster collapse epoch ([3]).

References

[1] A. C. Edge and G. C. Stewart, M.N.R.A.S 252 (1991) 414.
[2] N. Kaiser, M.N.R.A.S. 222 (1986) 323.
[3] K. Mitsuda and N. Ota, in preparation (2001).
[4] N. Ota, PhD thesis (2001) University of Tokyo, ISAS Research Note 727.
[5] N. Ota and K. Mitsuda, submitted to ApJ (2001).

The Dark Background Light

James OVERDUIN
Department of Physics, Waseda University, Okubo 3-4-1, Tokyo 169-8555, Japan

Abstract

A two-component galaxy model is used to obtain theoretical limits on the spectral intensity of the extragalactic background light (EBL) at near-optical wavelengths. The equivalent quantity is then calculated for cosmological models in which the effects of expansion are removed. Taking the ratio of the two shows that EBL intensity is determined primarily by the lifetime of the galaxies, and is reduced by only a factor of two to three by expansion, depending on assumptions about luminosity evolution and the makeup of the galaxy population.

Results

There has long been confusion over the relative importance of two factors in determining the brightness of the night sky: the lifetime of the galaxies (which limits the amount of light that has been pumped into intergalactic space), and expansion (which stretches and dims this light) [1]. Previous work using blackbody galaxy spectra with a simple evolution model has established that lifetime is the dominant factor, but has not set firm quantitative limits on the importance of expansion [2]. We do so here using spectral energy distributions (SEDs) which are now available down to wavelengths ~ 100Å, based on theoretical models of galaxy evolution for both normal and starburst-type galaxies [3]. We normalize these SEDs to data on total comoving luminosity density inferred from a photometric analysis of galaxies in the Hubble Deep Field (HDF) [4].

Total comoving luminosity density is taken as $\mathcal{L}(z) = n_{\rm N}(z)L_{\rm N} + n_{\rm S}(z)L_{\rm S}$ where $z =$ redshift, $n_{\rm N}(z) \equiv [1 - f(z)]n(z)$ and $n_{\rm S}(z) \equiv f(z)\,n(z)$ are comoving normal and starburst galaxy number densities, and $L_{\rm N}$ and $L_{\rm S}$ are normal and starburst galaxy B-band luminosities respectively. We match $\mathcal{L}(z)$ to fits to the HDF data by requiring that the starburst galaxy fraction $f(z)$ satisfy $f(z) = [r_{\rm L}(1+z)^{-\eta}\mathcal{N}(z) - 1]/(r_{\rm L} - 1)$ for $z \leq z_{\rm P}$ and $f(z) = 1$ for $z > z_{\rm P}$, while $n(z)$ is given by $n(z) = n_0(1+z)^{\eta}$ for $z \leq z_{\rm P}$ and $n(z) = n_0\mathcal{N}(z)$ for $z > z_{\rm P}$. Here $\mathcal{N}(z) \equiv [1/r_{\rm L} + (1 - 1/r_{\rm L})f_0]\mathcal{L}(z)/\mathcal{L}_0$, $\eta = \ln[\mathcal{N}(z_{\rm P})]/\ln(1+z_{\rm P})$ is the merger parameter, $f_0 \equiv f(0)$, $r_{\rm L} \equiv L_{\rm S}/L_{\rm N}$, $z_{\rm P}$ is the redshift of peak comoving luminosity

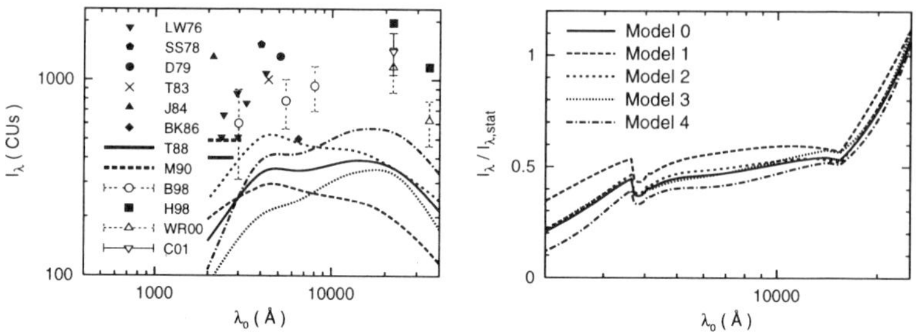

Fig. 1. (a) Spectral EBL intensity I_λ; (b) Ratio of $I_\lambda/I_{\lambda,\mathrm{stat}}$

density, $\mathcal{L}_0 = (2.0 \pm 0.2) \times 10^8 h_0 L_\odot$ Mpc^{-3} [5] and h_0 is Hubble's constant in units of 100 km s^{-1} Mpc^{-1} as usual. The free parameters of this model are the local starburst fraction f_0, the luminosity ratio r_L and the fit to the comoving luminosity density $\mathcal{L}(z)$ inferred from the HDF data.

Results for the intensity $I_\lambda(\lambda_0)$ of the integrated galaxy EBL at λ_0 are shown for the ΛCDM cosmology ($\Omega_m = 0.3$, $\Omega_\Lambda = 0.7$) in Fig. 1(a) along with existing observational upper limits (solid symbols) and reported detections (empty symbols) in continuum units (1 CU $\equiv$ 1 photon s^{-1} cm^{-2} Å^{-1} ster^{-1}). Model 0 is a best fit to $\mathcal{L}(z)$, with $f_0 = 0.05$ and $r_\mathrm{L} = 20$. Odd (even) numbered models have the weakest (strongest) luminosity density evolution compatible with the HDF data. Models 1 and 2 are maximally starburst-dominated, with $f_0 = 0.1$ and $r_\mathrm{L} = 1000$; while Models 3 and 4 are least, with $f_0 = 0.01$ and $r_\mathrm{L} = 10$.

Fig. 1(b) shows the ratio of $I_\lambda(\lambda_0)$ to $I_{\lambda,\mathrm{stat}}(\lambda_0)$, the quantity that would be measured in an equivalent universe with no Hubble expansion. Over the B-band, one sees that expansion reduces the intensity of the EBL by a factor of between 2 and 3 from what it would be in an equivalent static model, results which broaden only slightly over the full range of plausible Friedmann-Lemaître cosmologies. This confirms that the finite lifetime of the galaxies, rather than cosmic expansion, is responsible for the darkness of the night sky.

References

[1] P. S. Wesson, K. Valle and R. Stabell, Astrophys. J. 317 (1987) 601
[2] P. S. Wesson, Astrophys. J. 367 (1991) 399
[3] J. E. G. Devriendt et al., Astron. Astrophys. 350 (1999) 381.
[4] M. J. Sawicki, H. Lin and H. K. C. Yee, Astron. J. 113 (1997) 1
[5] M. Fukugita, C. J. Hogan and P. J. E. Peebles, Astrophys. J. 503 (1998) 518

Direct Measurement of the Acceleration of the Universe Using 0.1 Hz Band Laser Interferometer Gravitational Wave Antenna in Space

Naoki Seto
Department of Earth and Space Science, Osaka University, Toyonaka 560-0043, Japan
Seiji Kawamura
National Astronomical Observatory, Mitaka 181-8588, Japan
Takashi Nakamura
Yukawa Institute for Theoretical Physics, Kyoto University, Kyoto 606-8502, Japan

Abstract

It may be possible to construct a laser interferometer gravitational wave antenna in space with $h_{rms} \sim 10^{-27}$ at $f \sim 0.1$Hz in this century. We show possible specification of this antenna which we call DECIGO. Using this antenna we show that 1) typically 10^5 ($10^4 \sim 10^6$) chirp signals of coalescing binary neutron stars per year may be detected with S/N $\sim 10^4$. 2) We can directly measure the acceleration of the universe by ten years observation of binary neutron stars. 3) The stochastic gravitational waves of $\Omega_{GW} \sim 10^{-20}$ predicted by the inflation may be detected by correlation analysis for which effects of the recent cosmic acceleration would become highly important. Our formula for phase shift due to accelerating motion might be also applied for binary sources of LISA.

Direct Measurement of the Acceleration of the Universe

As shown in figure 1, ~ 0.1Hz is just between the ground-based detectors (TAMA300, LIGO, VIRGO, GEO600, etc) and the Laser Interferometer Space Antenna (LISA) [1]. It has been rarely discussed what kind of gravitational wave astronomy we can perform at this band. In Ref.[2] we proposed a new method to directly measure the acceleration of the Universe, more specifically the parameter $X \equiv (H_0 - H(z)/(1+z))/2$ where $H(z)$ is the Hubble parameter at redshift z (see also Ref.[3]). It is reasonable to assume that at the age of DECIGO we can estimate the redshift of binaries by simple inversion of the $d_L - z$ relation [4] *e.g.* determined from SNAP (http://lbl.gov).

Using the interferometer DECIGO with specification shown in figure 1,

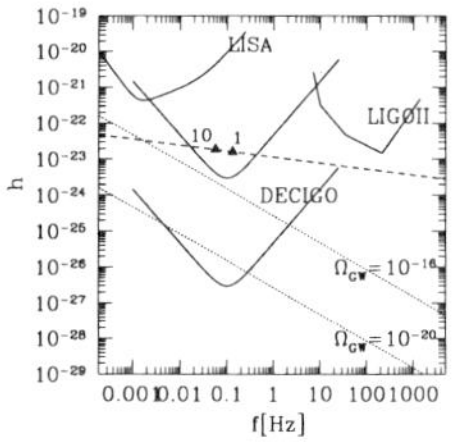

Fig. 1. Sensitivity (effectively S/N=1) for various detectors (LISA, DECIGO, LIGOII and a detector 10^3 times less sensitive than DECIGO) in the form of h_{rms} (solid lines). The dashed line represents evolution of the characteristic amplitude h_c for NS-NS binary at $z = 1$ (filled triangles; wave frequencies at 1yr and 10 yr before coalescence). The dotted lines represent the required sensitivity for detecting stochastic background with $\Omega_{GW} = 10^{-16}$ and $\Omega_{GW} = 10^{-20}$ by ten years correlation analysis (S/N=1).

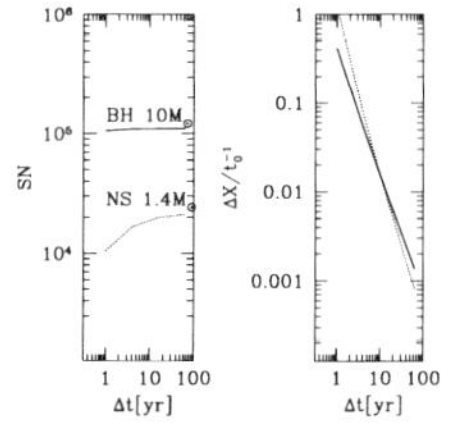

Fig. 2. S/N and the estimation error ΔX (normalized by the age of the universe; t_0) for NS-NS ($1.4M_\odot$) or BH-BH ($10M_\odot$) binaries at $z = 1$. We evaluate them for DECIGO (sensitivity: $h_{rms} \sim 10^{-27}$ at $f \sim 0.1$Hz).

the parameter X can be measured with accuracy $\Delta X \sim 10^{-2}$ by observing a double neutron star binary at $z \sim 1$ for about 10 yr. Considering the estimated coalescence rate of cosmological double neutron binaries $\sim 10^5 \text{yr}^{-1}$ it would be possible to determine X very accurately.

References

[1] P. L. Bender et al., *LISA Pre-Phase A Report, Second edition* (1998).
[2] N. Seto, S. Kawamura and T. Nakamura, Phys. Rev. Lett. (2001) 87, 21103.
[3] A. Loeb, Astrophys. J. 499 (1998) L111.
[4] B. F. Schutz, Nature, 323 (1986) 310

Oscillating scalar field dark matter

Osamu SETO, Kazunori KOHRI, and Takashi NAKAMURA
Yukawa Institute for Theoretical Physics, Kyoto University, Kyoto 606-8502, Japan

Abstract

We show that an oscillating scalar field in supergravity of mass of the order of $\sim$ TeV with a nonzero vacuum expectation value ($\sim 10^{10}$ GeV) can be a candidate of cold dark matter (CDM). To avoid the gravitino problem, we need a low reheating temperature after the primordial inflation. Then, the energy density of the oscillating scalar field satisfies all the requirements for CDM at present in the universe.

1. Introduction

It is widely believed that a significant fraction of energy density in the universe is in the form of cold dark matter (CDM). Recently it was reported that the contribution of CDM to density parameter is approximately $\Omega \simeq 0.3$. It is one of the most important problems in cosmology and particle physics to clarify the nature and the origin of dark matter.

In this paper, we consider stable scalar fields with an electroweak scale mass $\sim \mathcal{O}(1)$TeV and a large vacuum expectation value (VEV) $\sim \mathcal{O}(10^{10})$ GeV. These scalar fields naturally appear in supergravity. However, whenever we consider a model based on supergravity, we are faced with "gravitino problem". To avoid the gravitino problem, we need a low reheating temperature after the primordial inflation. In this situation, the energy density of the oscillation of the scalar field satisfies the requirements for CDM at present in the universe.

2. Model

In supergravity, we know there exist a lot of scalar fields with an almost flat potential. We consider a gauge singlet chiral superfield Φ which has a minimal Kähler potential $K = |\Phi|^2 M_G^{-2}$ and a superpotential,

$$W = \frac{\lambda}{n+3} \frac{\Phi^{n+3}}{M_G^n} + C, \tag{1}$$

where $\lambda \sim \mathcal{O}(1)$, and C is a constant. In addition, the scalar field acquires a soft mass ($\sim$ 1 TeV) through a SUSY breaking effect. When we assume that the cosmological constant vanishes at the VEV, it is natural to take $C \simeq m_{3/2}M_G^2$, where $m_{3/2}$ is the gravitino mass ($\sim$ 1 TeV). Then, we obtain the following low energy effective potential,

$$V(\phi) = V_0 - m_0^2|\phi|^2 + \frac{n}{n+3}\frac{\lambda C}{M_G^2}\frac{\phi^{n+3}+\phi^{*n+3}}{M_G^n} + \lambda^2\frac{|\phi|^{2n+4}}{M_G^{2n}}. \tag{2}$$

One finds that the VEV of the scalar field $M \equiv \langle\phi\rangle \sim \sqrt{m_0 M_G} \sim \mathcal{O}(10^{10})$ GeV in the case of $n = 1$ and the vacuum energy at $\phi = 0$ is $V_0 \sim M^2 m_0^2$.

While the Hubble expansion rate is large $H \gg m_0$, the scalar field ϕ would be trapped dynamically at the origin by an additional SUSY breaking effect. When the Hubble expansion rate becomes smaller than the mass of the scalar field, i.e., $H < m_0$, the additional SUSY breaking effect disappears and the scalar field begins to roll down to its VEV.

Hereafter we mainly consider the case of n = 1. When the reheating process finished, the ratio of the energy density of ϕ to the entropy density s is estimated as

$$\begin{aligned}\frac{\rho_\phi}{s} &= \frac{V_0 T_R}{4M_G^2 m_0^2}, \\ &\simeq 0.5\times10^{-9}\text{GeV}\left(\frac{T_R}{10^7\text{GeV}}\right)\left(\frac{M}{10^{10}\text{GeV}}\right)^2,\end{aligned} \tag{3}$$

where $s = 2\pi^2 g_* T_R^3/45$ and T_R is reheating temperature. The contribution of ρ_ϕ to density parameter Ω is estimated as

$$\Omega_\phi \simeq 0.28\times\left(\frac{T_R}{10^7\text{GeV}}\right)\left(\frac{M}{10^{10}\text{GeV}}\right)^2\left(\frac{0.7}{h}\right)^2. \tag{4}$$

From Eq. (4), we see that $\Omega_\phi \simeq 0.3$ for $T_R \simeq 10^7$ GeV.

3. Conclusion

In this paper, within the framework of inflationary cosmology we have shown that the stable scalar field with the electroweak scale mass and the large VEV is now oscillating, and the energy density of the oscillation significantly contributes to the density parameter Ω and satisfies the requirements for CDM at present in the universe.

References

O. Seto, K. Kohri and T. Nakamura, JHEP **0109**, 032 (2001).

Influence of Resolution of Timesteps on Making Merging Trees

Mamoru SHIMIZU
Department of Physics, University of Tokyo, 7-3-1 Hongo, Bunkyo-ku, Tokyo 113-0033, Japan
Tetsu KITAYAMA
Department of Physics, Toho University, 2-2-1 Miyama, Funabashi-shi, Chiba 274-8510, Japan
Shin SASAKI
Department of Physics, Tokyo Metropolitan University, 1-1 Minami-osawa, Hachioji-shi, Tokyo 192-0397, Japan
Yasushi SUTO
Department of Physics and Research Center for the Early Universe, University of Tokyo, 7-3-1 Hongo, Bunkyo-ku, Tokyo 113-0033, Japan

Abstract

We investigate effects of modeling of merging trees of dark matter halos on results of Monte-Carlo modeling of galaxy formation. Especially, we study effects of differences of timestep and mass resolution of merging trees of dark matter halos on gas cooling in the halos. We find that in order to obtain a suitable result on gas cooling, we should adopt smaller timestep and mass resolution of merging trees than previous studies in the modeling. The difference of these model parameters can be important to understand galaxy formation because cold gas fraction between our parameters and those of the previous studies differs by about factor 2.

1. Introduction

Semi-analytic models for galaxy formation (SAMs) are those which can be used for discussion of galaxy formation in a hierarchical universe like a cold dark matter model.

The most important ingredient in the SAMs is the conditional joint probability distribution function for a set of *progenitor* halos at a redshift z_2 which is a part of a *parent* halo at $z_1(< z_2)$. However, we cannot understand the conditional joint probability distribution function on the basis of the extended Press-Schechter theory. One needs to use the one-point probability distribution function and employ an additional *assumption* in generating realizations of merging trees of halos in general (e.g. [2]). Furthermore any numerical procedure to generate them necessarily involves several *ad hoc* parameters such as the finite timestep of computation.

We perform a systematic investigation of possible artificial effects of the above problems on merging tree realizations. In particular we focus on the extent to which the resulting merging trees reproduce the conditional one-point probability distribution function predicted by the extended Press-Schechter theory which directly changes the fraction of cold gas.

We adopt a conventional ΛCDM model with the cosmological parameters $\Omega_0 = 0.3$, $\lambda_0 = 0.7$, $h = 0.7$, $\sigma_8 = 1.0$, and $\Omega_B = 0.015\ h^{-2}$, and neglect star formation and feedback effect for definiteness.

2. Results

Our model of merging histories of dark matter halos is mainly based on that of Somerville and Kolatt (1999). We adopt the logarithmically equal timestep in redshift for $z_{\rm min} = 0$ and $z_{\rm max} = 15$. Then, We compare the result in case that the number of timestep, $N_{\rm step}$, is equal to 100 and 1000. The results with $N_{\rm step} = 1000$ better reproduces the prediction of extended Press-Schechter theory mainly because the progenitors with small scale are reproduced well.

Moreover, we consider how the resulting merger trees affect the cold gas fraction in an individual halo. We assume that the density profile of gas in the halos is that of the isothermal β-model with $\beta = 2/3$. The cold gas fraction with $N_{\rm step} = 1000$ is greater nearly by a factor of 2 than that of $N_{\rm step} = 100$ for large halo mass. Although we cannot understand effects of the factor 2 of cold gas fraction on the amount of stars precisely because of heating of gas by supernova feedback, the amount may vary by about factor 2. We cannot ignore the variation, when we compare calculation of the SAMs with observation, if it is true.

References

[1] Shimizu et al., submitted
[2] Somerville and Kolatt, 1999, MNRAS, 305, 946

Nucleosynthesis in Type II Supernovae and Abundances in Extremely Metal-poor Stars

Marii Shirouzu[1], Chiaki Kobayashi[1], Takayoshi Nakamura[1], Hideyuki Umeda,[1,2] and Ken'ichi Nomoto[1,2]
[1]*Department of Astronomy, School of Science, The University of Tokyo, 7-3-1 Hongo, Bunkyo-ku, Tokyo 113-0033, Japan*
[2]*Research Center for the Early Universe (RESCEU), School of Science, The University of Tokyo, 7-3-1 Hongo, Bunkyo-ku, Tokyo 113-0033, Japan*

Abstract

We calculate nucleosynthesis yields from Type II supernovae (SNe II) and hypernovae for very wide parameter ranges and then search for extremely metal-poor stars whose abundance patterns are in good agreement with these stellar yields. We find significant constraints on the model parameters. For example, stars with $M \gtrsim 20 M_\odot$ are likely to undergo hypernova-like explosions.

1. Introduction

Recent high resolution abundance surveys pay great attention to extremely metal-poor stars with [Fe/H] $\lesssim -2$. We assume that such a star formed from the interstellar matter that had been polluted by only a single supernova of a Population III star. We then expect that the abundance patterns of extremely metal-poor stars are in good agreement with one of the yields of explosion models and some constraints on model parameters can be found from direct comparisons between the abundances of very metal-poor stars and the stellar yields.

2. Models for SN II Explosions and Nucleosynthesis

Stars more massive than $8M_\odot$ explode as SNe II at the end of their life and nucleosynthesis proceeds explosively because of high temperatures and densities in the stellar interior. In spite of many efforts, the detailed explosion mechanism is still controversial. Therefore, our present theoretical models for explosions and nucleosynthesis yields in SNe II depend on the following parameters: 1) the mass of the progenitor, 2) metallicity of the progenitor (Z), 3) explosion energy, 4) mass cut, 5) Y_e (i.e., neutron excess), and 6) mixing.

3. Abundances in Extremely Metal-Poor Stars

The abundances of 80 extremely metal-poor stars are taken from [1] – [5]. The selective judgments are made by χ-square between the yields and the abundance of each stars. Parameter sets are determined from these comparisons,

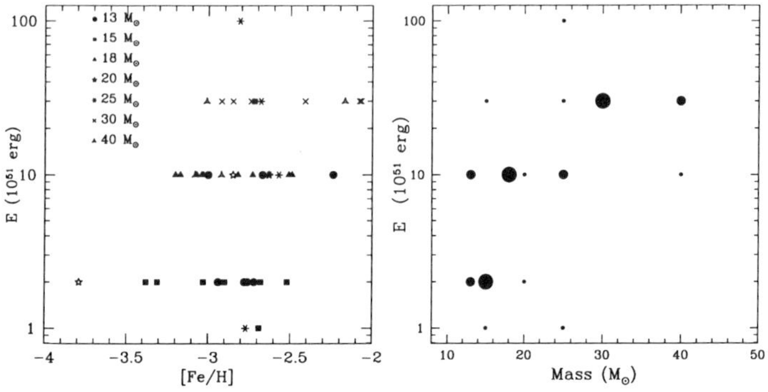

Fig. 1. Left: The [Fe/H] distribution of explosion energy and progenitor mass. Right: The progenitor mass - explosion energy relation of SNe that can reproduce the abundances of the extremely metal-poor stars. Large, middle, and small circles show the number of the selected stars as ≥ 5, $3-4$, and ≤ 2, respectively.

if abundance ratios in the models are similar to those observed in the programmed stars. We derive the following constraints on the parameters from most of the 45 selected stars; 1) $Y_e \simeq 0.4997 - 0.5$, and 2) the mixing must occur to bring up iron-peak elements that are produced in the deepest region of SNe. Our models can also explain the dispersion of the abundance ratios in the extremely metal-poor stars. A significant relation between the progenitor mass and explosion energy (the right panel of Fig.1.) is also found. This implies that most stars with $M \gtrsim 20 - 25 M_\odot$ undergo hypernova-like explosions.

References

[1] Aoki, W., Norris, J. E., Ryan, S. G., Beers, T. C., & Ando, H. 2001 preprint (astro-ph/0111297)

[2] Honda, S. 2002, PhD Thesis, The Graduate University for Advanced Studies

[3] McWilliam, A. et al. 1995, AJ, 109, 2757

[4] Norris, J. E., Ryan, S. G., & Beers, T. C. 1997b, ApJ, 488, 350; 1997c, ApJ, 489, L169; 2001, ApJ, 561, 1034; Norris, J. E., Ryan, S. G., Beers, T. C., & Deliyannis, C. P. 1997a, ApJ, 485, 370

[5] Ryan, S.G., Norris, J.E., & Beers, T.C. 1996, ApJ, 471, 254; 1998, ApJ, 506, 892

First Structure Formation ($12 > z > 7$) in a CHDM Model

Doru Marian SURAN and Nedelia Antonia POPESCU *Astronomical Institute of Romanian Academy, Cutitul de Argint,5, 75212 Bucharest 28, Romania*

Abstract

We describe the results of N-body+SPH simulations in a CHDM universe concerning the properties of object collapsing at high redshifts. We analyze both Large Scale structures (LSS) and halos properties and try to trace the relation between the evolution of the first order filamentary Web structures and the galactic and cluster structures.

1. Introduction

Our simulations have been made using PM models for DM+massive neutrinos+SPH (CHDM models, 2 massive neutrino species) at different redshifts: z = 11, 10, 9, 8. The simulations are made on the SGI platform of the Astronomical Institute of the Romanian Academy. The simulations are of $128^3 - 256^3/256^3 - 512^3$ points on an intermediate scale of 25 $h^{-1}Mpc$ with chosen parameters: $H_0 = 50kms^{-1}Mpc^{-1}$, $\Omega_m = 0.8$, $\Omega_b = 0.05$, $\Omega_\nu = 0.2$. The N-body simulations are supplied by a series of different proper models and algorithms for the interpretation of these simulations: halo identification; halo history and progenitor identification; shape of under and over dense regions identification and evolution of LSS objects (pancakes, filaments).

2. Results

The comparison between our halos catalogue and the Press-Schecter analytical formalism indicates a different distribution of our halos at small and intermediate scales ($M < 10^{12}M_\odot$), similar with the WDM calculations. In order to study the LSS structure we take into accont three different types of structures: a high density halos like structure ($\rho / < \rho >> 50$), a smooth intermediate background ($5 < \rho / < \rho >< 50$) and a low density component ($\rho / < \rho >< 5$). The time (redshift) variation of these components indicates a permanent transfer of matter between the intermediate and the low+high scales, with a bigger rate to the low scale. The correlation functions calculated for the three components indicate a different behaviour, i.e. different LSS spatial distribution of these. The correlation functions of the halos distribution carry both signals of galactic and cluster like objects. To take into account the matter exchange rates we have studied four LSS components: (1) unvirialized diffuse matter;(2) diffuse matter in host halos;(3) subhalos;(4) isolated halos. The results reveal that in time the unvirialized matter diminishes. The shapes (ellipticities) of intermediate LSS component show the persistence of the filamentary like structures at these redshifts (in comparison with the pancakes like ones). The shapes of the halos indicate the existence of more prolate and spherical shapes than the oblate ones. Radial profiles of the infall velocity distribution inside and in the vicinity of halos present shocks at the border of halos. Radial profiles of the Mach number is less than two with a mean in the range 1.3-1.4. This indicates mild supersonic velocities of the shocks. The mass concentration parameter shows a different distribution (function of redshifts) for higher and lower mass halos. The mass concentration parameter for host halos, subhalos and isolated halos (function of redshifts) presents different trends, suggesting the importance of the merging phenomena. The spin parameter (λ) shows a different distribution (with a mean distribution of ~ 0.1) in relation to the CDM model (with a mean distribution ~ 0.05). Also, the spin distribution is different for low and high mass halos, suggesting diferent mechanism of evolution. The angular momentum alignment with the minor axis of the best-fit ellipsoids of the halos is still present, but with much noise than in the CDM models. The local velocity fields and the deformation tensors can reconstruct the general LSS hydrodynamical flow involved in the simulations. The time evolution of general LSS flow and that of the halos movement 0 give a better description of the hydrodynamical evolution of the simulations.

On the Moduli Problem and Baryogenesis in Gauge-mediated SUSY Breaking Models

Shinta KASUYA, Masahiro KAWASAKI, and Fuminobu TAKAHASHI *Research Center for the Early Universe (RESCEU), School of Science, The University of Tokyo, 7-3-1 Hongo, Bunkyo-ku, Tokyo 113-0033, Japan*

Abstract

We investigate whether the Affleck-Dine mechanism can produce sufficient baryon number of the universe in the gauge-mediated SUSY breaking models, while evading the cosmological moduli problem by late-time entropy production. We find that the Q-ball formation renders the scenario very difficult to work, irrespective of the detail mechanism of the entropy production.

1. Introduction

In superstring theories, there generally exist various dilaton and modulus fields. These fields (we call them "moduli" in this paper) are expected to acquire masses of the order of the gravitino mass $m_{3/2}$. It is well known that the moduli cause serious cosmological problem [1] because they have only gravitationally suppressed interactions with other particles and hence have long lifetimes.

The gauge mediation models have attractive points that are absent in the hidden sector models; they can avoid the flavor problem and the mass pattern of the particles in the observable sector is predictable [2]. Therefore, in this paper we consider the moduli problem in the gauge-mediated SUSY breaking models.

In order to avoid the moduli problem, we need some huge entropy production process by which the moduli density is diluted. However, then the baryon asymmetry is also diluted. In this paper we study the cosmological moduli problem and baryogenesis in the gauge-mediated SUSY breaking models taking into account the Q-ball formation.

2. Affleck-Dine mechanism and Q-ball formation

In the Minimal Supersymmetric Standard Model (MSSM), there exist flat directions, along which there are no classical potentials. Since flat directions consist of squarks and/or sleptons, they carry baryon and/or lepton numbers, and can be identified as the Affleck-Dine field Φ. In the gauge-mediated SUSY breaking models, the potential of a flat direction is parabolic at the origin, and almost flat beyond the messenger scale [3, 4, 5]:

$$V_{gauge} \sim \begin{cases} m_\phi^2|\Phi|^2 & (\Phi \ll M_S) \\ M_F^4 \log^2 \dfrac{|\Phi|^2}{M_S^2} & (\Phi \gg M_S) \end{cases}, \tag{1}$$

where M_S is the messenger mass scale.

However, the AD field feels spatial instabilities during its coherent oscillation, and deforms into nontopological solitons, Q balls [4, 6, 7]. From numerical calculations [7, 5], Q balls absorb almost all the baryon charges which the AD field obtains. Hence the baryon asymmetry is suppressed by $\Delta Q/Q$, where ΔQ is the charge evaporated from Q balls.

The present baryon asymmetry can be related to the moduli density parameter as follows:

$$Y_B = \frac{n_B}{\rho_{mod}} \frac{\rho_{mod}}{\rho_c} \frac{\rho_c}{s_0} \times \frac{\Delta Q}{Q}, \tag{2}$$

where ρ_{mod} is the energy density of the moduli, ρ_c is the critical density, and s_0 is the entropy density at present.

3. Conclusion

We have investigated the possibility of the AD baryogenesis in the gauge-mediated SUSY breaking scenario, while evading the cosmological moduli problem by the late-time entropy production. In all the cases, the Q-ball formation makes the efficiency of the baryon number production considerably diminish.

References

[1] G.D. Coughlan, W. Fischler, E.W. Kolb, S.Raby, and G.G. Ross, Phys. Lett. B131 (1983) 59.
[2] For a review, see, e.g. G.F. Giudice and R. Rattazzi, Phys. Rep. 322 (1999) 419.
[3] A. de Gouvêa, T. Moroi, and H. Murayama, Phys. Rev. D56 (1997) 1281.
[4] A. Kusenko and M. Shaposhnikov, Phys. Lett. B418 (1998) 46.
[5] S. Kasuya and M. Kawasaki, to appear in Phys. Rev. D, hep-ph/0106119.
[6] K. Enqvist and J. McDonald, Nucl. Phys. B538 (1999) 321.
[7] S. Kasuya and M. Kawasaki, Phys. Rev. D61 (2000) 041301.

Earth Effects on Supernova Neutrino Spectra and their Implications for Neutrino Parameters

Keitaro TAKAHASHI *Department of Physics, The University of Tokyo, 7-3-1 Hongo, Bunkyo-ku, Tokyo 113-0033, Japan*
Katsuhiko SATO
Research Center for the Early Universe (RESCEU), School of Science, The University of Tokyo, 7-3-1 Hongo, Bunkyo-ku, Tokyo 113-0033, Japan

Abstract

We perform a detailed study of the Earth matter effects on supernova neutrinos with neutrino oscillation parameter LMA and small θ_{13}. The Earth effects show significant dependences on the distance which neutrinos travel in the Earth and the value of Δm^2.We show that making use of these dependences, we can obtain implication for the value of Δm^2_{12} by comparing the observed energy spectrum to the predicted one. In much of the neutrino-detection-time-Δm^2_{12} plane, Δm^2_{12} can be determined with an accuracy equal to or better than $\pm 0.5 \times 10^{-5}\mathrm{eV}^2$.

1. Supernova Neutrino Spectrum and Δm^2_{12}

Neutrino oscillation can affect the energy spectrum of supernova neutrino drastically. Neutrinos of all flavor are produced in the high dense region of the iron core [1] and interact with matter before emerging from the supernova. The presence of non-zero masses and mixing in vacuum among various neutrino flavors results in strong matter dependent effects, including conversion from one flavor to another. Hence, the observed neutrino flux in the detectors may be dramatically different for certain neutrino flavors and for certain values of mixing parameters, due to neutrino oscillation [2, 3]. Similar matter effect occurs in the Earth, too and the length of neutrino path in the Earth depends on the arrival time as seen in Top-Left of Fig. 1. [4]. In our previous paper, we found that the spectrum of ν_e change significantly due to the Earth matter effect when the solution of the solar neutrino problem is LMA and θ_{13} is small[5]. In this contribution, we study the Earth matter effect more carefully.

The Top-Right and Bottom-Left of Fig. 1. shows the nadir angle and Δm^2_{12} dependence of the spectrum at SNO, respectively. The dependence of the spectrum on Δm^2_{12} is so significant that we can determine Δm^2_{12} from the shape

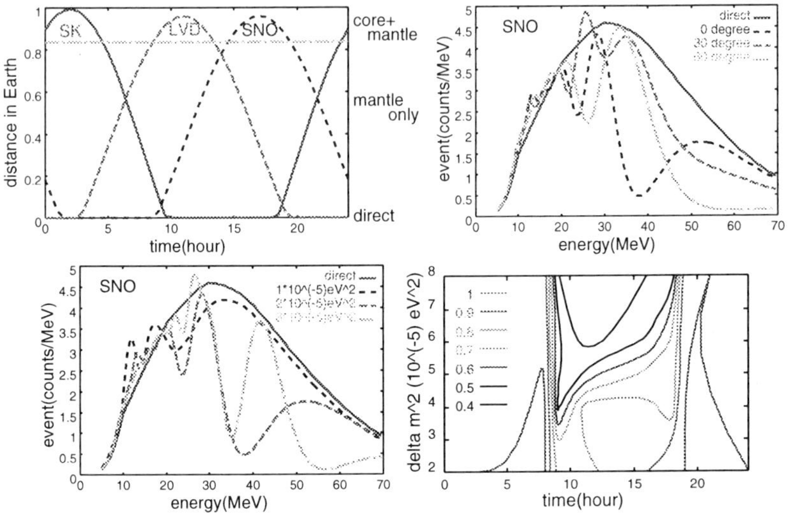

Fig. 1. Top-Left:The distance which neutrinos travel in the Earth as functions of the arrival time of neutrino burst [4]. We assume a supernova which located in the Galactic center. Top-Right:Nadir angle dependence of the spectrum at SNO. Bottom-Left:Δm_{12}^2 dependence of the spectrum at SNO. Bottom-Right:Contour plot of the probability that Δm_{12}^2 is determined with an accuracy of $\pm 0.5 \times 10^{-5}$.

of the observed spectrum. We perform Monte Carlo simulation to obtain the expected observed spectra, assuming a certain value of Δm_{12}^2 and compare the simulated spectrum to the theoretical spectra with various Δm_{12}^2 values with a χ^2 method (for more detail, see [6]). Bottom-Right of Fig. 1. is a contour plot of the probability that Δm_{12}^2 is determined with an accuracy of $\pm 0.5 \times 10^{-5}$. From this figure, we can see that the probability is $> 90\%$, when SK detects neutrinos from the Earth and that the probability is $> 80\%$ in 2/3 of time-Δm_{12}^2 plain.

References

[1] T.Totani, K.Sato, H.E.Dalhed and J.R.Wilson, Astrophys. J. 496 (1998) 216.

[2] G. Dutta, D. Indumathi, M. V. N. Murthy and G. Rajasekaran, Phys. Rev. D61 (2000) 013009.

[3] K. Takahashi, M. Watanabe, K. Sato and T. Totani, Phys. Rev. D64 (2001) 093004.

[4] C. Lunardini and A. Y. Smirnov, Nucl. Phys. B616 (2001) 307.

[5] K. Takahashi, M. Watanabe and K. Sato, Phys. Lett. B510 (2001) 189.

[6] K. Takahashi and K. Sato, hep-ph/0110105, submitted to Phys. Rev. D.

A Bridge from Optical to Infrared Galaxies: Explaining Local Properties, Predicting Galaxy Counts and the Cosmic Background Radiation

Tsutomu T. TAKEUCHI
National Astronomical Observatory, Mitaka, Tokyo 181-8588, Japan
Tomonori TOTANI
Princeton University Observatory, Princeton, NJ 08544-1001, USA

1. Introduction

To understand when and how stars and galaxies formed in the universe is one of the most fundamental issues in modern astronomy. We construct a new model for infrared galaxy counts and cosmic background radiation (CBR) to tackle the hidden star formation history in the Universe.

2. Model

We construct a model of infrared galaxy counts and CBR based on the backwards approach for luminosity evolution of galaxies, by extending the model for optical/NIR galaxy counts (Totani & Yoshii 2000, hereafter TY00) to include dust emission component in infrared bands. In our model, The SED is constructed by empirical way, but dust temperature is determined by an energy balance condition. Details are presented in our paper Totani & Takeuchi (2001).

3. High Redshift Infrared Galaxies

We make predictions for high-z infrared galaxies such as faint galaxy counts and the cosmic background radiation. Dust temperature evolution is much stronger than expected from the empirical $L_{\rm IR}$-$T_{\rm dust}$ relation that have been widely used in the previous works. Especially, the peak dust temperature during the starburst phase of elliptical galaxies in the screen model is $\sim$ 60–70K. This must have significant effect on the infrared SED of primordial elliptical galaxies, and hence on faint counts and CBR. Nearby spiral galaxies are dominant in the brightest counts, while contribution from elliptical galaxies rapidly appears when the flux is decreased down to a certain value, because of the strong dusty starburst of these galaxies at their formation stage. The model also succeed in reproducing the overall shape of the observed CBR.

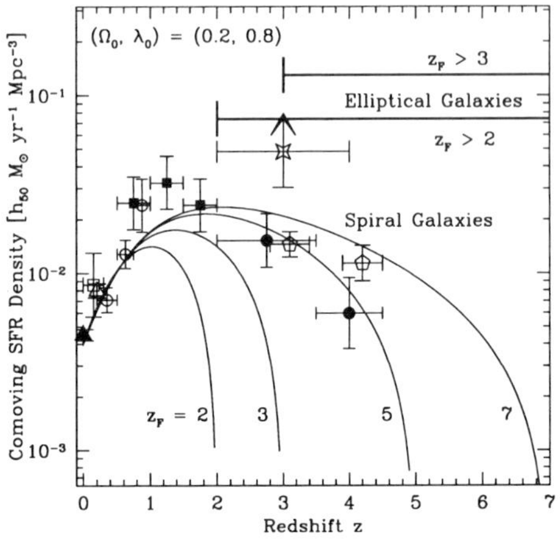

Fig. 1. The cosmic star formation history derived from our model.

4. Implications and Discussion

Figure 1. shows the cosmic history of global star formation rate as a function of redshift. Elliptical galaxies in our model have much higher global SFR at $z > 3$ than the peak at $z \sim 1$ suggested by the optical observations and the model is still consistent with the FIR background detected by *COBE*/FIRAS. This is in very sharp contrast to a number of previous studies which claimed that such high SFR beyond $z \sim 1$ is disfavored from the *COBE*/FIRAS data. This is because the dust temperature of forming elliptical galaxies is much higher than expected from the local empirical relation. The SFR expected from elliptical galaxies is close to the lower limit from the submillimeter observation, and this suggests that a considerable part of forming elliptical galaxies has been detected by submillimeter observations. On the other hand, SFR expected from spiral galaxies is in reasonable agreement with those obtained by optical observations.

References

[1] Totani, T., & Yoshii, Y. 2000, ApJ, 540, 81 (TY00)
[2] Totani, T., & Takeuchi, T. T. 2001, ApJ, submitted

Have We Already Detected Astrophysical Symptoms Of Space-time Noncommutativity ?

Takashi TAMAKI, Tomohiro HARADA, Umpei Miyamoto
Department of Physics, Waseda University, Ohkubo, Shinjuku, Tokyo 169-8555, Japan
and Takashi Torii
Research Center for the Early Universe (RESCEU), School of Science, The University of Tokyo, 7-3-1 Hongo, Bunkyo-ku, Tokyo 113-0033, Japan

Abstract

We discuss astrophysical implications of κ-Minkowski space-time. We first derive a velocity formula which shows that a massless particle moves at a constant speed as in the usual Minkowski space-time. We also analyze reaction processes in κ-Minkowski space-time which are related to the detections of the super Greisen-Zatsepin-Kuzmin (GZK) events and of high-energy ($\sim$20 TeV) γ-rays from Markarian (Mk) 501. We can naturally account for them.

1. Introduction

Recently, since super GZK events has been reported, some authors argue the Lorentz invariance (LI) violation to solve them[1], which may also explain detections of γ-rays above 20 TeV [2]. Since it is pointed out that an arrival time analysis by γ-rays from Mk 421 gives a severe constraint on LI violation [3], it may be impossible [4]. Here, we use a model called κ-Minkowski space-time [5] to consider them [6]. We use the units in which $c = \hbar = 1$.

2. κ-Minkowski space-time

The basic commutation relations are $[x^i, t] = i\lambda x^i$, $[x^i, x^j] = 0$, where the indices i, j run 1, 2, 3. If we denote the additive operator in κ-Minkowski space-time by $\hat{+}$ to distinguish from the conventional one, we can write as

$$(E_1 \hat{+} E_2, \boldsymbol{p}_1 \hat{+} \boldsymbol{p}_2) := (E_1, \boldsymbol{p}_1)(E_2, \boldsymbol{p}_2) = (E_1 + E_2, \boldsymbol{p}_1 + e^{\lambda E_1}\boldsymbol{p}_2) \ , \tag{1}$$

where E and $\boldsymbol{p}$ are the energy and the momentum, respectively.

We describe a plane wave as $\psi_{(E,\boldsymbol{p})} = e^{i\boldsymbol{p}\cdot\boldsymbol{x}}e^{iEt}$, and place the t generator to the right of x generator [5]. Then, the property $\psi_{(E_1,\boldsymbol{p}_1)(E_2,\boldsymbol{p}_2)} = \psi_{\boldsymbol{p}_1,E_1}\psi_{\boldsymbol{p}_2,E_2}$ is found. We can also define $\psi_{(E,\boldsymbol{p})^{-1}} := e^{-i\boldsymbol{p}e^{-\lambda E}\cdot\boldsymbol{x}}e^{-iEt} = e^{-iEt}e^{-i\boldsymbol{p}\cdot\boldsymbol{x}}$, which implies that $(E,\boldsymbol{p})^{-1}$ is an inversion of $(E,\boldsymbol{p})$. Because of these structures, the dispersion relation is altered as (m is a mass of the particle.)

$$\lambda^{-2}(e^{\lambda E} + e^{-\lambda E} - 2) - \boldsymbol{p}^2 e^{-\lambda E} = \lambda^{-2}(e^{\lambda m} + e^{-\lambda m} - 2)\ . \tag{2}$$

To derive the group velocity formula, we consider infinitesimal changes ΔE and $\Delta\boldsymbol{p}$ as a result of adding $(\Delta E', \Delta\boldsymbol{p}')$ as $(E\hat{+}\Delta E', \boldsymbol{p}\hat{+}\Delta\boldsymbol{p}') = (E+\Delta E, \boldsymbol{p}+\Delta\boldsymbol{p})$. We can express as $(\Delta E', \Delta\boldsymbol{p}') \cong (\Delta E, \frac{\Delta\boldsymbol{p}}{e^{\lambda E}})$, where we keep only terms in first order in ΔE and $\Delta\boldsymbol{p}$. Then, we make a wave packet as follows

$$\begin{aligned} I &= \psi_{(E-\Delta E,\boldsymbol{p}-\Delta\boldsymbol{p})} + \psi_{(E+\Delta E,\boldsymbol{p}+\Delta\boldsymbol{p})} = \psi_{(E,\boldsymbol{p})(-\Delta E',-\Delta\boldsymbol{p}')} + \psi_{(E,\boldsymbol{p})(\Delta E',\Delta\boldsymbol{p}')} \\ &= \psi_{(E,\boldsymbol{p})}\psi_{(-\Delta E',-\Delta\boldsymbol{p}')} + \psi_{(E,\boldsymbol{p})}\psi_{(\Delta E',\Delta\boldsymbol{p}')} = \psi_{(E,\boldsymbol{p})}[e^{-i\Delta\boldsymbol{p}'\cdot\boldsymbol{x}}e^{-i\Delta E't} + e^{i\Delta\boldsymbol{p}'\cdot\boldsymbol{x}}e^{i\Delta E't}] \\ &\cong 2e^{i\boldsymbol{p}\cdot\boldsymbol{x}}e^{iEt}\cos\left[\frac{\Delta\boldsymbol{p}}{e^{\lambda E}}\cdot\left(\boldsymbol{x} + \frac{e^{\lambda E}\Delta E t}{\Delta\boldsymbol{p}}\right)\right]. \end{aligned} \tag{3}$$

By considering $|I|^2$, the group velocity $\boldsymbol{v}_l$ can be written as $\boldsymbol{v}_l := e^{\lambda E}\frac{dE}{d\boldsymbol{p}}$. We also consider a similar relation $(\Delta E'\hat{+}E, \Delta\boldsymbol{p}'\hat{+}\boldsymbol{p}) = (E+\Delta E, \boldsymbol{p}+\Delta\boldsymbol{p})$, where the corresponding group velocity $\boldsymbol{v}_r$ is $\boldsymbol{v}_r := \frac{dE}{d\boldsymbol{p}}\left(1 - \lambda\boldsymbol{p}\cdot\frac{dE}{d\boldsymbol{p}}\right)^{1}$. Using (2), we obtain the important conclusion that *massless particles move in a speed* $|\boldsymbol{v}_l| = |\boldsymbol{v}_r| = 1$ *as in the usual Minkowski space-time.*

3. Conclusion

Based on the above results, we considered threshold values for reactions $\gamma + \gamma \to e^+ + e^-$ and $p + \gamma \to p + \pi_0$ in κ-Minkowski space-time and analyzed their relavance to ~20 TeV photons and the super GZK events. We found that we can naturally account for such high energy events. If this is the case, we might have already detected symptoms of the space-time noncommutativity.

References

[1] E.g., S. Coleman and S. L. Glashow, Phys. Rev. D **59**, 116008 (1999).
[2] E.g., T. Kifune, Astrophys. J. Lett. **518**, L21 (1999).
[3] E.g., G. Amelino-Camelia et al., Nature **393**, 763 (1998).
[4] G. Amelino-Camelia and T. Piran, Phys. Rev. D **64**, 036005 (2001).
[5] G. Amelino-Camelia and S. Majid, Int. J. Mod. Phys. A **15**, 4301 (2000).
[6] T. Tamaki, T. Harada, U. Miyamoto and T. Torii, gr-qc/0111056.

Lognormal Property of Weak Lensing Fields

Atsushi TARUYA
Research Center for the Early Universe (RESCEU), School of Science, The University of Tokyo, 7-3-1 Hongo, Bunkyo-ku, Tokyo 113-0033, Japan

Abstract

Using the ray-tracing simulations, non-Gaussian property of weak lensing fields is studied quantitatively. In particular, validity of the lognormal model prediction to the weak lensing statistics is discussed. Comparing the lognormal predictions of one-point probability distribution function and Minkowski functionals of convergence field with those obtained from the simulation data, we find that the empirical lognormal model is safely applicable as a useful cosmological tool, as long as we are concerned with the non-Gaussianity of $\nu \lesssim 5$ for low-z_s samples.

1. Introduction

Cosmic shear, coherent distortions in galaxy images caused by the gravitational field of intervening large-scale structure, has now been recognized as a powerful cosmological tool. Recently, measurements of the cosmic shear have been independently performed by several groups [1]. The estimated shear variance from the different observational data set quantitatively reconciles with each other, which is in good agreement with theoretical predictions based on the cold dark matter (CDM) model of structure formation.

To obtain a strong constraint on the cosmological parameters and/or the dark matter distribution, quantifying the non-Gaussian features of weak lensing field is important. Here, based on the empirical lognormal model, we discuss a statistical description for the local convergence, κ, i.e., the surface mass density projected along a line-of-sight on the image plane. The usefulness of logrnomal models in characiterizing the non-Gaussianity of κ is investigated in detail using the ray-tracing simulations.

2. Testing lognormal models against ray-tracing simulations

From the ray-tracing simulations, we first construct the weak lensing map that covers a field-of-view of $5^\circ \times 5^\circ$. We then compute the smoothed convergence fields by convolving the top-hat smoothing kernel and the various statistical quantities characitering the non-Gaussianity are evaluated: (i) *probability distribution function*(PDF) of local convergence, $P(\kappa)$; (ii) *Minkowski functionals* $v_i(\nu)$ $(i = 0, 1, 2)$; (iii) *skewness* and *kurtosis*. Note that the lognormal model prescription gives simple analytic expressions to these quantities, which are basically characterized by the three parameters, i.e., minimum value of the convergence, $\kappa_{\rm min}$, RMS of the convegence and its gradient fields, $\langle\kappa^2\rangle^{1/2}$ and $\langle(\nabla\kappa)^2\rangle^{1/2}$ [2].

As an illustrative example, in Fig.1, we show the one-point PDF as a function of local convergence for various smoothing angles in Lambda CDM model. In each panel, solid and dashed lines respectively indicate the lognormal prediction and the Gaussian PDF. Clearly, the significant deviation from Gaussian distribution is found and the lognormal PDF remarkably reproduces the simulation results in a good accuracy. The detailed analysis for other statistics is presented in Ref.[2].

Now, we briefly summarize the main results of our analysis. The lognormal models of local convergence accurately describe the simulation results of one-point PDF and Minkowski functionals on both small and large smoothing angular scales in the case of the low-$z_{\rm s}$ data. On the other hand, the detailed comparison revealed that the lognormal model does not provide an accurate prediction for the statistics sensitive to the rare events such as the skewness and kurtosis of the convergence. We therefore conclude that, as long as we are concerned with the appropriate range of the convergence, $\nu \lesssim 5$ $(\nu \equiv \kappa/\langle\kappa^2\rangle^{1/2})$, the lognormal model empirically but quantitatively gives a useful approximation characterizing the non-Gaussianity features on the convergence field.

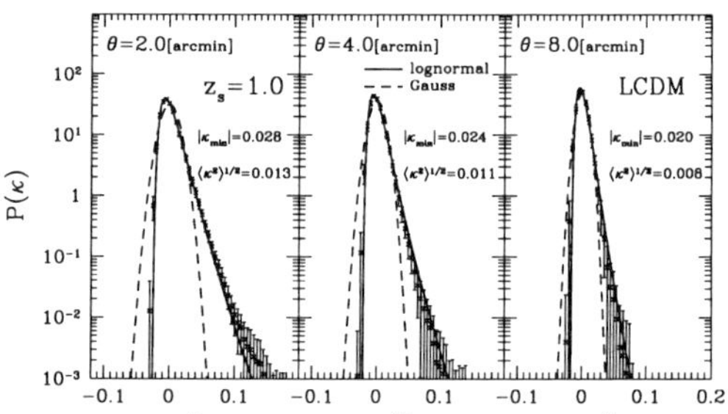

Fig. 1. One-point PDF as a function of convergene field, $P(\kappa)$ in Lambda CDM model with the smoothing angle, $\theta = 2'$(*left*), $4'$ (*middle*) and $8'$(*right*). In each panel, solid(dashed) lines indicate lognormal(Gaussian) predictions.

References

[1] e.g., D. Bacon, A. Refregier, and R. Ellis, MNRAS 318 (2000) 625; L. Van Waerbeke et al., A & A 358 (2000) 30.

[2] A. Taruya, M. Takada, T. Hamana, I. Kayo, and T. Futamase, Astrophys. J. (2001), submitted.

Brane World and Holography

Takashi TORII, Tetsuya SHIROMIZU, and Daisuke IDA
Research Center for the Early Universe (RESCEU), School of Science, The University of Tokyo, 7-3-1 Hongo, Bunkyo-ku, Tokyo 113-0033, Japan

Abstract

We consider the brane-world in the holographic point of view. First of all, we find the constraint on the coupling of the scalar fields with the matter (not holographic CFT) on the brane. We show that the traceless part of the energy-momentum tensor of holographic CFT is a part of the bulk Weyl tensor. The trace part which comes from the trace-anomaly is corresponding to the ρ^2-term appeared in the generalized FRW equation in the brane-world.

The Randall-Sundrum brane-world is the attractive and simple model which describes the stringy cosmology, where the 4-dimensional spacetime is regarded as a 4-dimensional Minkowski spacetime embedded in the 5-dimensional anti-de Sitter (AdS) spacetime. In this model, it has been checked that the Newton gravity is recovered in the linearized theory.

We shall consider that the gravitational equations on the brane from the holographic point of view. We will compare the gravitational equations, which are obtained via the purely geometrical reduction without any approximations, with the effective equations derived by using the AdS/CFT correspondence [1]. This direction was initiated by Witten, and then formulated by others. They have shown that the correction to the Newton gravity can be calculated by CFT living on the brane. Here we will consider what kind of brane-world is compatible to the AdS/CFT correspondence. There are two ways to obtain the effective equation on the brane. By comparing them, our purpose should be attained.

We begin with the action with the 5-dim gravity, bulk scalar field and matters on the brane. From the Gauss equation and the Israel's junction condition, the gravitational equation on the brane is given by

$$\begin{aligned} {}^{(4)}G_{\mu\nu} &= \frac{2{\kappa_5}^2}{3}\hat{T}_{\mu\nu}(\phi) + g_{\mu\nu}\Big[-\Lambda_4 + \frac{{\kappa_5}^2}{16}\Big(2\frac{d\lambda}{d\phi} + \frac{\partial\mathcal{L}_m}{\partial\phi}\Big)\frac{\partial\mathcal{L}_m}{\partial\phi}\Big] \\ &\quad + 8\pi G_4(\phi)\tau_{\mu\nu} + {\kappa_5}^4\pi_{\mu\nu} - E_{\mu\nu}. \end{aligned} \tag{1}$$

See Ref.[2] for details. $\tau_{\mu\nu}$ is the energy-momentum tensor of $\mathcal{L}_m$ on the brane.

The second-rank trace-free symmetric tensor $E_{\mu\nu}$ is the so-called "electric" part of the 5-dimensional Weyl tensor, ${}^{(5)}C_{KLMN}$. Note that we did not use any approximation to derive the gravitational equations. In this sense, Eq. (1) is exact under the assumption.

To derive the effective equation on the brane from the holographic point of view, we follow the Giddings, Katz and Randall's argument[3], where they have used the AdS/CFT correspondence[1]. The counter term S_{ct} is assumed to have the local form in terms of the 4-dimensional quantities, which can be easily determined by using the Hamilton-Jacobi equation [4]. Hamilton-Jacobi equation will be solved by expanding on shell action S with respect to the order of the derivatives, that is, we perform the derivative expansion. We obtain the effective gravitational equation on the brane:

$$ {}^{(4)}G_{\mu\nu} = 8\pi G_4 \tau_{\mu\nu} + D_\mu\phi D_\nu\phi - \frac{1}{2}q_{\mu\nu}(D\phi)^2 + \frac{4}{\ell\sqrt{-q}}\frac{\delta S^{(4)}}{\delta q_{\alpha\beta}}q_{\mu\alpha}q_{\nu\beta} + \frac{4}{\ell}\langle T_{\mu\nu}\rangle_{\rm CFT}. \quad (2) $$

Comparing Eq. (1) with Eq. (2), we obtain the relation between $E_{\mu\nu}$ and the energy-momentum tensor of CFT in the 2nd derivative order or $O(T_{\mu\nu})$. On the other hand, from the trace of the effective equation, we obtain $\pi^\mu{}_\mu = 4\langle T^\mu{}_\mu\rangle_{\rm CFT}/\ell$. It is already well-known that the term, $\pi^\mu{}_\mu$, implies the modification of the FRW models: $H^2 = 8\pi G_4\rho/3 + \rho^2/36$. Hence we can conclude that ρ^2-term comes from the trace-anomaly from the holographic point of view.

We stress that the brane-world models with a correct AdS/CFT interpretation will belong to a rather limited class in general. As an example, we considered the bulk massless scalar field and showed that the coupling of the scalar field with the matters on the brane cannot be admitted. This is also one of our main results. On the other hand, it is reminded that holographic CFT is coupled to the bulk scalar on the branes.

References

[1] J. Maldacena, Adv. Theor. Math. Phys. 2 (1998) 231; O. Aharony, S. S. Gubser, J. Maldacena, H. Ooguri and Y. Oz, Phys. Rep. 323 (2000) 183.

[2] T. Shiromizu, K. Maeda and M. Sasaki, Phys. Rev. D62 (2000) 024012.

[3] S. B. Giddings, E. Katz, and L. Randall, JHEP 0003 (2000) 023.

[4] P. Kraus, F. Larsen and R. Siebelink, Nucl. Phys B563 (1999) 259.

Gravitational Particle Production from Metric Perturbations

Shinji TSUJIKAWA
Research Center for the Early Universe (RESCEU), School of Science, The University of Tokyo, 7-3-1 Hongo, Bunkyo-ku, Tokyo 113-0033, Japan
Bruce A. BASSETT
Relativity and Cosmology Group (RCG), University of Portsmouth, Mercantile House, Portsmouth, PO1 2EG, England
Marco PELOSO
Physikalisches Institut, Universität Bonn Nussallee 12, D-53115 Bonn, Germany
Lorenzo SORBO
SISSA/ISAS, via Beirut 2-4, I-34013 Trieste, Italy, and INFN, Sezione di Trieste, via Valerio 2, I-34127 Trieste, Italy

Abstract

We study gravitational creation of light fermions in the presence of classical scalar metric perturbations about a flat Friedmann-Lemaitre- Robertson-Walker (FLRW) background. These perturbations can be large during preheating, breaking the conformal flatness of the background spacetime. We compute numerically the total number of particles generated by the modes of the metric perturbations which have grown sufficiently to become classical. In the absence of inhomogeneities massless fermions are not gravitationally produced, and then this effect may be relevant for abundance estimates of light gravitational relics.

1. Introduction

It is well known that particles are created in an expanding Universe [1]. In the presence of inhomogeneous metric perturbations, the number of particles created by inhomogeneities can be computed by a perturbative evaluation of the S-matrix [2]. In this case the occupation number is composed by three parts, *viz*, the zeroth-order contribution due to the homogeneous expansion, a first-order (in the perturbation amplitude) part arising from the interference between 0- and 2- particle states, and a second-order contribution which comes from the interaction between nonzero particle states. In the massless and conformally coupled cases, the first two terms vanish in rigid, exactly FLRW backgrounds, and only inhomogeneity contributes to the gravitational particle production. In what folllows we shall consider the gravitatinal creation of massless fermions from inhomogeneities.

2. Results

Let us consider perturbations around the FLRW background, $ds^2 = a^2(1+2\Phi)d\eta^2 - a^2(1-2\Phi)\delta_{ij}dx^i dx^j$, with a and Φ are the scale factor and the gravitational potential, respectively. In the interaction picture, the evolution of the initial vacuum (zero particle) state $|0\rangle$ is determined by the interaction Lagrangian $\mathcal{L}_I(\Phi)$. In the case of the massless fermion field ψ, the following second order term contributes to the particle production [3]:

$$N = \frac{1}{15\,(2\,\pi)^2\,a^3} \int_0^\infty dp_0 \int_{|\mathbf{p}|<p_0} d^3\mathbf{p}\,|\mathbf{p}|^4 \left| \int_{\eta_i}^{\eta_f} d\eta\, e^{i\,p_0\,\eta}\,\tilde{\Phi}\,(\mathbf{p},\eta) \right|^2 , \qquad (1)$$

where $\tilde{\Phi}$ is the Fourier component of Φ. We apply this formalism to the preheating phase after inflation with the potential $V = \lambda\,\phi^4/4 + g^2\phi^2\,\chi^2/2$ (ϕ is an inflaton field). In particular, the longwavelength modes of the second field χ are parametrically amplified whenever $g^2/\lambda \simeq 2\,n^2$ with integer n. At linear order in field perturbations, these modes are coupled to metric perturbations through a term proportional to $\dot{\chi}\delta\chi_k$. Metric perturbations are amplified by this term, and particle production is strongly enhanced. In the center of the first resonance band ($g^2/\lambda = 2$), the occupation number of fermions is found to be $\tilde{n}_\psi \equiv a^3 n_\psi/(\sqrt{\lambda}\,\phi_0)^3 \simeq 5 \times 10^{-10}$ (see ref. [3] for details).

3. Conclusion

We discussed gravitational creation of fermions by inhomogeneous perturbations. The gravitational production of massless particles is solely induced by metric perturbations which break the conformal flatness of the background. Although in the cases considered the final abundance is still below the nucleosynthesis, this effect may be important for the abundance estimates of light particles such as the modulus field.

References

[1] L. Parker, Phys. Rev. Lett. 21 (1968) 562; Phys. Rev. 183, (1969) 1057.

[2] J. A. Frieman, Phys. Rev. D39 (1989) 389; J. Cespedes and E. Verdaguer, Phys. Rev. D41 (1990) 1022; A. Campos and E. Verdaguer, Phys. Rev. D45 (1992) 4428.

[3] B. A. Bassett, M. Peloso, L. Sorbo, and S. Tsujikawa, Nucl. Phys. B to appear, hep-ph/0109176.

A Graph-Theoretical Approach of Large-Scale Structure in the Universe

Haruhiko UEDA
Faculty of Education and Human Studies, Akita University, 1-1 Tegata-gakuen, Akita-shi, Akita 010-8502, Japan
Tsutomu T. TAKEUCHI
National Astronomical Observatory of Japan, 2-21-1 Osawa, Mitaka Tokyo 181-8588, Japan
Makoto ITOH
Faculty of Economics and Information, Gifu Shotoku Gakuen University, 1-38 Nakauzura, Gifu-shi, Gifu 500-8288, Japan

Abstract

With the help of a graph-theoretical approach, we have quantified two-dimensional galaxy distributions in observation and Cold Dark Matter (CDM) simulations. We compared these distributions and found that the CDM model with a density parameter of less than one is preferable to reproduce the galaxy distributions in the observations.

1. Introduction

One of the most important goals of cosmology is to determine a desirable theoretical model that suitably describes our universe. Distributions of our nearby galaxies play an important role for determine suitable theoretical model. In this paper, we use a graph–theoretical approach, and quantify two-dimensional galaxy distributions based on observations. We have prepared a two-dimensional galaxy sample produced from the the Lyon–Meudon Extragalactic Database (LEDA: see Paturel et al. 1997). We extracted a magnitude-limited subsample from the original LEDA galaxies with limiting magnitudes up to 14.5. We, hereinafter, refer to our two-dimensional magnitude limited sample as 'the LEDA2d subsample'.

To compare the LEDA2d subsample with a theoretical model, we used cosmological N-body simulations (Ueda, Itoh 1999). Three CDM models with $(\Omega_0, \lambda_0) = (0.1, 0.9), (0.5, 0.5), (1.0, 0.0)$ were adopted. Hereafter, we call these three models LCDM (Low-density CDM), MCDM (Middle-density CDM), and HCDM (High-density CDM), respectively. We here construct four artificial two-dimensional surveys from these CDM simulations.

2. Analysis

We use following algorithm for qunatifying the galaxy distributions. We first make constellation graphs from the galaxy distributions. A constellation graph is constructed by connecting nearest-neighbor pairs of galaxies. We next assign an adjacency matrix of each constellation graph. In our analysis, we use six-type weighted constellation graphs by assigning the edge angle, θ^j, with $j = 0, 0.5, 1.0, 1.5, 2.0, 2.5$ to each edge. We finally estimate the eigenvalues of the adjacency matrices. In this procedure, we obtain the six-type distribution functions of the eigenvalues ($0 \leq j \leq 2.5$). To treat the galaxy distributions in a statistical way, we estimate the mean absolute deviation, $D(j)$, of the distribution function of the eigenvalue as a function of the weight index, j.

3. Comparison and Conculsions

We now compare the galaxy distributions in the LEDA2d subsample and the CDM artificial samples. Fig1 is the result of the relative mean absolute deviation, D/D_{HCDM} as a function of the weight index, j ($0 \leq j \leq 2.5$).

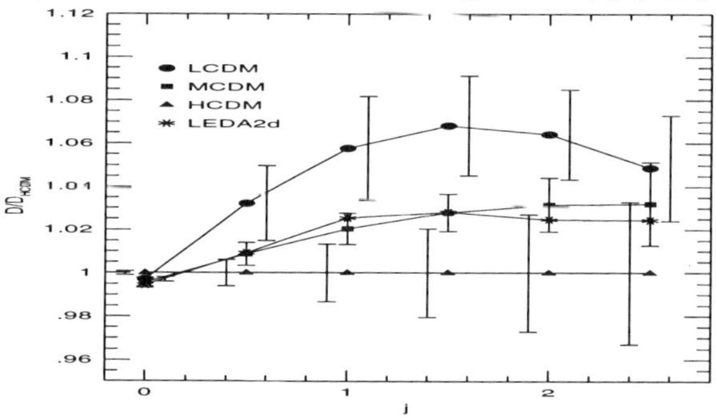

Fig. 1. Relative mean absolute deviation of the distribution function of the eigenvalue of the LCDM, MCDM, HCDM, and LEDA2d subsample.

From above figure, we found that the CDM model with a density parameter of less than one is preferable to reproduce the galaxy distributions in the observation (LEDA2d).

References

[1] H. Ueda, T.T. Takeuchi, M. Itoh, Publ. Astron. Soc. Japan 53 (2001) 381.

Nucleosynthesis of Zinc and Iron-Peak Elements in Massive Pop III Supernovae: Abundances of Very Metal-Poor Halo Stars

Hideyuki UMEDA and Ken'ichi NOMOTO
Research Center for the Early Universe (RESCEU), and
Department of Astronomy, School of Science, The University of Tokyo,
7-3-1 Hongo, Bunkyo-ku, Tokyo 113-0033, Japan

Abstract

We calculate evolution and nucleosynthesis in massive Pop III stars with $M = 13 \sim 270 M_{\odot}$, and compare the results with abundances of very metal-poor halo stars. The observed abundances can be explained by the energetic core-collapse supernovae with $M \lesssim 100 M_{\odot}$ ("hypernovae") but not by pair-instabillity supernovae (PISNe) with $M \sim 150 - 270 M_{\odot}$. This result constrain the IMF for the Pop III and very metal-poor Pop II stars.

The observed abundances of metal-poor halo stars show quite interesting trends. There are significant differences between the abundance patterns in the iron-peak elements below and above [Fe/H]~ -2.5. This is because the Galaxy was not well mixed in such a early stage, and the abundance pattern of each supernova (SN) may be kept in the very metal-poor stars. For [Fe/H]$\lesssim -2.5$, the mean values of [Cr/Fe] and [Mn/Fe] decrease toward smaller metallicity, while [Co/Fe] and [Zn/Fe] increases.

These trends are explained simultaneously in the context of Fe core-collapse SNe, if the mass from complete Si-burning is relatively large compared with that from incomplete Si-burning (Nakamura et al. 1999). However, all previous model calculations under-produced the absolute value of [Zn/Fe] significantly (Fig. 1 left). We find that [Zn/Fe] is larger for deeper mass-cuts, smaller neutron excess, and larger explosion energies. Among them the large explosion energy is most important to realize the large [Zn/Fe] ratio (Fig. 1 right; Umeda & Nomoto 2002). The observed trends of the abundance ratios among the iron-peak elements are well explained with this high energy ("Hypernova") models rather than the simple "deep" mass-cut effect, because the overabundance of Ni can be avoided in the hypernova models.

We also investigate the yields of pair-instability supernova explosions of $M \simeq 130 - 300 M_{\odot}$ stars. In Fig.2 we compare the abundance pattern of a

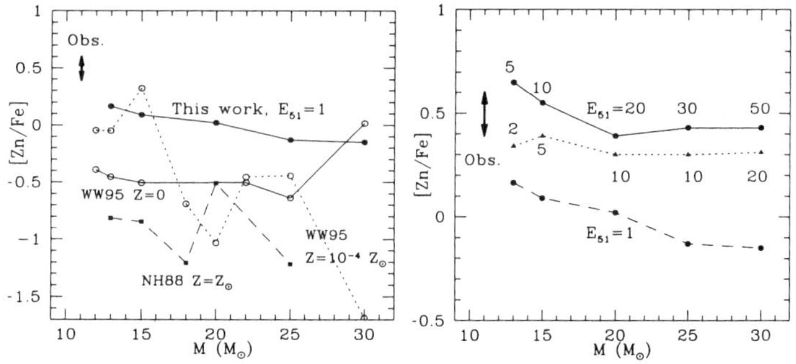

Fig. 1. Observed range of [Zn/Fe] for the very metal-poor stars ([Fe/H] $\lesssim -3$) compared with the yields of theoretical models with low explosion energy $E_{51} = E_{exp}/10^{51}$ erg =1 (left) and with high energies $E_{51} = 2 \sim 50$ (right panel). The observed large [Zn/Fe] value and also other unexpected trends in [Co, Mn, Cr/Fe] can be explained with this high energy ("Hypernova") models.

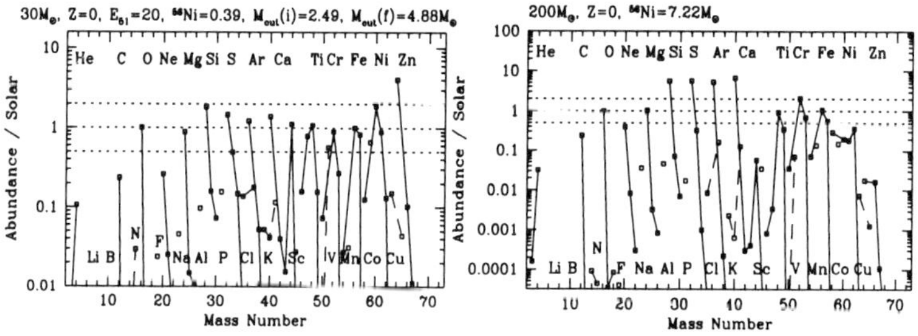

Fig. 2. The abundance pattern of a hypernova (left) and a PISN model (right panel), normalize to the solar abundance. For PISNe with initial masses $M \simeq 130-300M_{\odot}$ [Zn/Fe] is small, so that the abundance features of very metal-poor halo stars cannot be explained by these SNe.

hypernova and a PISN model As can be seen [Zn/Fe] of PISNe are always small, because of relatively large incomplete Si-burning region for the explosion (Umeda & Nomoto 2002). Therefore, the abundance features of very metal-poor stars cannot be explained by pair-instability supernovae.

References

[1] Nakamura, T., Umeda, H., Iwamoto, K., Nomoto, K., Hashimoto, M., Hix, W.R., Thielemann, F.-K. 2001, ApJ, 550, 880

[2] Umeda, H., Nomoto, K. (2002), ApJ, 565, in press (astro-ph/0103241)

Weak Lensing Study of the Galaxy Cluster MS 1054-03

Keiichi Umetsu
Academia Sinica Institute of Astronomy and Astrophysics (ASIAA), P.O. Box 23-141, Taipei 106, Taiwan, R.O.C.

Abstract

We present a weak lensing analysis of the rich, X-ray luminous galaxy cluster MS 1054-03 at redshift $z = 0.83$ on the basis of deep, wide-field Suprime-Cam images of faint background galaxies taken with the Subaru telescope. We apply the Schneider's aperture mass technique to the Suprime-Cam R_c-band data to obtain the distribution of the weak lensing signal-to-noise ratio over the data field. We demonstrate that wide-field deep Suprime-Cam data are powerful, efficient tools for weak lensing cluster survey out to redshfits of order unity.

1. Introduction

The weak but coherent image distortions of faint background galaxies (FBG) induced by intervening mass inhomogeneity have been used to reconstruct the projected mass distributions of galaxy clusters (Bartelmann & Schneider 2001). Schneider (1996) pointed out that the coherent distortions can also be used as a tool to search for mass concentrations on the sky independent of their luminosities. The paper presents our weak lensing analysis of the MS 1054-03 field on the basis of a mosaic of deep Suprime-Cam images taken with the Subaru telescope in R_c-band; the details of our observations and data reduction will be present elsewhere. Throughout the paper, we will use $\Omega_m = 0.3$, $\Omega_\Lambda = 0.7$.

2. Weak Lensing Analysis

We apply the aperture mass technique of Schneider (1996) to the Suprime-Cam R_c-band data for measuring the local weak lensing signal-to-noise ratio (S/N) over the data field. To integrate the shear-induced galaxy ellipticities, we use the filter function sensitive to the structure with radius $1.5'$ ($0.5h^{-1}$ Mpc at $z = 0.83$). Figure 1 shows the resulting weak lensing S/N map from the FBG sample. We find two S/N peaks above a threshold of $\nu_c = 4$: the first S/N peak corresponds to MS 1054-03, and the second one coincides well with another galaxy concentration

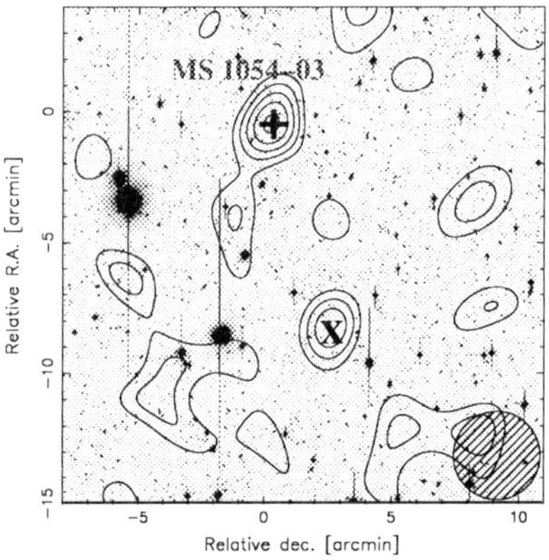

Fig. 1. Contour plots of the weak lensing S/N distribution obtained from the FBG catalog overlayed on the R_c-band image. The contour lines indicate levels of 1σ, 2σ, 3σ, etc. North is to the right, and east is to the top.

located about $8'$ northwest of MS 1054-03. The statistical significance of the detected S/N peaks can be assessed analytically according to the peak theory of random Gaussian field (Bond & Efstashiou 1987). The probability of finding a random noise peak with $\nu > \nu_c$ is calculated to be 5×10^{-4}.

3. Conclusions

We have performed a weak lensing analysis of the MS 1054-03 field on the basis of Suprime-Cam images taken with the Subaru telescope. We detected two significant mass concentrations in the weak lensing S/N map, which coincide well with galaxy concentrations in the R_c-band image. Our results demonstrate that the wide-field deep Subaru/Suprime-Cam images are powerful, efficient tools for weak lensing cluster survey out to redshifts of order unity.

This work is based on the collaboration with Toshifumi Futamase, Toru Yamada, Masaru Kajisawa, Yanick Mellier, and Takashi Hamana..*

References

[1] M. Bartelmann & P. Schneider, Phys. Rep. 340 (2001) 291
[2] J.R. Bond & G. Efstashiou, MNRAS 226 (1987) 655
[3] P. Schneider, MNRAS 283 (1996) 837

*Based on data collected at Subaru Telescope, which is operated by the NAOJ.

Acceleration of the Universe as a Consequence of Gravity Properties

Leonid Verozub
Kharkov National University, Kharkov 61077, Ukraine

Abstract

A simple model of the selfgraviting expending dust-ball is considered in view of our gravitation equations proposed earlier. It is shown that at specific conditions the acceleration of the ball surface can be positive.

Bimetric gravitation equations that have no the physical singularity in the center of the spherically-symmetric field were proposed in [1]. The equations can be considered as a realization of the idea of dependency of space-time properties on the reference frame being used [2] The equations do not contradict experimental data [2] . In particular, they were tested by the binary pulsar PSR1913+16. [3]. Furthermore, the energy-momentum tensor that follows from these equation leads to the finite proper gravitational energy of a point mass [2]. For this reason we have to be careful about predictions of these equations.

To demonstrate which consequences follow from the above gravitation equations for the problem of the Universe acceleration consider in flat space-time an expanding selfgraviting spherically symmetric homogeneous dust-ball with the parameters similar to the ones of the observed Universe.

The spherically symmetric solution of the bimetric equations differ very little from the ones of the Einstein equations. However they are completely different at the small distances from the center. They have no horizon events at the Schwarzschshild radius r_g and no physical singularity in the center.

At the distance $r/r_g \simeq 2$ the gravitation force, affecting free moving test particle, changes its sign and becomes repulsive. It is well known that the Schwarzshild radius $GM \ / \ c^2$ for the observable mass M of the Universe matter is of the order of its radius R . Therefore, we can expect the negative sign of the magnitude d^2R/dt^2 for objects with large redshifts.

To consider the problem more rigorously we must proceed from the Lagrangian that describes the motion of the particles of the sphere surface.

The radial velocity $\dot{R} = dR/dt$ of the particles of the ball surface with the

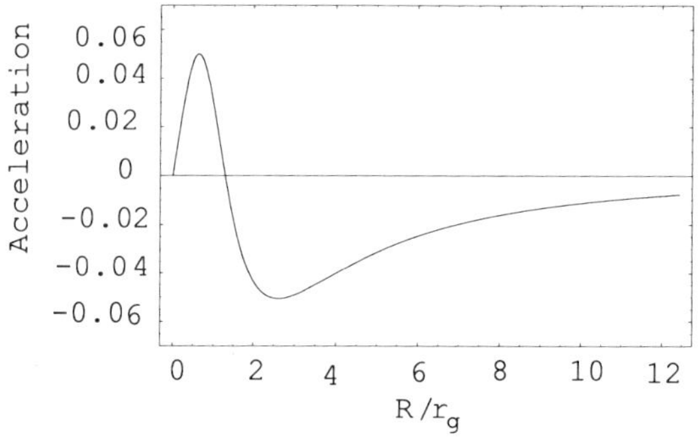

Fig. 1. Aceleration of the expanding ball vs. its radius

radius R are given by

$$\dot{R}^2 = \frac{c^2 C}{A}\left[1 - \frac{C}{\bar{E}^2}\right], \tag{1}$$

where $\bar{E} = E/mc$ and E is the energy of the specks of dust, C and A are some functions of R.

We assume here, for definiteness, that at the moment the radial velocity is $V_0 = H_0 R_0$ where $H_0 = 0.25 \cdot 10^{-17} s^{-1}$ is the Habble constant, $R_0 = 3 \cdot 10^{27} cm$, the constant $\bar{E}^2$ is equal to 0.60 . Fig. 1 shows the dependence of the acceleration $d\dot{R}/dt = (d\dot{R}/dR)\dot{R}$ (arbitrary units) upon R .

It follows from the figures that the acceleration is negative at f $R/r_g \gg 1$, and is positive if R/r_g is of the order of r_g or less than that. For example, if the radius $R = R_U$ and the matter density $\rho = 2 \cdot 10^{-28} g/cm^3$, the value of $R_U/r_g = 0.9 < 1$ and the acceleration is equal to $1 \cdot 10^{-8} cm/s^2$ which is half as large as this magnitude resulting from the value of $q_0 = -1$ that was found in [5].

In paper [4] the Riess et al these results were studied in detail in view of the model above. A good compliance was found.

References

[1] L. Verozub, Phys. Lett A156 (1991) 404

[2] L. Verozub *Gravitation, Cosmology and Relativistic Astrophysics*, (2001) Kharkov National University 44

[3] L. Verozub and A. Kochetov, Grav. and Cosmol. 6 (2000) 246.

[4] L. Verozub and A. Kochetov, Radio Phys. and Radio Astron 6 (2001) 93.

[5] A. Riess, A. Filipenko, P. Challis, et al., AJ 116 (1998) 1009.

Microscopic Study of Nuclear "Pasta" by Quantum Molecular Dynamics

Gentaro WATANABE[a,b], Katsuhiko SATO[a,c], Kenji YASUOKA[d], Toshikazu EBISUZAKI[b]
[a]*Department of Physics, University of Tokyo, Tokyo 113-0033*
[b]*Division of Computational Science, RIKEN, Saitama 351-0198*
[c]*RESCEU, University of Tokyo, Bunkyo, Tokyo 113-0033*
[d]*Department of Mechanical Engineering, Keio University, Yokohama 223-8522*

Abstract

Structure of cold dense stellar matter at subnuclear densities is investigated with quantum molecular dynamics (QMD). We succeeded in showing that the phases with slab-like and rod-like nuclei etc. can be formed dynamically without any assumptions on the nuclear shape. Our result suggests the existence of these kinds of phases in neutron star crusts.

1. Introduction

At subnuclear densities in NS crusts and SN cores, nuclei are expected to have exotic structures such as rod-like and slab-like shape etc., which are called nuclear "pasta" [1]. This prediction has been obtained by previous studies [2, 3] based on free energy calculations assuming specific nuclear shapes.

However, the formation process of the "pasta" phases has not been discussed except for some perturbative approaches [1, 6]. It is important to adopt microscopic and dynamical approach which allows arbitrary nuclear structures in order to understand the formation of the non-spherical nuclei. We have started studying dense matter by QMD [4] simulation whose final goal is to obtain a realistic picture about NS crusts and SN cores. As the first step of our quest, we will answer the following question in this study: "Can the 'pasta' phases be formed dynamically?"

2. Simulation and results

We have performed QMD simulations of (n,p,e) system with proton fraction $x = 0.3$ and 0.5 for various nucleon densities ρ. We set a cubic box which is imposed periodic boundary conditions. Total nucleon numbers are 1372 and

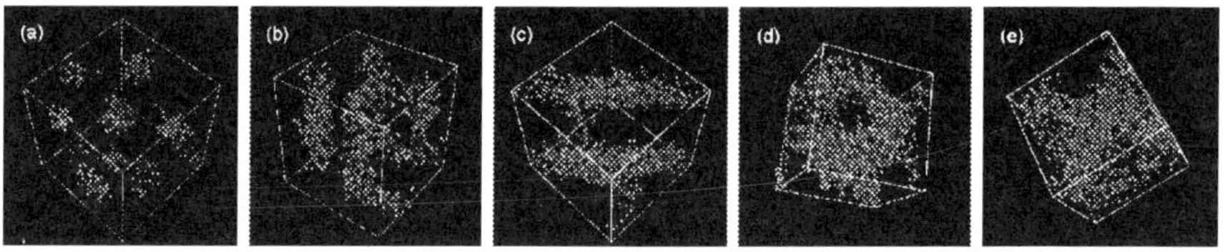

Fig. 1. Nucleon distributions of cold symmetric nuclear matter at various densities; (a) sphere phase, $0.1\rho_0$; (b) cylinder phase, $0.25\rho_0$; (c) slab phase, $0.4\rho_0$; (d) cylindrical hole phase, $0.5\rho_0$ and (e) spherical hole phase, $0.6\rho_0$. The darker particles represent protons and brighter ones represent neutrons. Temperatures are $O(0.1)$ MeV.

2048. The Coulomb interaction is calculated by the Ewald method and relativistic electrons are treated as a uniform background. For the nucleon interaction, we use a model Hamiltonian developed by Maruyama et al. [5].

We first prepare uniform hot nucleon gas at $k_B T \sim 20$ MeV as an initial condition, and then cool it down slowly for $O(10^3 - 10^4)$ fm/c until the temperature gets ~ 0.1 MeV. Note that any artificial fluctuations are not given during the simulation. Shown in Fig. 1 is the resultant nucleon distributions of cold symmetric matter for $\rho = 0.25 - 0.6\rho_0$ (ρ_0 : the normal nuclear density). We can see from Fig. 1 that the phases with rod-like and slab-like nuclei, cylindrical and spherical bubbles are reproduced. Also for $x = 0.3$, those phases are reproduced at about $0.05\rho_0$ lower than the symmetric matter case. The density dependence of the nuclear shape is quite sensitive and various intermediate phases with complicated structures are observed. Whole phase diagrams of cold matter and properties of the nuclear shape changes will be discussed elsewhere [7]. It is concluded that our results suggest the existence of the "pasta" phases in NS crusts which cool down keeping the local thermal equilibrium after the proto-NS is formed, and their cooling time scale is much longer than that of our simulations.

References

[1] C.J. Pethick and D.G. Ravenhall, Ann. Rev. Nucl. Part. Sci. 45 (1995) 429.
[2] D.G. Ravenhall, C.J. Pethick and J.R. Wilson, PRL 50 (1983) 2066.
[3] M. Hashimoto, H. Seki and M. Yamada, PTP 71 (1984) 320.
[4] J. Aichelin and H. Stöcker, PLB 176 (1986) 14.
[5] T. Maruyama et al., PRC 57 (1998) 655.
[6] K. Iida, G. Watanabe and K. Sato, PTP 106 (2001) 551.
[7] G. Watanabe, K. Sato, K. Yasuoka, T. Ebisuzaki, in preparation.

The Most Stringent Limit on Solar Axions

Satoru WATANABE and Hiromoto SHIBAHASHI
Department of Astronomy, The University of Tokyo, Tokyo 113-0033, Japan

The axion is a light pseudoscalar particle introduced to solve the strong CP problem [1] [2] and one of the most likely candidates for the dark matter. In the interior of the Sun, the axion can be produced through the Primakoff effect. The axion interacts so weakly with other particles that it escapes freely from the Sun once it is produced. Therefore it functions as an energy-loss mechanism. We reexamine the theoretical limit on the axion coupling to two photons, $g_{a\gamma\gamma}$, in the light of a new solar model and the latest solar neutrino observations.

Helioseismology has been successful in determining precisely the sound-speed profile, $c(r)$, and the density profile, $\rho(r)$, in the Sun based on precise observations. By imposing the constraints of $c_{\rm obs}(r)$ and $\rho_{\rm obs}(r)$, we can solve the basic equations governing the radiative core of the Sun (the continuity equation, the hydrostatic equation, the energy equation, and the energy transfer equation) directly without assuming and following the evolutionary history of the Sun. We call this model as the seismic solar model (hereafter SeiSM) [3]. SeiSM has advantages over the conventional evolutionary solar models in the sense that it is more faithful to the seismic data.

We calculate a series of SeiSMs taking account of the axionic energy-loss with varying $g_{a\gamma\gamma}$. The axionic energy-loss rate is proportional to the square of $g_{a\gamma\gamma}$ [4]. In this procedure, SeiSMs are constructed so as to be always consistent with $c_{\rm obs}(r)$ and $\rho_{\rm obs}(r)$. An increase in axionic energy-loss must be compensated by an increase in the nuclear energy generation, which leads to an increase in the theoretically expected neutrino fluxes.

The Sudbury Neutrino Observatory (SNO) has recently announced the first results on solar neutrinos [5]. SNO detected the ^{8}B-neutrino via the charged current (CC) reaction on deuterium and by the elastic scattering (ES) of electrons. The CC reaction is sensitive exclusively to ν_e, while the ES reaction, the same reaction as the Super-Kamiokande (SK), is sensitive to all active neutrino flavors (ν_e, ν_μ, ν_τ), but with weak sensitivity to ν_μ and ν_τ. If there are flavor transformations between active neutrino flavors (ν_e, ν_μ, ν_τ), comparison of the neutrino flux deduced from the ES reaction assuming no neutrino oscillations, $\phi^{\rm ES}(\nu_x)$, with that measured by the CC reaction, $\phi^{\rm CC}(\nu_e)$, can provide the flux of non-electron flavor active neutrinos, $\phi(\nu_{\mu,\ \tau})$. By comparing $\phi^{\rm CC}(\nu_e)$ with the

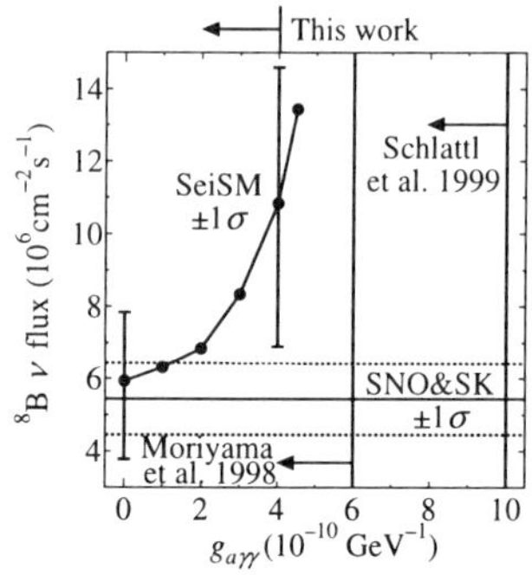

Fig. 1. The theoretically expected ^{8}B-neutrino fluxes of SeiSM with different $g_{a\gamma\gamma}$. For a comparison, the observationally determined active neutrino flux is also shown by the horizontal solid line with the $\pm 1\ \sigma$ error band (the dotted lines).

Super-Kamiokande's precise value of $\phi^{\rm ES}(\nu_x)$ [6], the total flux of the active ^{8}B-neutrino, $\phi(\nu_{e,\ \mu,\ \tau})$, is determined to be $5.44{\pm}0.99{\times}10^6$cm^{-2} s^{-1} [5].

The theoretically expected neutrino flux based on SeiSM with $g_{a\gamma\gamma} = 0$ is consistent with the value experimentally derived in this way. Because a higher $g_{a\gamma\gamma}$ than a critical value makes the model's neutrino flux too high to be consistent with that determined from SNO and SK, we can limit $g_{a\gamma\gamma}$ (Fig. 1). From the comparison of these neutrino fluxes, we set a limit $g_{a\gamma\gamma} < 4.0 \times 10^{-10}$ GeV^{-1}[7]. This limit based on a new procedure is about a factor of 3 improvement over the previous theoretical limit [4] and a more severe limit than the latest solar axion experiment [8]. Therefore this limit is the most stringent limit on solar axions.

References

[1] R.D. Peccei, and H.R. Quinn, Phys. Rev. Lett. 38 (1977) 1440; Phys. Rev. D16 (1977) 1791.

[2] S. Weinberg, Phys. Rev. Lett. 40 (1978) 223; F. Wilczek, Phys. Rev. Lett. 40 (1978) 279.

[3] S. Watanabe, and H. Shibahashi, submitted to Publ. Astron. Soc. Japan

[4] H. Schlattl, A. Weiss, and G. Raffelt, Astrop. Phys. 10 (1999) 353.

[5] SNO Collaboration, Q.R. Ahmad et al., Phys. Rev. Lett. 87 (2001) 071301.

[6] Super-Kamiokande Collaboration, S. Fukuda et al., Phys. Rev. Lett. 86 (2001) 5651.

[7] S. Watanabe, and H. Shibahashi, hep-ph/0112012.

[8] S. Moriyama et al., Phys. Lett. B434 (1998) 147.

Parallelization and Vectorization of the AMR N-body Code

Hideki Yahagi and Yuzuru Yoshii
Institute of Astronomy, University of Tokyo
2-21-1 Osawa, Mitaka, Tokyo 181-0015, Japan

Abstract

Particle-mesh method is a fast N-body method and appropriate for periodic boundary problems. However, its force resolution is restricted by the mesh spacing of the structured grid. This drawback is resolved by the adaptive mesh refinement (AMR) technique which introduces finer meshes where higher resolution is required [1, 3]. On the other hand, the new supercomputer system at National Astronomical Observatory of Japan began its operation with Fujitsu's vector parallel processors VPP5000. We vectorized and parallelized our N-body code with AMR to draw its performance. In this paper, we describe our vectorization and parallelization schemes briefly. Details of our AMR N-body code and optimization schemes are given in [2, 3].

1. Vectorization

It is trivial to vectorize loops which treat cell data and particle data separately, such as the Poisson solver and the integration of the equation of motion. However, it is not straightforward to vectorize loops which need cell data and particle data simultaneously. Such loops exist only in three functions: force interpolation, mass assignment, and particle sieve. Among them, force interpolation is vectorizable, because it only updates particle data just once per call. The other two functions, especially mass assignment, has been thought to be unvectorizable. However, these functions are also vectorizable by the scheme described below. This vectorization scheme is also applicable to parallelize these functions for shared memory processors.

Accessing particles in cells by rolling up each linked list in depth-first order cannot be vectorized (*zigzag arrow* in the left panel of the figure), because the length of each loop and which particles are in the linked list are not known beforehand. However, changing the particle sweep order from depth-first order to width-first order, this particle sweeping is vectorized (*horizontal arrows*). On the other hand, the mass assignment schemes cannot be vectorized because those schemes update more than one cells data per particle during a scan of the loops. Such data accesses are prohibited for vectorization. This difficulty is also avoid-

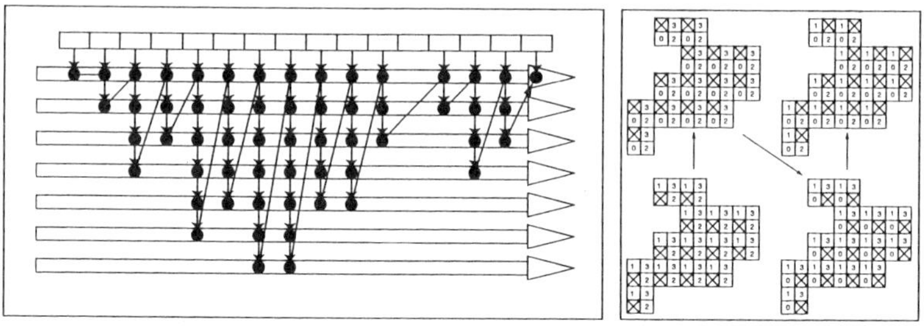

able by splitting loops into 2^{dim} loops. An example for a two dimensional case is shown in the right panel of the figure.

The only remained function to be vectorized is particle sieve. This function is split into two steps, stay particle sieve and leaked particle sieve. First, stay particle sieve divides particles into two groups, particles which remain in the same cell from the previous step and those which do not. Then, remained particles in the same cell are connected by a linked list. Although the above processes are vectorizable by simply applying the width-first sweeping, leaked particle sieve cannot be vectorized. However, this does not bother at all because time steps must be controlled so that particles move much shorter distance than the cell separation. This confirms that the number of leaked particles is much smaller than that of stay particles.

2. Parallelization

Cell data are sorted in each level by the Morton order. The Morton order is a recursive N-shaped ordering, and lines up three dimensionally distributed cells in a row. Cell data at each level are distributed among processors simply cutting this sequence. With this data structure, parallelization of our code is only developing following three communication functions: 1) Communication between neighboring cells in the same level. 2) Communication between a cell and its children, and vice versa. 3) Communication between particles and cells.

Our code is developed partially on Fujitsu's VPP5000 installed at the Astronomical Data Analysis Center, National Astronomical Observatory of Japan.

References

[1] Kravtsov, A. V., Klypin, A. A., & Khokhlov, A. M. 1997, ApJS, 111, 73
[2] Yahagi, H. 2002, Doctor Thesis, University of Tokyo, submitted
[3] Yahagi, H., & Yoshii, Y. 2001, ApJ, 558, 463

Constraints on Cosmic Quintessence

Masanobu Yahiro
Department of Physics and Earth Sciences, University of the Ryukyus, Nishihara-chou, Okinawa 903-0213, Japan
Grant James Mathews
Center for Astrophysics, Department of Physics, University of Notre Dame, Notre Dame, IN 46556, USA
Kiyotomo Ichiki, Toshitaka Kajino, and Manabu Orito
National Astronomical Observatory, 2-21-1, Osawa, Mitaka, Tokyo 181-8588, Japan

Abstract

The "quintessence" models are constrained by primordial nucleosynthesis, cosmic microwave background and Type Ia supernovae.

1. Introduction

Recent observations [1, 2] of Type Ia supernovae indicate that the universe may be presently accelerating due to the influence of a dominant dark energy with a negative pressure. The simplest interpretation of this dark energy is a cosmological constant. A second possibility is derived from the so-called "quintessence" models. Observational constraints on such quintessence models have been of considerable recent interest. In the present work, we consider the power-law effective potential [3] and its supergravity(SUGRA)-corrected form [4]. We construct a detailed mapping of the allowed parameter space for these quintessence models.

2. Constraints

The primordial light-element abundances are deduced from observations. Adding energy density from the Q field tends to increase the universal expansion rate and change the abundances. In the present work we deduce an absolute upper limit of 5.6% of the background radiation energy density allowed in the quintessence field. Another constraint is Type Ia supernovae. A comprehensive study of the observational constraints on the present value of $w_Q \equiv P_Q/\rho_Q$ has been recently summarized by Wang et al. [5]. We adopt the same constraints.

A third constraint on the quintessence field arises from its effect on the epoch of photon decoupling. There are two effects to be considered: look-back effect and increase of energy by Q field. The look-back constraint is already satisfied by demanding that $\Omega_Q = 0.7$ at the present time. The second constraint is deduced by [6], that is, $\Omega_Q \leq 0.39$ during the epoch of photon decoupling.

3. Results and Conclusions

For the bare Ratra power-law potential, $V(Q) = M^{(4+\alpha)}Q^{-\alpha}$, the main constraint is simply the requirement that the equation of state be sufficiently negative by the present time. The allowed values for the parameter α is < 2, if the most stringent constraint $w_Q \leq -0.6$ is adopted. For such small α, the parameter M becomes a very small fraction of the Planck mass, and the fine tuning problem is reintroduced. For the SUGRA-corrected Q fields, on the other hand, the constraint from w_Q is greatly relaxed. In fact, w_Q is sufficiently negative ($w_Q < -0.6$) for all $\alpha < 10^4$.

We conclude that for both the bare Rata inverse power-law potential and its SUGRA-corrected form, the main constraints for models which achieve the tracker solution by the nucleosynthesis epoch is from the requirement that the equation of state parameter becomes sufficiently negative by the present epoch. The main constraint from nucleosynthesis is for models in which the Q field has not yet obtained the tracker solution by the time of nucleosynthesis. All solutions consistent with the primordial nucleosynthesis constraints are also consistent with our adopted CMB Q-energy constraint. The SUGRA-corrected potential is the least constrained and avoids the fine tuning problem for M. Therefore, it may be the preferred candidate for the quintessence field.

Our full analyses concerning this issue will be published in Phys. Rev. D.

References

[1] P. M. Garnavich, et al., Astrophys. J., **509**, 74 (1998); Riess, et al., Astron. J, **116**, 1009 (1998).

[2] S. Perlmutter, et al., Nature **391**, 51 (1998); Astrophys. J., **517**, 565 (1998).

[3] B. Ratra and P.J.E. Peebles, Phys. Rev. D37, 3406 (1988); P.J.E. Peebles and B. Ratra, Astrophys. J. Lett., 325, L17 (1988).

[4] P. Brax and J. Martin, Phys. Lett., **B468**, 40 (1999); Phys.Rev. D61 (2000) 103502; P. Brax, J. Martin, A. Riazuelo, Phys. Rev. **D62**, 103505 (2000).

[5] L. Wang, R.R. Caldwell, J.P. Ostriker and P. J. Steinhardt, Astrophys. J., 530, 17 (2000).

[6] R. Bean, S. H. Hansen, and A. Melchiorri, astro-ph/0104162, in press (2001).

Dynamics of Stellar Orbits in Disk Galaxies

Taihei YANO, Toshiyuki KAN-YA & Naoteru GOUDA
National Astronomical Observatory, Mitaka, Tokyo 181-8588, Japan

Abstract

We have investigated the dynamics of stellar motions in galactic potentials with bar-shaped potentials in order to clarify whether the structures in galaxies are formed. In this investigation, we examine whether the axis of an oval orbit align with that of a bar-shaped potential. When the orbital axis is aligned with the bar, the structures will be formed, and when the axis is aligned perpendicular to the bar, no structures are formed. Earn & Lynden-Bell(1996) introduce the condition that an axis of an oval orbit is aligned with a bar around the Lindblad resonance. We consider the more general case, i.e., we discuss how the orbit behaves in all regions of a galaxy. Our results are raughly summarized as follows; When $r/r_{ILR} > 1$, the axis of the oval orbit is aligned with that of the perturbed potential. On the other hand, when $r/r_{ILR} < 1$, the axis of the oval orbit is aligned perpendicular to that of the perturbed potential. Therefore the structures in galaxies appear in the outer region of inner Llindblad resonance. These results are consistent with the linear analysis of eauations of motion.

1. Introduction

Spiral structures or bar structures are generally observed in disk galaxies. It is important to investigate how these structures are formed on the basis of the orbital analysis, i.e., the gravitational interactions of orbits of individual stars with the disk potential.

Lynden-Belli1979jexamine the dynamics of stellar motions in galactic potentials with bar-shaped perturbation, and consider how bar and spiral structures are formed. Furthermore, Earn & Lynden-Bell (1996) investigate more generally the dynamics of orbits in galaxies and introduce the condition that an axis of an oval orbit is aligned with that of a bar potential based on the motion of orbits in the galactic potential.

It is very important for the formation of bar or spiral structures that an orbit tends to align with the potential or tends to be aligned perpendicular to the potential. When each oval orbit tends to align with each other, the potential caused by orbits becomes deeper and deeper. On the other hand, when each oval orbit tends to be aligned perpendicular to each other, the garavity of an orbit tend to cancel the potential trough caused by the other orbits, i.e., no structures are formed in this case.

Earn & Lynden-Bell introduce the condition that orbits align with each other under the assumption that the difference between each angular velocity is small against its angular velocity. This assumption is restrict regions in galactic disks to around the resonance. So, when we consider the formation of bar structures according to Lynden-Bell(1979), only around the inner Lindblad resonance is worked. Consequently, in the regions where resonance is not occurred, in almost all regions except for the resonance radius, we do not know how these orbits behave.

Here, we show how oval orbits behave by the dynamical interactions of orbits when $|\Omega_i - \Omega_p|/\Omega \not\ll 1$, where Ω_p is the pattern speed, Ω is an angular velocity of a star, and $\Omega_i = \Omega - \kappa/2$.

2. Summary

We generalize the investigation of Earn & Lynden-Bell(1996). They obtained the condition whether the axis of an oval orbit is aligned with that of the bar-shaped potential, which is a function of the galactic potential parameterized by β ($\zeta = \zeta(\beta)$), where β is defined as $v \propto r^\beta$, (v: circular velocity). On the other hand, we obtain the condition whether the axis of an oval orbit is aligned with that of the perturbed potential is a function of the galactic potential parameterized by β and radius r of the galaxy $\zeta = \zeta(\beta, r)$. However, as shown above, the condition is almost only the function of radius ($\zeta = \zeta(\beta, r) \simeq \zeta(r)$). Earn & Lynden-Bell's condition is satisfied only around the region of Lindblad resonance.

References

[1] Earn, D. J. & Lynden-Bell, D., 1996, MARAS, 278, 395
[2] Lynden-Bell, D., 1776, MNRAS, 187, 101

Beyond H_0 Determination

Atsunori YONEHARA
Center for Computational Physics, University of Tsukuba, Tennoudai 1-1-1, Tsukuba, Ibaraki 305-8577, Japan
Shin MINESHIGE
Yukawa Institute for Theoretical Physics, Kyoto University, Sakyo-ku, Kyoto, 606-8502, Japan
George CHARTAS
Department of Astronomy and Astrophysics, Pennsylvania State University, University Park, PA 16802, U.S.A.
Edwin L. TURNER
Princeton University Observatory, Peyton Hall, Ivy Lane, Princeton University, Princeton, NJ08544-1001, U.S.A.

Abstract

We can directly measure H_0 from time delay between multiple images of a gravitational lens system. In this paper, we investigate two new ideas beyond this research. One is to measure other cosmological parameters, $\Omega_{\rm m}$ and Ω_Λ. Another is to probe physics of the central region of lensed quasars, accretion disk. Former method does work, if monitoring with sub-day time resolution becomes available. Latter method does work, if monitoring with $\sim$ second time resolution becomes available.

1. Basic Concept and Estimations

Since redshift-distance relation is determined by H_0, $\Omega_{\rm m}$ and Ω_Λ, a method proposed by Refsdal (1964) should be useful to probe not only H_0 but also $\Omega_{\rm m}$ and Ω_Λ. However, $\Omega_{\rm m}$ and Ω_Λ dependence of distance and/or time delay is weaker than H_0 dependence. Additionally, by using a single lens system, the effect of $\Omega_{\rm m}$ and Ω_Λ to the distance is degenerate with H_0. Therefore, we compile the lens and the source redshift, measured time delay for 6 lens systems, assume a simple lens potential, perform χ^2 fitting procedure, and constrain H_0, $\Omega_{\rm m}$ and Ω_Λ. Contour of the best fit H_0 values are plotted with dotted lines in $\Omega_{\rm m}$-Ω_Λ plane (figure 1).

Unfortunately, with current time resolution ($\geq$ 1day), we can never obtain useful constrain on $\Omega_{\rm m}$ and Ω_Λ. However, the time resolution is expected to be

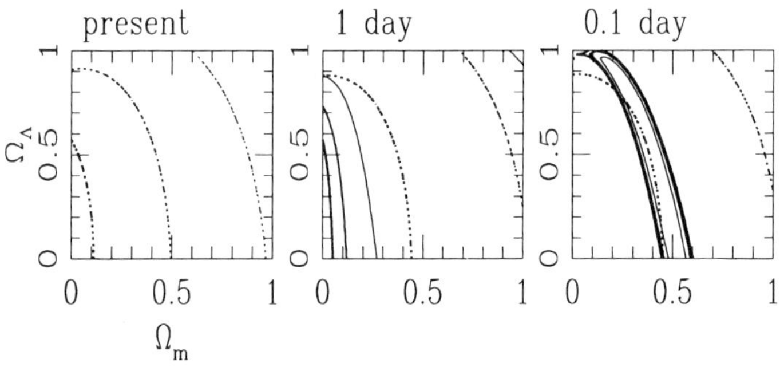

Fig. 1. Current status is shown in left panel. H_0 =90, 85, and 80 km s^{-1} Mpc^{-1} contours are presented from left to right (dotted lines), respectively. Middle and right panel are expected results for 1 day and 0.1 day time resolution, respectively. In these panels, H_0 =65, and 60 km s^{-1} Mpc^{-1} contours are presented from left to right, respectively. Additionally, 1-, 2-, and 3-σ level confidence regions are shown in thin, middle and thick real lines, respectively.

fine in near future. If the time resolution become smaller than 0.1day, constrain on Ω_m and Ω_Λ will become tight and we will be able to determine Ω_m and Ω_Λ with fairly narrow error bar (see figure 1).

In contrast, location of flares on the quasar accretion disk can be measured by time delay difference between different flares. As same as the estimation performed by Yonehara (1999), this relation is derived from fundamental equation of gravitational lensing and has following form;

$$\frac{\delta r}{r_g(M_{\mathbf{BH}} = 10^9 M_\odot)} \sim 10 \left(\frac{\delta\tau}{1\mathrm{sec}}\right) \left(\frac{1+z_{\mathrm{ol}}}{2}\right)^{-1} \left(\frac{D_{\mathrm{ls}}}{D_{\mathrm{ol}}}\right) \left(\frac{\Delta\theta}{1\mathrm{arcsec}}\right)^{-1}, \quad (1)$$

where δr is distance between flares, $\delta\tau$ is time delay difference between flares, z_{ol} is the redshift of the lens, D_{ls} and D_{ol} are angular diameter distances from the lens to the source, and from observer to the source, respectively, and $\Delta\theta$ is separation between lensed quasar images. Thus, if the 1sec time resolution is achieved, we will be able to know locations of flares on the accretion disk around super-massive black hole.

This work was supported in part by the JSPS, 9514 (AY).

References

[1] S. Refsdal MNRAS 128 (1964) 307.
[2] A. Yonehara ApJ 519 (1999) L31.

Propagation of Ultra-High Energy Cosmic Rays from Sources in the Super-Galactic Plane

Hiroyuki YOSHIGUCHI
Department of Physics, School of Science, The University of Tokyo, 7-3-1 Hongo, Bunkyo-ku, Tokyo 113-0033, Japan, hiroyuki@utap.phys.s.u-tokyo.ac.jp

Abstract

We have performed the detailed numerical simulations on the propagation of the UHE protons in the energy range $E = (10^{19.5} - 10^{22.0})$ eV in the relatively strong extragalactic magnetic field with strength $B = (10, 100)$ nG within about 40 Mpc. In this case, the deflection angles of UHECRs become so large that the no counterparts problem is simply solved. As for the source distribution, we assumed that it is proportional to the number distribution of galaxies. We have found many clusters, which mean the small-scale anisotropy, in our simulations. Compared with the AGASA data, the significant anisotropy on the arrival directions of UHECRs are found. This may originate from the incompleteness of the ORS database. This problem may be also solved if the source distribution is slightly changed. For example, this problem may be solved if we assume that UHECRs come from some of the galaxies such as AGNs and radio galaxies.

1. Introduction

The observed differential cosmic ray spectrum is remarkably featureless and extends beyond 10^{20} eV [1]. Moreover, there is a puzzling problem on the distribution of the arrival directions of UHECRs [2]. It is reported that there is no significant large-scale anisotropy of arrival direction distribution of UHECRs. On the other hand, there are one triplet and three doublets within a separation angle of 2.5 degrees for the 47 cosmic rays above 4×10^{19} eV, and the probability of observing these clusters by a chance coincidence under an isotropic distribution is smaller than 1% [1]. We have to give a natural explanation for these puzzling problems.

In this study, arrival directions of UHECRs are calculated numerically taking the effects of the extramagnetic fields into consideration. Optical Redshift Survey (ORS) data [3] is used as the source distribution of UHECRs. Detailed explanation of calculation method and another numerical results such as energy spectrum are presented in our paper [4]

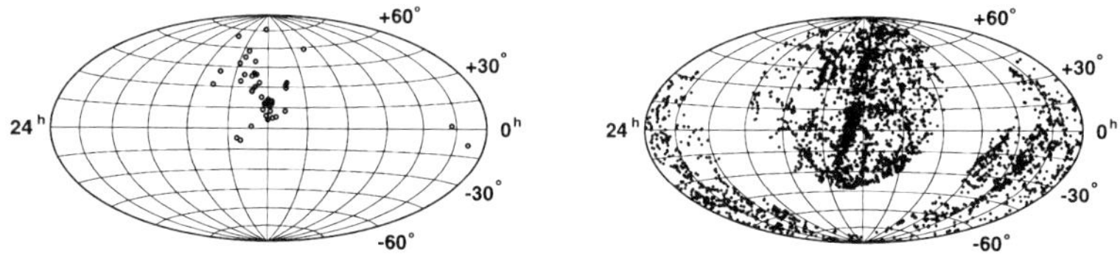

Fig. 1. (a) Arrival directions of UHECRs above $4\times$ in equatorial coordinates for the case of $B = 10$ nG and $l_c = 1$ Mpc. (b) Angular distribution of galaxies in ORS data within 40 Mpc in equatorial coordinates. It is noted that this database does not contain any data within $|b| \leq 20°$.

2. Results and Conclusion

In Fig.1a, we show one realization of the arrival directions of UHECRs above 4×10^{19} eV in equatorial coordinates for the case of B = 10nG and l_c = 1Mpc. For comparison, we show the distribution of the galaxies within 40 Mpc in ORS data catalogue in equatorial coordinates in Fig.1b. In order to compare our results with the AGASA data [1], the number of events is chosen to be 47 and the arrival directions are restricted in the range $-15° \leq b \leq 80°$. As easily seen, there is the significant anisotropy on the arrival directions of UHECRs. This may be due to the incompleteness of the ORS database. If we assume that UHECRs come from some of the galaxies such as AGNs and radio galaxies, this problem may be solved. We will investigate in detail the dependence of our conclusion presented in this study on the source distribution in the forthcoming paper.

References

[1] M. Takeda, et al. ApJ 522 (1999) L225.
[2] P. Blasi, A.V. Olinto PRD 59 (2000) 023001.
[3] B.X. Santiago, et al. ApJ 446 (1995) L457.
[4] Y. Ide, S. Nagataki, S. Tsubaki, H. Yoshiguchi, K. Sato PASJ 53 (2001) 1153.

Stochastic Biasing of Galaxies and Dark Halos in Cosmological Hydrodynamic Simulations

Kohji YOSHIKAWA[1], Atsushi TARUYA[2], Y. P. JING,[3] and Yasushi SUTO[2]
[1]*Department of Astronomy, Kyoto University Kitashirakawa-Oiwake-cho, Sakyo-ku, Kyoto 606-8502, Japan*
[2]*Department of Physics and Research Center for the Early Universe (RESCEU), School of Science, The University of Tokyo, 7-3-1 Hongo, Bunkyo-ku, Tokyo 113-0033, Japan*
[3]*Shanghai Astronomical Observatory, the Partner Group of MPI für Astrophysik, Nandan Road 80, 200030 Shanghai China*

Abstract

We investigate nonlinear and stochastic biasing of galaxies dark matter cores (DM cores) and dark halos using cosmological hydrodynamic simulations. We compare their biasing properties with the halo biasing model of Taruya & Suto. Dark halos in our simulation exhibit reasonable agreement with the predictions only on the scales larger than $\sim 10h^{-1}$Mpc, while the volume exclusion effect due to their finite size underpopulated them on smaller scales. Interestingly, the biasing of galaxies and DM cores are in relatively good agreement with the theoretical predictions for halo biasing. We also find the clear dependence of galaxy biasing on their formation epoch. The distribution of old population of galaxies tightly correlate with underlying dark matter density field, while that of the young population is slightly more stochastic and anti-biased. Furthermore, the old population of galaxies selectively reside within massive halos, and the young galaxies are formed preferentially in smaller halos. Assuming that early and late-type galaxies correspond to the old and young population of galaxies in our simulation, respectively, all these segregation of galaxies are consistent with the observational results on the morphology dependent clustering of galaxies.

1. Introduction

In this work, we use the cosmological hydrodynamic simulations[1] of galaxy formation in cold dark matter (CDM) universe and investigate the biasing properties of galaxies dark matter cores (DM cores) and dark halos. Particularly, we focus on the comparison of their biasing with the theoretical prediction of halo biasing by Taruya & Suto [2].

2. Numerical Simulations

Our simulation code is a hybrid of Particle-Particle-Particle-Mesh (P^3M) Poisson solver and smoothed particle hydrodynamics (SPH) algorithm[1]. The simulation presented in this paper adopts $N_{DM} = 128^3$ dark matter particles and the same number of gas particles. We consider the spatially-flat low-density CDM (LCDM) universe with $\Omega_0 = 0.3$, $\lambda_0 = 0.7$, $\sigma_8 = 1.0$ and $h = 0.7$, where Ω_0 is the mean mass density parameter, λ_0 the dimensionless cosmological constant, σ_8 the rms density fluctuation on a scale of $8h^{-1}$Mpc, and h the Hubble constant in the units of 100 [km/s/Mpc]. We assume the mean baryon density parameter to be $\Omega_b = 0.015h^{-2}$. The simulation is carried out with a periodic cube of $(75h^{-1}\mathrm{Mpc})^3$, with the gas and dark matter mass per particles being $2.4 \times 10^9 M_\odot$ and $2.2 \times 10^{10} M_\odot$, respectively.

3. Conclusion

Using a cosmological SPH simulation, we directly simulate the formation of galaxies and identify galaxies as isolated and distinct objects. We find that the biasing of dark halos suffers from the volume exclusion effect due to their finite size. On the other hand, biasing of galaxies and DM cores are in good agreement with the theoretical prediction by Taruya & Suto[2]. The biasing of galaxies clearly depend on their formation epoch; the old galaxies shows tight correlation with dark matter and the young ones avoid the dense environment. Considering the observed morphology dependent clustering of galaxies, the formation epoch is the major parameter in determining the morphological type of galaxies.

References

[1] Yoshikawa, K., Jing, Y.P. & Suto, Y. 2000,
[2] Taruya, A. & Suto, Y. 2000, ApJ, 542, 559 (TS)

The 5th RESCEU International Symposium on "New Trends in Theoretical and Observational Cosmology"

Sanjo Conference Hall,
University of Tokyo (Tokyo, Japan)
November 13 – 16, 2001

Program: ORAL

Monday, November 12

16:00 Open of the registration desk
17:00 Welcome Party

Tuesday, November 13

• Morning Session

Chair: M. Sasaki (9:20–11:00)

9:20–9:25	Opening address K. Sato	
9:25–9:30	Congratulatory address J. Kawamura	
9:30–10:15	Primordial perturbation spectra from inflation D. Wands	(40+5)
10:15–11:00	Baryogenesis T. Yanagida	(40+5)

11:00–11:15 Tea Break

Chair: K. Maeda (11:15–12:45)

11:15–12:00	Brane world cosmology J. Garriga	(40+5)
12:00–12:45	Why does Inflation start at the top of the hill? S. W. Hawking	(40+5)

12:45–13:45 Lunch

• Afternoon Session

Chair: N. Sugiyama (13:45–16:20)

13:45–14:30	LISA as a tool for cosmology B. Schutz	(40+5)
14:30–15:15	Measurements of the CMB: Current Status and Future Prospects L. Page	(40+5)
15:15–16:00	Cosmology with Supernovae and Supernova Acceleration Probe (SNAP) S. Perlmutter	(40+5)
16:00–16:20	First distant supernova search with SUBARU M. Doi	(15+5)

16:20–16:50 Tea Break

Chair: M. Kawasaki (16:50–18:50)

16:50–17:10	Cosmic Microwave Background Anisotropy with Late Time Entropy Production Scenario T. Takahashi	(15+5)
17:10–17:30	A cosmological constant from degenerate vacua J. Yokoyama	(15+5)
17:30–17:50	Cosmological aspects of Affleck-Dine mechanism and Q-ball formation in the gauge-mediated SUSY breaking models S. Kasuya	(15+5)
17:50–18:10	Global structure of the Brane Universe H. Ishihara	(15+5)
18:10–18:30	Bulk Gravitational Field and Cosmological Perturbations on the Brane K. Koyama	(15+5)
18:30–18:50	Brane World, Mass Hierarchy and the Cosmological Constant S. Ichinose	(15+5)

Wednesday, November 14

• Morning Session

Chair: J. Yokoyama (9:20–10:50)

9:20–9:40	Primordial black hole formation in natural double inflation in supergravity M. Yamaguchi	(15+5)
9:40–10:00	The Origin of Inflation:Quantum Effects and AdS/CFT in Cosmology T. Hertog	(15+5)
10:00–10:20	Galaxy distribution and type Ia supernova data in the accelerating universe with a local void K. Tomita	(15+5)

10:20–10:50	Axion Telescope Experiment M. Minowa	(25+5)

10:50–11:05 Tea Break

Chair: H. Kodama (11:05–12:25)

11:05–11:35	Implications from recent X-ray studies of clusters of galaxies K. Makishima	(25+5)
11:35–11:55	Deep X-ray Surveys with Chandra and XMM-Newton R. Griffiths	(15+5)
11:55–12:25	Observational Approach to Molecular Cloud Formation S. Yamamoto	(25+5)

12:25–13:25 Lunch

• Afternoon Session

Chair: K. Nomoto (13:25–15:45)

13:25–13:55	Low energy antiprotons in cosmic rays during a half solar cycle period T. Maeno	(25+5)
13:55–14:40	Results from the Super-Kamiokande Experiment K. Nishijima	(40+5)
14:40–15:25	Extremely High Energy Cosmic Rays and Neutrinos by the Telescope Array M. Sasaki	(40+5)
15:25–15:45	Diversity of Type Ia Supernovae: Rapidly Rotating Progenitors T. Uenishi	(15+5)

15:45–16:15 Tea Break

Chair: K. Makishima (16:15–17:35)

16:15–16:35	Neutrino Burst from Supernovae and neutrino Oscillation K. Sato	(15+5)
16:35–16:55	Probing the Formation and Evolution of Galaxies and Clusters with Gamma-rays and Cosmic Rays S. Inoue	(15+5)
16:55–17:15	Iron line emission in the Afterglow of GRB000214 and the environment around the GRB S. Nagataki	(15+5)
17:15–17:35	Implications of $\bar{\nu}_e e^- \to W^- \gamma$ for high-energy $\bar{\nu}_e$ observation G. L. Lin	(15+5)

18:15– Banquet

Thursday, November 15

• Morning Session

Chair: Y. Suto (9:20–10:50)

9:20–10:05	Cosmological parameters and galaxy properties from the 2dF galaxy redshift survey O. Lahav	(40+5)
10:05–10:50	Science from the Sloan Digital Sky Survey M. Fukugita	(40+5)

10:50–11:05 Tea Break

Chair: T. Shigeyama (11:05–12:30)

11:05–11:50	Clustering and density profile of dark matter halos Y. Suto	(40+5)
11:50–12:10	Making galaxies with high-resolution N-body simulations Y. Jing	(15+5)
12:10–12:30	Structure of Dark Matter Halos From Hierarchical Clustering T. Fukushige	(15+5)

12:30–13:30 Lunch

• Afternoon Session

Chair: M. Fukugita (13:30–15:45)

13:30–14:15	Galaxies and Intergalactic Material at High Redshift K. Adelberger	(40+5)
14:15–15:00	Observational Tests of Hierarchical Galaxy Formation R. Ellis	(40+5)
15:00–15:45	Photometric properties of galaxies at $z \sim 4$–5 in the Subaru Deep Fields K. Shimasaku	(40+5)

15:45–16:15 Tea Break

Chair: N. Gouda (16:15–18:15)

16:15–16:35	Clustering of Galaxies at redshift 4 and 5 in the Subaru Deep Fields M. Ouchi	(15+5)
16:35–16:55	Constraining Cosmological Parameters from Clusterings of intermediate-redshift Galaxies T. Matsubara	(15+5)
16:55–17:15	Imprint of Large Scale Structure on CMB anisotropies A. Cooray	(15+5)
17:15–17:35	H_2, D/H and the CMBR temperature at $z = 3.025$ toward QSO 0347–3819 S. Levshakov	(15+5)
17:35–17:55	Galaxy Alignments from Cosmic Tidal Forces J. Lee	(15+5)
17:55–18:15	Large scale structure in the SDSS and 2dF galaxy surveys A. Doroshkevich	(15+5)

Friday, November 16

- Morning Session **Chair: T. Futamase (9:30–12:00)**

9:30–10:15	Cosmic Shear: Advances and Prospects A. Refregier	(40+5)
10:15–10:55	Numerical study of weak lensing T. Hamana	(35+5)
10:55–11:40	Weak lensing and large scale structure using Suprime-Cam S. Miyazaki	(40+5)
11:40–12:00	Gravitational lensing of SN1997ff C. Gunnarsson	(15+5)

12:00–13:00 Lunch

- Afternoon Session **Chair: S. Ikeuchi (13:00–15:30)**

13:00–13:20	Is the luminosity function of galaxies universal? S. Okamura	(15+5)
13:20–14:05	The MAGNUM Project: AGN variability as a new technique for distance determination Y. Yoshii	(40+5)
14:05–14:50	Nucleosynthesis in Pop III stars and Cosmic/Galactic Chemical Evolution K. Nomoto	(40+5)
14:50–15:10	Environmental variation of photometric properties of galaxies along the large scale structure around a z=0.4 cluster T. Kodama	(15+5)
15:10–15:30	Closing Address K. Makishima	(20)

- Special Lecture

17:00–	Prof. Stephen W. Hawking "Brane new world" Yasuda Lecture Hall, The University of Tokyo

Program: Poster

The size of a poster board is 120 cm (height) × 90 cm (width).

Ando	Shinichiro	Determining the Supernova Direction by its Neutrinos
Asada	Hideki	Astrometric Microlensing due to an Extrasolar Planet
Bristow	Paul	Theory, Observation and Selection: New trends in modelling selection effects in galaxy surveys?"
Deng	Jinsong	Light curve models for the hypernova SN 1999as
Enoki	Motohiro	Effects of Formation Epoch Distribution on X-ray Luminosity and Temperature Functions of Galaxy Clusters
Fukui	Takao	A Non-Separable Solution of the Vacuum 5D STM Universe
Harada	Tomohiro	Stability criterion for self-similar solutions in general relativity
Hikage	Chiaki	Non-linear stochastic biasing effects in genus statistics for dark halo distributions
Hirai	Shiro	Influence of Inflation on Parametric Resonance during Preheating
Ibrahim	Arbab	Cosmic acceleration with a positive cosmological term
Ida	Daisuke	Stability of AdS × S × S
Iguchi	Hideo	Is dark energy the only solution to the apparent acceleration of the present universe?
Inoue	Kaiki T.	Non-Gaussianity as a probe of topology of the space
Iwata	Ikuru	Peculiar Velocities of the Galaxies beyond the Local Void
Kawasaki	Wataru	Strong Correlation between the High-latitude Steady Unidentified EGRET Sources and Possible Merging Clusters of Galacxies
Kayo	Issha	Log-normal property of the cosmological density fluctuations
Kobayashi	Chiaki	Chemodynamical Evolution of Elliptical Galaxies
Kohri	Kazunori	Equation of state of electron in strong magnetic fields and neutrino driven wind in supernovae
Kotake	Kei	Specifying the environments around GRB, Explaining the Fe line in the X-Ray afterglow of GRB000214
Kudoh	Hideaki	Nonlinear gravitational interactions in the stabilized Randall-Sundall two branes model
Kurita	Yasunari	Quasi normal modes of D-brane black holes
Kuwabara	Takeshi	The psirwise Peculiar Velocity Distribution Function from the Halo Density Profile
Lee	Chul H.	Structure of the force-free black hole magnetosphere
Maeda	Keiichi	Aspherical hypernova explosions and the early galactic chemical evolution
Mizuno	Shuntaro	Quintessence in a brane world

Murata	Tsukasa	Fluctuation-Dissipation Relation of the Vacuum Noise in Curved spacetime
Nakamura	Kouji	Hoop conjecture and brane world gravity
Nagasawa	Masahiro	Dynamical Evolution of Monopoles and Anti-Monopoles
Nagashima	Michiyasu	Galaxy Number Counts in the Subaru Deep Field: A Constraint on a Hierarchical Galaxy Formation Model
Narita	Makoto	Global foliatoins of Gowdy symmetric spacetimes with stringy matter
Nishioka	Hiroaki	Measuring the time evolution of the redshift-distortion parameter
Ochiai	Hirotaka	Creation of the Randall-Sundrum Bubble
Oguri	Masamune	Time Delay Statistics of Strong Gravitational Lensing and the Density Profile of Dark Halos
Okoshi	Katsuya	Chemical Enrichment in Damped Lyman alpha Systems on Hierarchical Galaxy Formation
Orito	Manabu	Cosmological Constraints on Neutrino Degeneracy
Ota	Naomi	Two Classes of Cluster Size and the Connection to the L–T Relation
Overduin	James	Dark Matter and the Background Light
Potter	Frank	Quantum Celestial Mechanics: Galaxy Rotation Curves without requiring Dark Matter
Seto	Naoki	Direct measurement of the acceleration of the universe using 0.1 Hz band laser interferometer gravitational wave antenna in space
Seto	Osamu	Oscillating scalar field dark matter
Shimizu	Mamoru	Influence of resolution of timesteps on making merging trees
Shimizu	Mamoru	Numerical solution of AdS x S x S and its stability
Shirouzu	Marii	Nucleosynthesis in Type II Supernovae and Abundances of Pop III Stars
Suran	Marian D.	First structure formation ($12 > z > 7$) in a CHDM model
Takahashi	Fuminobu	On the Moduli Problem and Baryogenesis in Gauge-mediated SUSY Breaking Models
Takahashi	Keitaro	Effects of neutrino oscillation on supernova neutrino
Takeuchi	Tsutomu T.	A Bridge from Optical to Infrared Galaxies: Explaining Infrared Local Properties, Predicting Galaxy Counts and the Cosmic Background Radiation
Tamaki	Takashi	Have we already detected astrophysical symptoms of space-time noncommutativity?
Taruya	Atsushi	Lognormal property of weak lensing fields
Torii	Takashi	Brane world vs. Holographic Renormalization Group
Tsujikawa	Shinji	Gravitational particle production from metric perturbations
Ueda	Haruhiko	A Graph–Theoretical Approach of large-scale structure in the Universe

Umeda	Hideyuki	Nucleosynthesis of Zn and Iron-Peak Elements in Pop III Type II Supernovae:Abundances of Very Metal-Poor Halo Stars
Umetsu	Keiichi	Weak Lensing Analysis of the high redshift cluster MS 1054-03
Verozub	Leonid	Acceleration of the Univers as a Consequence of Gravity Properties
Watanabe	Gentaro	Microscopic Study of Nuclear "Pasta" by Quantum Molecular Dynamics
Watanabe	Satoru	The most stringent limit on solar axions
Yahagi	Hideki	Parallelization and Vectorization of the AMR N-body code
Yahiro	Masanobu	Constraints on Cosmic Quintessence
Yano	Taihei	Dynamics of Stellar Orbits in Disk Galaxies
Yonehara	Atsunori	Beyond H_0 determination
Yoshiguchi	Hiroyuki	Propagation of UHECRs from the Sources in the Super-Galactic Plane
Yoshikawa	Kohji	Stochastic Biasing of Galaxies and Dark Halos in Cosmological Hydrodynamic Simulations
Zhang	Zhiqiang	Feeding Back Theory of the Universe

List of Participants

Adelberger, K.
Harvard-Smithsonian Center for Astrophysics
kadelber@diocletian.harvard.edu

Akiyama, R.
Ibaraki University
tofu@goif.sci.ibaraki.ac.jp

Ando, S.
University of Tokyo
ando@utap.phys.s.u-tokyo.ac.jp

Arafune, J.
Institute of Cosmic Ray Research, University of Tokyo
arafune@icrr.u-tokyo.ac.jp

Asada, H.
Fuculty of Science and Technology, Hirosaki University
asada@phys.hirosaki-u.ac.jp

Bristow, P.
European Southern Observatory
bristowp@eso.org

Burdyuzha, V.
Lebedev Physical Institute of Russian AS
burdyuzh@asc.rssi.ru

Chiba, M.
NationalAstronomicalObservatory
chibams@gala.mtk.nao.ac.jp

Cooray, A.
California Institute of Technology
asante@hyde.uchicago.edu

Deng, J.
University of Tokyo
deng@astron.s.u-tokyo.ac.jp

Doi, M.
University of Tokyo
doi@ioa.s.u-tokyo.ac.jp

Doroshkevich, A.
Theoretical Astrophysics Center
dorr@tac.dk

Ellis, R.
California Institute of Technology
rse@astro.caltech.edu

Enoki, M.
National Astronomical Observatory
enoki.motohiro@nao.ac.jp

Eriksson, M.
Stockholm University
mate@physto.se

Fukugita, M.
ICRR, University of Tokyo
fukugita@sdss1.icrr.u-tokyo.ac.jp

Fukui, T.
Dokkyo University
tfukui@dokkyo.ac.jp

Fukushige, T.
College of Arts and Sciences, University of Tokyo
fukushig@provence.c.u-tokyo.ac.jp

Furihata, Y.
Asahikawa National College of Technology
furi@asahikawa-nct.ac.jp

Futamase, T.
Tohoku University
tof@venus.astr.tohoku.ac.jp

Garriga, J.
IFAE, Department de Fisica, Universitat Autonoma de Barcelona
garriga@ifae.es

Gouda, N.
National Astronomical Observatory of Japan
naoteru.gouda@nao.ac.jp

Griffiths, R.
Carnegie Mellon University
griffith@astro.phys.cmu.edu

Gunnarsson, C.
Stockholm University
cg@physto.se

Hamana, T.
National Astronomical Observatory Japan
hamana@yukawa.kyoto-u.ac.jp

Harada, T.
Waseda University
harada@gravity.phys.waseda.ac.jp

Hawking, S.W.
DAMTP, CMS, University of Cambridge
S.W.Hawking@damtp.cam.ac.uk

Hertog, T.
University of Cambridge
T.Hertog@damtp.cam.ac.uk

Hikage, C.
University of Tokyo
hikage@utap.phys.s.u-tokyo.ac.jp

Hirai, S.
Osaka Electro-Communication Junior College
hirai@isc.osakac.ac.jp

Hirayama, T.
KEK
thirayam@kek.jp

Ibrahim, A.
Omdurman Ahlia University
arbab64@yahoo.com

Ichikawa, K.
University of Tokyo
ichikawa@resceu.s.u-tokyo.ac.jp

Ichinose, S.
University of Shizuoka
ichinose@u-shizuoka-ken.ac.jp

Ida, D.
Research Center for the Early Universe
ida@resceu.s.u-tokyo.ac.jp

Igi, K.
Kanagawa University
igi@info.kanagawa-u.ac.jp

Iguchi, H.
Tokyo Institute of Technology
iguchi@th.phys.titech.ac.jp

Ikeuchi, S.
Nagoya University
ikeuchi@a.phys.nagoya-u.ac.jp

Inoue, S.
National Astronomical Observatory
inoue@th.nao.ac.jp

Inoue, K.T.
National Astronomical Observatory
tinoue@th.nao.ac.jp

Ishihara, H.
Osaka City University
ishihara@sci.osaka-cu.ac.jp

Iwata, I.
Kyoto University
iwata@kusastro.kyoto-u.ac.jp

Jean-Michel, Alimi
CNRS, Observatoire de Paris-Meudon
Jean-Michel.Alimi@obspm.fr

Jing, Y.
Shanghai Observatory,
Chinese Academy of Sciences
ypjing@center.shao.ac.cn

Kakuhari, Y.
Ibaraki University
kakuhari@golf.sci.ibaraki.ac.jp

Kanai, T.
University of Tokyo
kanai@resceu.s.u-tokyo.ac.jp

Kang, X.
Shanghai astronomy observatory,
Chinese Academy of Sciences
kangx@center.shao.ac.cn

Kao, W.F.
Chiao Tung University
wfgore@cc.nctu.edu.tw

Kasai, M.
Hirosaki University
kasai@phys.hirosaki-u.ac.jp

Kasuya, S.
University of Tokyo
kasuya@resceu.s.u-tokyo.ac.jp

Kawasaki, M.
University of Tokyo
kawasaki@resceu.s.u-tokyo.ac.jp

Kawasaki, W.
Department of Astronomy,
The University of Tokyo
kawasaki@astron.s.u-tokyo.ac.jp

Kayo, I.
University of Tokyo
kayo@utap.phys.s.u-tokyo.ac.jp

Kobayashi, C.
Department of Astronomy,
The University of Tokyo
chiaki@astron.s.u-tokyo.ac.jp

Kobayashi, Y.
NAOJ
yuki@merope.mtk.nao.ac.jp

Kodama, H.
Yukawa Institute for Theoretical Physics
kodama@yukawa.kyoto-u.ac.jp

Kodama, T.
University of Tokyo
kodama@astron.s.u-tokyo.ac.jp

Kohri, K.
Yukawa Institute for Theoretical Physics (YITP)
kohri@yukawa.kyoto-u.ac.jp

Kotake, K.
University of Tokyo
kkotake@utap.phys.s.u-tokyo.ac.jp

Koyama, K.
Kyoto University
kazuya@phys.h.kyoto-u.ac.jp

Kudoh, H.
Yukawa institute for theoretical physics
kudoh@yukawa.kyoto-u.ac.jp

Kurita, Y.
Kyoto University
kurita@phys.h.kyoto-u.ac.jp

Kuwabara, T.
University of Tokyo
kuwabara@utap.phys.s.u-tokyo.ac.jp

Lahav, O.
Institute of Astronomy,
University of Cambridge
lahav@ast.cam.ac.uk

Lee, C.H.
University of Alberta
chlee@hepth.hanyang.ac.kr

Lee, J.
Academia Sinica
taiji@asiaa.sinica.edu.tw

Levshakov, S.
National Astronomical Observatory
lev@th.nao.ac.jp

Lin, G.-L.
National Chiao-Tung University
glin@cc.nctu.edu.tw

Lin, W.
Shanghai Observatory
linwp@center.shao.ac.cn

Liu, G.-C.
ASSIA, Taiwan
gcliu@asiaa.sinica.edu.tw

Maeda, H.
Waseda university
hideki@gravity.phys.waseda.ac.jp

Maeda, K.
Waseda University
maeda@mse.waseda.ac.jp

Maeda, K.
Department of Astronomy,
The University of Tokyo
maeda@astron.s.u-tokyo.ac.jp

Maeno, T.
Kobe University
maeno@phys.sci.kobe-u.ac.jp

Makishima, K.
University of Tokyo
maxima@phys.s.u-tokyo.ac.jp

Matsubara, T.
Nagoya University
taka@a.phys.nagoya-u.ac.jp

Matsui, S.
Tohoku University
matsui@tuhep.phys.tohoku.ac.jp

Mellier, Y.
Institut d'Astrophysique de Paris
mellier@iap.fr

Minezaki, T.
University of Tokyo
minezaki@mtk.ioa.s.u-tokyo.ac.jp

Minowa, M.
University of Tokyo
minowa@phys.s.u-tokyo.ac.jp

Miyamoto, U.
Waseda University
umpei@gravity.phys.waseda.ac.jp

Miyazaki, A.
Nagasaki Prefectural University
miyazaki@nagasakipu.ac.jp

Miyazaki, S.
NationalAstronomicalObservatory
satoshi@anela.mtk.nao.ac.jp

Mizuno, S.
Waseda University
shuntaro@gravity.phys.waseda.ac.jp

Mizutani, K.
Saitama University
mizutani@cr.phy.saitama-u.ac.jp

Moroi, T.
Tohoku University
moroi@phys.tohoku.ac.jp

Motoyama, K.
Ibaraki University
motoyama@golf.sci.ibaraki.ac.jp

Murata, T.
Hiroshima University
murata@hiroshima-u.ac.jp

Nagasawa, M.
Kanagawa University
nagasawa@info.kanagawa-u.ac.jp

Nagashima, M.
National Astronomical Observatory
masa@th.nao.ac.jp

Nagataki, S.
University of Tokyo
nagataki@utap.phys.s.u-tokyo.ac.jp

Nakamichi, A.
Gunma Astronomical Observatory
akika@astron.pref.gunma.jp

Nakamura, K.
National Astronomical Observatory
kouchan@th.nao.ac.jp

Narita, M.
Waseda University
narita@gravity.phys.waseda.ac.jp

Nishijima, K.
Tokai University
kyoshi@tkikam.sp.u-tokai.ac.jp

Nishioka, H.
Hiroshima University
nishioka@theo.phys.sci.hiroshima-u.ac.jp

Nobili, S.
University of Stockholm
serena@physto.se

Nomoto, K.
Department of Astronomy,
The University of Tokyo
nomoto@astron.s.u-tokyo.ac.jp

Ochiai, H.
University of Tokyo
ochiai@utap.phys.s.u-tokyo.ac.jp

Oguri, M.
University of Tokyo
oguri@utap.phys.s.u-tokyo.ac.jp

Okamura, H.
Department of Physics,
Kogakuin University
okamura@cc.kogakuin.ac.jp

Okamura, S.
University of Tokyo
okamura@astron.s.u-tokyo.ac.jp

Okoshi, K.
National Astronomical Observatory Japan
okoshi@th.nao.ac.jp

Orito, M.
National Astronomical Observatory Japan
orito@th.nao.ac.jp

Ota, N.
Tokyo Metropolitan University
naomi@phys.metro-u.ac.jp

Ouchi, M.
University of Tokyo
ouchi@astron.s.u-tokyo.ac.jp

Overduin, J.
Waseda University
overduin@gravity.phys.waseda.ac.jp

Page, L.
Princeton University
page@pupgg.princeton.edu

Perez, J.
Ecole Nationale Superieure de Technique Avancees
perez@ensta.fr

Perlmutter, S.
Lawrence Berkeley National Laboratory,
University of California
saul@lbl.gov

Potter, F.
University of California, Irvine
drpotter@lycos.com

Refregier, A.
University of Cambridge
ar@ast.cam.ac.uk

Sanuki, T.
University of Tokyo
sanuki@phys.s.u-tokyo.ac.jp

Sasaki, M.
ICRR, University of Tokyo
sasakim@icrr.u-tokyo.ac.jp

Sasaki, M.
Osaka University
misao@vega.ess.sci.osaka-u.ac.jp

Sato, F.
Konan University
sato@tap.scphys.kyoto-u.ac.jp

Sato, K.
University of Tokyo
sato@phys.s.u-tokyo.ac.jp

Schutz, B.
MPI fur Gravitationsphysik (AEI)
schutz@aei-potsdam.mpg.de

Seto, N.
Osaka University
seto@vega.ess.sci.osaka-u.ac.jp

Seto, O.
Yukawa Institute for Theoretical Physics
oseto@yukawa.kyoto-u.ac.jp

Shibata, M.
University of Tokyo
shibara@provence.c.u-ac.jp

Shigeyama, T.
University of Tokyo
shigeyama@astron.s.u-tokyo.ac.jp

Shimasaku, K.
University of Tokyo
shimasaku@astron.s.u-tokyo.ac.jp

Shimizu, M.
University of Tokyo
mshimizu@utap.phys.s.u-tokyo.ac.jp

Shimizu, Y.
University of Tokyo
yshimizu@utap.phys.s.u-tokyo.ac.jp

Shinkai, H.
RIKEN
shinkai@atlas.riken.go.jp

Shiromizu, T.
University of Tokyo
siromizu@utap.phys.s.u-tokyo.ac.jp

Shirouzu, M.
University of Tokyo
shirouzu@astron.s.u-tokyo.ac.jp

Sugiyama, N.
National Astronomical Observatory
naoshi@th.nao.ac.jp

Suran, M.D.
Astronomical Institute of Romanian Academy
suran@roastro.astro.ro

Susa, H.
Tsukuba University
susa@rccp.tsukuba-ac.jp

Susukita, R.
RIKEN
susukita@postman.riken.go.jp

Suto, Y.
University of Tokyo
suto@phys.s.u-tokyo.ac.jp

Takahara, M.
Doshisha Women's College of Liberal Arts, University of Tokyo
mtakahar@dwc.doshisha.ac.jp
takahara@utap.phys.s.u-tokyo.ac.jp

Takahashi, F.
University of Tokyo
fumi@resceu.s.u-tokyo.ac.jp

Takahashi, K.
University of Tokyo
ktaro@utap.phys.s.u-tokyo.ac.jp

Takahashi, T.
Tohoku University
tomo@tuhep.phys.tohoku.ac.jp

Takeuchi, T.T.
National Astronomical Observatory of Japan
takeuchi@optik.mtk.nao.ac.jp

Tamaki, T.
Waseda University
tamaki@gravity.phys.waseda.ac.jp

Tanabe, Y.
ytanbe@highway.ne.jp

Tanaka, I.
Osaka City University
itanaka@sci.osaka-cu.ac.jp

Tanaka, T.
Yukawa Institute for Theoretical Physics (YITP)
tanaka@yukawa.kyoto-u.ac.jp

Taruya, A.
University of Tokyo
ataruya@phys.s.u-tokyo.ac.jp

Tatekawa, T.
Waseda University
tatekawa@gravity.phys.waseda.ac.jp

Terasawa, T.
University of Tokyo
terasawa@eps.s.u-tokyo.ac.jp

Tomita, K.
Yukawa Institute for Theoretical Physics
tomita@yukawa.kyoto-u.ac.jp

Torii, T.
University of Tokyo
torii@resceu.s.u-tokyo.ac.jp

Tsujikawa, S.
University of Tokyo
shinji@resceu.s.u-tokyo.ac.jp

Uchida, S.
Fukui Medical University
uchida@physics.fukui-med.ac.jp

Ueda, H.
Akita University
ueda@quartet.ipc.akita-u.ac.jp

Uematsu, S.
Ibaraki University
sachi@golf.sci.ibaraki.ac.jp

Uenishi, T.
University of Tokyo
uenishi@astron.s.u-tokyo.ac.jp

Umeda, H.
University of Tokyo
umeda@astron.s.u-tokyo.ac.jp

Umetsu, K.
Institute of Astronomy and Astrophysics, Academia Sinica (ASIAA)
keiichi@asiaa.sinica.edu.tw

Verozub, L.
Kharkov National University
verozub@gravit.kharkov.ua

Wakamatsu, K.
Gifu University
waka@info.gifu-u.ac.jp

Wands, D.
University of Portsmouth
david.wands@port.ac.uk

Watabe, H.
Waseda University
watabe@gravity.phys.waseda.ac.jp

Watanabe, G.
University of Tokyo
gentaro@utap.phys.s.u-tokyo.ac.jp

Watanabe, S.
University of Tokyo
watanabe@astron.s.u-tokyo.ac.jp

Yahagi, H.
University of Tokyo
hyahagi@ioa.s.u-tokyo.ac.jp

Yahiro, M.
University of the Ryukyu
yahiro@sci.u-ryukyu.ac.jp

Yamaguchi, M.
University of Tokyo
gucci@resceu.s.u-tokyo.ac.jp

Yamamoto, A.
KEK
akira.yamamoto@kek.jp

Yamamoto, K.
Kyoto University
yamamoto@nucleng.kyoto-u.ac.jp

Yamamoto, S.
University of Tokyo
yamamoto@phys.s.u-tokyo.ac.jp

Yanagida, T.
University of Tokyo
yanagida@phys.s.u-tokyo.ac.jp

Yano, T.
National Astronomical Observatory
yano@pluto.mtk.nao.ac.jp

Yasuda, N.
NAOJ
naoki.yasuda@nao.ac.jp

Yokoi, N.
University of Tokyo
yokoi@icrr.u-tokyo.ac.jp

Yokoyama, J.
Osaka University
yokoyama@vega.ess.sci.osaka-u.ac.jp

Yonehara, A.
Tsukuba University
yonehara@rccp.tsukuba.ac.jp

Yoshiguchi, H.
University of Tokyo
hiroyuki@utap.phys.s.u-tokyo.ac.jp

Yoshii, Y.
IOA,University of Tokyo
yoshii@ioa.s.u-tokyo.ac.jp

Yoshikawa, K.
Kyoto Univeristy
kohji@kusastro.kyoto-u.ac.jp

Yoshioka, S.
Tokyo University of Mercantile Marine
yoshioka@phys.tosho-u.ac.jp

Zhang, Z.
CAAC, China
zzq7@mail.dlptt.ln.cn

Author Index

Int Background Light 359

BB
21
MAR 24